T0257747

Recent Update on Organic Pollutants

Recent Update on Organic Pollutants

Edited by **Bruce Horak**

New York

Published by Callisto Reference,
106 Park Avenue, Suite 200,
New York, NY 10016, USA
www.callistoreference.com

Recent Update on Organic Pollutants
Edited by Bruce Horak

© 2015 Callisto Reference

International Standard Book Number: 978-1-63239-547-4 (Hardback)

This book contains information obtained from authentic and highly regarded sources. Copyright for all individual chapters remain with the respective authors as indicated. A wide variety of references are listed. Permission and sources are indicated; for detailed attributions, please refer to the permissions page. Reasonable efforts have been made to publish reliable data and information, but the authors, editors and publisher cannot assume any responsibility for the validity of all materials or the consequences of their use.

The publisher's policy is to use permanent paper from mills that operate a sustainable forestry policy. Furthermore, the publisher ensures that the text paper and cover boards used have met acceptable environmental accreditation standards.

Trademark Notice: Registered trademark of products or corporate names are used only for explanation and identification without intent to infringe.

Printed in the United States of America.

Contents

Permissions

List of Contributors

Preface

This book aims to highlight the current researches and provides a platform to further the scope of innovations in this area. This book is a product of the combined efforts of many researchers and scientists, after going through thorough studies and analysis from different parts of the world. The objective of this book is to provide the readers with the latest information of the field.

Even a decade after the Stockholm Convention on Persistent Organic Pollutants (POPs) was signed, a large variety of organic chemicals still pose ecological hazards of top most priority. The expansion of information base on organic pollutants (OPs), environmental fate and effects, as well as the cleansing methods, is accompanied by a rise in importance of certain pollution sources, associated with possible generation of new hazards for humans and nature. This book provides an analytical and environmental update, and can be specifically important for engineers and experts.

I would like to express my sincere thanks to the authors for their dedicated efforts in the completion of this book. I acknowledge the efforts of the publisher for providing constant support. Lastly, I would like to thank my family for their support in all academic endeavors.

Editor

Part 1

High Concern Sources of Organic Pollutants

Textile Finishing Industry as an Important Source of Organic Pollutants

Alenka Majcen Le Marechal[1], Boštjan Križanec[2],
Simona Vajnhandl[1] and Julija Volmajer Valh[1]
[1]University of Maribor, Faculty of Mechanical Engineering, Maribor,
[2]Environmental Protection Institute, Maribor,
Slovenia

1. Introduction

The textile finishing industry is, among all industries in Europe, the greatest consumer of high quality fresh water per kg of treated material and with the natures of their production processes significantly contributing to pollution. Wastewater from the textile industry is also a significant environmental pollution source of persistent organic pollutants.

Not only textile wastewater but also textile products often contain chemicals such as formaldehyde, azo-dyes, dioxins, pesticides and heavy metals, that might pose a risk to humans and the environment. Some of these chemicals found in finished products are there as residues from the production of dyes and auxiliary chemicals (the synthesis of dyes involves a large variety of chemicals with complex synthesis paths, during which toxic, carcinogenic and persistent organic compounds can be formed, such as dioxins, and traces can be found in commercial dyes), others are added to give certain characteristics to the products (colour, flame retardancy, anti wrinkling properties *etc.*) (Križanec & Majcen Le Marechal, 2006), or are already present in the raw textile material. The mentioned compounds have been found in wastewater after home washing, in organic solvent after dry-cleaning and also in the atmosphere after incineration. Possible sources of organic pollutants are also wastewater treatment methods and the incineration of textile materials.

The formation of dioxins can occur via dyeing and textile finishing processes with conditions favourable for their generation (high temperature, alkaline conditions, ultraviolet (UV) radiation, and other radical initiators). Textile dyes are designed to be resistant to microbial, chemical, thermal and photolytic degradation. After the dyeing process, a lot of non-bonded dyes are released into the wastewater, which can also be treated by Advanced Oxidation Processes (AOPs) in order to destroy the dye molecule and to decolourise the wastewater and reduce organic pollution. It is well-known that under the experimental conditions of such methods, which can be very useful because of the short-time of treatment, hazardous compounds can be formed due to very powerful oxidizing agents such as hydroxyl radicals ($OH^{\bullet}$).

In line with the improvement of people's living standard and the growing awareness and need to preserve the environment several regulations were introduced also in the textile industry in order to control the use of chemicals in textile processes. Under REACH regulation (REACH regulation controlled the quality of fabric, apparels, and shoes and other textile materials, so as to protect human health and the environment) the following main groups of compounds in textiles are under control: Azo Dyes, Phthalates, Formaldehyde, Flame-retardants, Pentachlorophenol, Carcinogenic dyes, Sensitizing disperse dyes, Hexavalent chromium, Polychlorinated biphenyls, Heavy metals, Nickel release, Total lead content, Organic tin compounds, Total cadmium content, Organic chlorine carrier, Nonylphenol, Octylphenol, and Nonyl phenol ethoxylate, and several established directives (e.g. Azo dyes – Directive 2002/61/EC, Pentachlorophenol (PCP)-Directive 94/783/EC)) regulated/banned the use of these substances throughout the textile production chain (www.cirs-reach.com/textile/).

Market pressure, on the other hand, leads to the introduction of an extensive range of new products, especially dyes, and for many of them environmental and health impact data are still lacking. The quantification of chemicals within the environment is leading to the development of sensitive analytical methods in order to effectively detect and control pollution by organic pollutants.

2. Textile raw material

2.1 Fibres

Two general categories of fibres are used in the textile industry: natural and chemical (man-made) fibres. Man-made fibres encompass both purely synthetic materials of petrochemical origin, and regenerated cellulose material from wood fibres. A detailed classification of fibres is presented in Table 1.

Natural fibre		
Animal origin	*Vegetable origin*	*Mineral*
Raw wool	Cotton	Asbetos
Silk fibres	Flax	Glass
Other hair fibres	Jute	Metalic
Alpaca	Linen	Copper
Camel	Ramie	Steel
Cashmere	Hemp	
Horse		
Llama		
Mohair		
Rabbit		
Vicuna		
Chemical (man-made) fibres		
Natural polymers fibres	*Synthetic polymer fibres*	
Viscose, Cupro, Lyocell	Inorganic polymer	
Cellulose Acetate	Glass for fibre glass	
Triacetate	Metal for metal fibre	
	Organic polymers	
	Polyester (PES)	
	Polyamide (PA)	
	Polyacrylonitrile (PAC)	
	Polypropylene (PP)	
	Elastane (EL)	

Table 1. Classification of fibres

2.2 Organic pollutants in textile raw materials

Different kinds and amounts of organic pollutants (contaminants) can already be present in fibres before they arrive at the textile mill. The potential contaminants may be released from the raw materials into the water or air during processing, when the fabric is heated or scoured. Due to the large amount of the fibre used during textile manufacturing, even trace contaminants can produce large amounts of pollutants. Many textile operations lack an incoming quality-control system for fibres. Testing of the natural or chemical fibres for organic pollutants is very rarely done in textile mills.

Textile raw materials contain: natural impurities from cotton, wool, silk, etc., fibre solvents (when chemical fibres are produced by dry-spinning or solvent-spinning processes), monomers (caprolactam ex polyamide 6), catalysts (antimony trioxide in polyesters' fibres), sizing agents (woven textiles esp. cotton and cotton blends), and preparation agents (esp. woven and knitted textiles made of man-made fibres). (Lacasse & Baumann, 2004)

2.3 Natural fibres

Natural fibres are acquired from animal, mineral, and plant sources. Several types of organic pollutants are found in natural fibre and all have the potential to create significant pollution problems. For example, waxes, oils, fats, and grease from animal fibre can contribute to biochemical oxygen demand (BOD), and chemical oxygen demand (COD). Pesticide residues from plant fibre can contribute to aquatic toxicity. Metals can accumulate in sludge or during the treatment system itself, thus causing potential long-term problems.

Cotton is the most significant natural fibre. Cotton fibre contains 88-96 % of cellulose and the residue is pectin substances, wax, proteins, ash, and other organic components (less than 1%). Chemicals such as pesticides, herbicides, and defoliants can be used during the production of cotton. These chemicals may remain as a residue on raw cotton fibres that reach the textile mill. Tests of cotton samples from growing regions worldwide, performed from 1991 to 1993, reported levels of pesticides below the threshold limit value for foodstuffs (US EPA 1995).

Although the content of pesticides in raw cotton fibres was negligible, almost half of the insecticides used in agriculture were used in cotton production. For this reason the reduction of insecticides for economic, environmental, and human health reasons was necessary. Insect-resistant (Bt) cotton (GM IR) started being used in 1996. So, the production of Bt cotton during the first ten years (1996-2005) reduced the total volume of those active insecticide ingredients by 94.5 million kilograms, representing a 19 % reduction in insecticides (Naranjo, 2009).

In 2001 it was reported that cotton was contaminated by pentachlorophenol (PCP) when being used, not only as a defoliant, but also as a fungicide during transportation and storage. (UK, 2001).

Wool is another significant commercial natural fibre. It is an animal hair from the bodies of sheep, sheared once or sometimes twice a year, and its quality and quantity vary widely, depending on the breed of sheep and their environment. Raw wool contains natural impurities (wool grease, suint, dirt) and residues of pesticides. Pesticides are applied onto sheep in order to control external parasites such as lice, blowflies, mites etc.

The used pesticides generally fall into four main groups (BAT, 2003):

- **organochlorine insecticides (OCs):** γ-hexachlorocyclohexane (lindane), pentachlorophenol (PCP), dieldrin, DDT,
- **organophosphorous insecticides (OPs):** diazinon, propetamphos, chlorfenvinphos, chloryriphos, dichlorfenthion,

- **synthetic pyrethroids insecticides (SPs):** cypermethrin, deltamethrin, fenvalerate, flumethrin, cyhalothrin and
- **insect-growth regulators (IGRs):** cyromazine, dicyclanil, diflubenzuron, triflumuron.

Biocides, organohalogen and organophosphorus compounds are among the priority substances listed for emission-control in the IPPC Directive. Diazinon (OP), propetamphos (OP), cypermethrin (SP) and cyromazine (IGR) are the most commonly-used ectoparasiticides for treating sheep. Insect-growth regulators such as dicyclanil, diflubenzuron and triflumuron are registered only in Australia and New Zealand. Lindane, which is the most toxic of the hexachlorocyclohexane isomers and also the most active as a pesticide, is still found in wool coming from the former Soviet Union, the Middle East and some South American countries (BAT, 2003). Wools from South America exhibited the highest levels of organochlorine insecticides, whilst wools from Australia and New Zealand exhibited the lowest levels of OCs (Shaw, 1989).

Pesticides such as OCs, OPs, and SPs have a lipophilic nature, so they associate strongly with the natural oils within the wool and as removed using these oils during wool scouring operations. The chemical stability of organochlorine pesticides is reflected in their resistance to microbial degradation. The lipophilic nature, hydrophobicity, and low chemical and biological degradation rates of organochlorine pesticides have led to their accumulation in biological tissues, and subsequent magnification of concentrations in organisms progressing up the food-chain.

The fate of the Ectoparasiticides (antiparasitic drug) in the wool scouring process is following. 96 % of the pesticides are removed from the wool (4 % is retained on the fibre after scouring). Of this 96 % around 30 % or less is retained on-site in recovered grease and the remaining fraction is discharged into the effluent and submitted to wastewater treatment.

Levels of organochlorine insecticides pentachlorophenol (PCP) in textile products are usually too low to be quantified accurately by traditional methods. PCP has been found at levels as high as 100 ppm in consumer products such as wool carpets (Wimbush, 1989). Wimbush tested 140 wool carpets and found that 88 % of them had a PCP content below 5 ppm. Only 3 carpets contained more than 50 ppm of PCP (US EPA, 1996).

Silk accounts for only 0.2 % of total fibre production. It is a protein fibre like wool and derived from the silkworm. The silk fibre is composed of fibroin filaments wrapped with sericine (silk gum), which has to be removed during pretreatment. (BAT, 2003)

2.4 Chemical fibres

Synthetic fibres may contain several types of impurities existing in the fibres before they reach the textile mill. These kinds of impurities are imparted onto the fibres during fibre manufacturing and fall into the categories of finishes, polymer synthetic by-products, and additives. Impurities associated with synthetic fibres are:

- Finishes: antistatic, lubricant,
- Polymer synthesis by-products: non-reactive monomers, low-molecular-weight oligomers, residual catalyst, and
- Additives to facilitate processing: antistatic agents, lubricants, humectants, and others.

Table 2 presents those compounds typical of synthetic fibre extracts. The compounds in Table 2 were not only found in the process wastewater from synthetic fibre dyeing and finishing operations but were also detected in synthetic fibre extracts.

Fibre	Compounds
Polyester	*tetra* hydro-2,5-dimethyl *cis* furan, ketones (methyl isobutyl ketone, 3-methyl cycle pentanone, hexanone, diethyl ketone), dodecanol, alcohols (C_{14} and C_{18}), esters of carboxylic acids (C_{14}-C_{24}), hydrocarbons (C_{14}-C_{18}), carboxylic acids (C_{16}-C_{24}), phthalate esters
Acrylic	hydrocarbons (C_{15}-C_{18}), esters of carboxylic acids (C_{17}-C_{22}), alcohols, phthalate esters, N, N-dimethyl acetamide
Nylon 6	diphenyl ether, hydrocarbons (C_{16}-C_{20}), carboxylic acids (C_{14}-C_{18}) and dicarboxylic acids, esters of carboxylic acids (C_{10}-C_{18}), alcohols (C_{20}-C_{22})

Table 2. Compounds typical of synthetic fibre extracts

Some of these impurities are produced during the polymerization of synthetic fibres and some are added to control the surface and electrical properties of the fibre. These impurities can create pollution problems.

The above review of possible organic pollutants in textile raw materials indicates the necessity for quality control regarding incoming fibre. Some textile companies have endorsed standard testing methods for fibres and have actively worked to ensure that the fibre user and the producer exchange information about quality control. It is important to set up good incoming fibre quality-control based on performance-testing, statistical sampling, and the analysis of extractable materials, in order to identify potential pollution problems before they arise.

3. Finishing processes within the textile industry

3.1 Basic process within the textile industry

The textile chain begins with the production or harvesting of raw fibres. These basic processes are schematically presented in Fig. 1. Treatments that are broadly referred to as 'finishing processes' are pretreatment, printing, dyeing, finishing, and coating, including washing and drying.

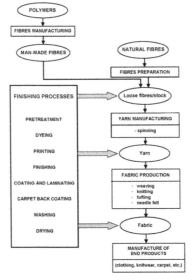

Fig. 1. Schematic presentation of textile processes

3.2 Possible organic pollutants in textile finishing processes

Most textile finishing operational units use chemical specialties. The major specialty consumption operations are pretreatment processes (desizing, scouring, bleaching), dyeing, printing, and finishing. (Mattioli, et al, 2002)

So, this chapter focuses on those textile finishing processes that might produce organic pollutants.

3.2.1 Pretreatment processes

Pretreatment processes depend on the kind and form of the treated fibre, as well as the amount of material to be treated. These processes should ensure the removal of foreign materials from the fibres in order to improve their uniformity, hydrophilic characteristics, and affinity for dyestuffs and finishing treatments. The pretreatment processes are desizing, scouring, bleaching, and mercerizing.

3.2.1.1 Desizing

Desizing is the process for removing size-chemicals from textiles. The possible organic pollutants in effluents after desizing, are presented in Table 3.

Fibres	Organic substances
Cotton	carboxymethyl cellulose, enzymes, fats, hemicelluloses, modified starches, non-ionic surfactants, oils, starch, waxes
Linen	
Viscose	
Silk	carboxymethyl cellulose, enzymes, fats, gelatine, oils, polymeric sizes, polyvinyl alcohol, starch, waxes
Acetates	
Synthetics	

Table 3. Possible organic pollutants in effluents after the desizing process

The washing water from desizing may contain up to 70 % of the total COD in the final effluent. Synthetic esters' oils are less problematic because they are emulsified or soluble in water and easily biodegraded. More problems are caused by compounds such as silicon oils, because they are difficult to emulsify, and poorly biodegradable. Silicon oils are found in elastane blends with cotton or polyamide.

3.2.1.2 Scouring

Scouring is the cleaning process for removing impurities from both natural and synthetic materials. In natural fibres, impurities can be present such as oils, fats, waxes, minerals, and plant-matter. Synthetic fibres can contain spinning, finishing, and knitting oils.

Scouring is performed in an alkali medium together with auxiliaries that include:

- non-ionic surfactants (alcohol ethoxylates, alkyl phenol ethoxylates) and anionic surfactants (alkyl sulphonates, phosphates, carboxylates),
- compounds for removing metal ions (nitrilotriacetic acid (NTA), ethylenediaminetetraacetate (EDTA), diethylene triamine pentaacetate (DTPA), gluconic acid, phosphonic acids as complexing agents),
- polyacrylates and phosphonates as special surfactant-free dispersing agents and
- sulphite and hydrosulphite as reducing agents (to avoid the risk of the formatting of oxycellulose when bleaching with hydrogen peroxide).

Table 4 presents those possible organic pollutants derived from scouring.

Fibers	Organic substances
Cotton	anionic surfactants, cotton waxes, fats, glycerol, hemicelluloses, non-ionic surfactants, peptic matter, sizes, soaps, starch
Viscose	anionic detergents, fats, non-ionic detergents, oils, sizes, soaps, waxes
Acetates	
Synthetics	anionic surfactants, anti static agents, fats, non-ionic surfactants, oils, petroleum spirit, sizes, soaps, waxes
Wool (yarn and fabric)	anionic detergents, glycol, mineral oils, non-ionic detergents, soaps
Wool (loose fiber)	acetate, anionic surfactants, formate, nitrogenous matter, soaps, suint, wool grease, wool wax

Table 4. Possible organic pollutants in waste-effluents after the scouring process

3.2.1.3 Bleaching

Bleaching is a chemical process for removing unwanted coloured matter from materials. Several different types of chemicals are used as bleaching agents and selection depends on the type of fibre. The most frequently used oxidizing agents for cellulose fibres are hydrogen peroxide (H_2O_2), sodium chlorate(I) (NaClO), sodium chlorate(III) ($NaClO_2$). Peracetic acid and optical brightening agents are also applicable.

Bleaching with H_2O_2 is connecting with the use of H_2O_2 stabilisers. During the bleaching process, hydroxyl radicals attack the cellulose fibre starting with oxidation of the hydroxyl groups, and eventually ending with cleavage of the cellulose molecules, thus decreasing the degree of polymerisation. Reaction is catalysed by heavy metals such as iron, copper, and cobalt. H_2O_2 stabilisers (EDTA, NTA, DTPA, gluconates, phosphonates and polyacrylates) inhibit these reactions. NTA, EDTA and DTPA form very stable metal complexes. EDTA and DTPA are also poorly eliminated compounds, and they could pass non-degraded through the common wastewater treatment system. Their ability to form a very stable complex with metal makes the problem even more serious because they can mobilise those heavy metals present in the effluent, and release them into the receiving water. Other auxiliaries used in hydrogen peroxide bleaching are surfactants with emulsifying, dispersing, and wetting properties.

Sodium chlorate(I) was, for a long time, one of the more widely-used bleaching agents throughout the textile finishing industry. Nowadays bleaching with sodium chlorate(I) is limited in Europe for ecological reason. NaClO leads to secondary reactions that form a number of organic halogen compounds, such as carcinogenic trichloromethane.

3.2.1.4 Mercerizing

Mercerizing is a chemical process that improves the strengths, lustres and affinities of dyes for cotton fabrics. Several organic pollutants can be used. Possible organic pollutants in mercerizing effluents are alcohol sulphates, anionic surfactants, and cyclohexanol.

3.2.2 Dyeing

Dyeing is used to add colour to textile materials. Textiles may be dyed at various stages of production.

Dyeing can be carried out in a batch or in a continuous mode. The choice between the two processes depends on the type of make-up, the chosen class of dye, the equipment available, and the cost involved. Both continuous and discontinuous dyeing involve preparation of the dye, dyeing, fixation, washing, and drying. Quite a large amount of the non-fixed dye leaves the dyeing units.

Besides dyes, other auxiliary substances can be added to the dyeing process, and may give rise to water pollution. Possible pollutants are: fatty amine ethoxylates (levelling agent), alkylphenol ethoxylates (levelling agent), quaternary ammonium compounds (retarders for cationic dyes), cyanamide-ammonia salt condensation products (auxiliaries for fastness improvement), acrylic acid-maleic acid copolymers (dispersing agent), ethylenediamine tetraacetate (EDTA), diethylenetriaminepentaacetate (DTPA), ethylenediaminetetra-(methylenephosphonic acid) (EDTMP), diethylenetriaminepenta(methylenphosphonic acid) (DTPMP).

All these are water soluble and non-biodegradable compounds, which can pass non-transformed or partially-degraded through a wastewater treatment system. Some of them are toxic (quaternary amines) or can give rise to metabolites that may affect the reproduction-chain within an aquatic environment (BAT, 2003).

Organic dyes are presented in Chapter 4.

3.2.3 Printing

Printing, like dyeing, is a process for applying colour to a substrate. The printing techniques are: rotary screen, direct, discharge, resist, flat screen, and roller printing. Pigments cover about 75-85 % of all printing operations, do not require washing steps and generate a small amount of waste. Compared to dyes, pigments are typically insoluble and have a high affinity for fibres. Printing paste residues, wastewater from wash-off and cleaning operations, and volatile organic compounds from drying and fixing, are typical emission sources from printing processes.

Printing process wastewater is small in volume, but the concentration of pollutants is higher than that in wastewater from dyeing. Wastewater after printing contains the following organic pollutants: urea, dyes or pigments, and organic solvents.

These pollutants are likely to be encountered in wastewater after printing, and are presented in Table 5 (BAT, 2003).

Additionally, organic pollutants after the finishing processes, such as aliphatic hydrocarbons (C_{10}-C_{20}) from binders, monomers (acrylates, vinylacetates, styrene, acrylonitrile, acrylamide, butadiene), methanol from fixation agents, other alcohols, esters, polyglycols from emulsifiers, formaldehyde from fixation agents, ammonia, N-methylpyrrolidone from emulsifiers, phosphoric acid esters, phenylcyclohexene from thickeners and binders, might also be distributed in the exhaust air.

3.2.4 Finishing

The term 'finishing' covers all those treatments that improve certain properties or the serviceability of the fibre. Finishing may involve those mechanical/physical and chemical treatments performed on fibre, yarn, or fabric, in order to improve appearance, texture, or performance.

Organic pollutants found during finishing processes are:

- cross-linking agents in easy-care finishing,
- flame-retardant agents,
- softening agents,
- antistatic agents,
- hydrophobic/oleophobic agents and
- biocides.

Pollutant	Source
Organic dyestuff	Non-fixed dye
Urea	Hydrotropic agent
Ammonia	In pigment printing pastes
Sulphates and sulphites	Reducing agents by-products
Polysaccharides	Thickeners
CMC derivates	Thickeners
Polyacrylates	Thickeners Binder in pigment printing
Glycerin and polyols	Anti-freeze additives in dye formulation Solubilising agents in printing pastes
m-nitrobenzene sulphonate and its corresponding amino derivative	In discharge printing of vat dyes as oxidising agent Direct printing with reactive dyes inhibits the chemical reduction of the dyes
Polyvinyl alcohol	Blanket adhesive
Multiple substituted aromatic amines	Reductive cleavage of azo dyestuff in discharge printing
Mineral oils/aliphatic hydrocarbons	Printing-paste thickeners (half-emulsion pigment printing pastes occasionally)

Table 5. Pollutants in wastewater after the printing process.

Cross-linking agents in easy-care finishing are mainly based on formaldehyde. They can be formaldehyde-rich, formaldehyde-poor or formaldehyde-very poor. Formaldehyde-rich cross-linking agents are: "self-crosslinking "agents such as hydroxymethyl urea, bis(methoxymethyl)urea, hydroxymethyl melamine, and bis(methoxymethyl)melamine. 'Reactant cross-linking' agents are formaldehyde-poor or have very poor cross-linking based on the derivatives of the bis(hydroxymethyl)-dihydroxyethylene urea. All these products may potentially produce emissions of free formaldehyde and methanol. Formaldehyde's presence in these finishing agents represents a potential risk, not only for wastewater and exhausted air, but also to workers in the textile mills, and the final users of the textile as well. Formaldehyde is also suspected of carcinogenity. These are the reasons for much effort being put into the production of free formaldehyde cross-linking agents. One cross-linking agent that is formaldehyde-free is also available on the market and is based on dimethyl urea and glyoxal [BASF, 2000]. Formaldehyde-free cross-linkers are considerably more expensive than formaldehyde cross-linkers, so this is the reason that formaldehyde-free cross-linkers have never been widely used in the textile industry.

The flame-retardant agents most commonly-used within the textile sector, and belong to the organic flame-retardant agents, are both halogenated organic and organo-phosphorus compounds.

As halogenated organic compounds only, brominated and chlorinated flame-retardant agents are used in practice. Brominated compounds are the most effective. Bromine can be bound aliphatically or aromatically, the aromatic derivatives are widely-used because of their high thermal stability. Chlorinated flame-retardant agents include chlorinated aliphatic and cycloaliphatic compounds. They are less expensive, less stable, and more corrosive to the equipment when compared to the brominated compounds. Polybrominated flame-retardants include the following compounds:

-   polybrominated diphenyl ethers (PBDE); pentabromodiphenyl ether (penta-BDE), octabromodiphenyl ether (octa-BDE), decabromodiphenyl ether (deca-BDE),
-   polybromo biphenyls (PBB); decabromobiphenyl and
-   tetrabromobisphenol A (TBBA).

A PDBE, which is of major use as a flame-retardant agent, is a deca-BDE. Deca-BDE and octa–BDE could break down into penta–BDE and tetra–BDE after release into the environment. Penta-BDE is a persistent substance liable to biocumulate.

Organo-phosphorus flame-retardant is represented by molecule phosphonic acid, (2-((hydroxylmethyl)carbamyl)ethyl)-dimethyl ester. Phosphorus containing flame-retardant agents is non readily biodegradable and water-soluble. According to one source, this product is not toxic or harmful to aquatic organisms and shows no potential to bioaccumulate, whilst another source concludes that knowledge about the toxicology of this compound is insufficient (BAT, 2003).

Softening agents are water-based emulsions or dispersions of water-insoluble active materials. There are four groups of softeners:

-   non-ionic surfactants (fatty acids, fatty esters, and fatty amides),
-   cationic surfactants (quaternary ammonium compounds, amido amines, imidazolines),
-   paraffin and polyethylene waxes and
-   organo-modified silicones.

Softening agents are molecules with high molecular weight but their volatility is low. The performance of each type of softener varies and each has advantages and disadvantages. Non-ionic surfactants are biodegradable, whilst paraffin and polyethylene waxes are non-biodegradable. Cationic surfactants have high aquatic toxicity. Paraffin waxes are still used although these types of softeners emit smoke when heated and producing air emissions from dryers.

Antistatic agents are applied as functional finishes to selected textile materials for use within static-sensitive environments. From the chemical point of view, antistatic agents are based on quaternary ammonium compounds and phosphoric acid ester derivatives.

Hydrophobic/Oleophobic agents fall under the following categories:

-   wax-based repellents (paraffin-metal salt formulations),
-   resin-based repellents (fatty modified melamine resins),
-   silocone repellents and
-   fluorochemical repellents (copolymers of fluoroalkyl acrylates and methacrylates).

Organic pollutants arise from silicone repellents, fluorochemical repellents and from resin-based repellents.

Silicone repellents contain polysiloxane-active substances (dimethylpolysiloxane and modified derivates), emulsifiers, hydrotropic agents (glycols), and water.

Fluorochemical repellents are copolymers of fluoroalkyl acrylates and methacrylates. Market formulations contain active agents together with emulsifiers (ethoxylated fatty

alcohols and acids, but also fatty amines and alkylphenols) and other by-products that are often solvents, such as:
- acetic acid esters (butyl/ethyl acetate),
- ketones (methylethyl keton and methylisobutyl ketone),
- diols (ethandiol, propandiol), and
- glycolethers.

Resin-based repellents are produced by condensing fatty compounds (acids, alcohols or amines) with methylolated melamines.

Biocides are used for antimicrobial finishes regarding hospital textile material or as odour suppressants for socks. For this purpose, the following active substancesare used:
- zinc organic compounds,
- tin organic compounds,
- dichlorophenyl(ester) compounds,
- benzimidazol derivatives,
- triclosane, and
- isothiazolinones (the most commonly used today).

Biocides are also applied in the carpet sector to impart wool fibre lifetime protection against a range of textile pests. These auxiliaries are usually known as mothproofing agents. As mothproofing agents are used, such as permethrin (synthetic pyrethoid), gyfluthrin (synthetic pyrethroid), and sulcofuron (halogenated diphenylurea derivative).

All biocides give rise to environmental concerns when they are discharged in wastewater, because of their toxicity to aquatic life.

4. Organic dyes

Dyes make the world more beautiful through coloured products, but cause a lot of problems. Organic dyes contain a majority of substituted aromatic and heteroaromatic groups. The colour results from the conjugated chains or rings can absorb different wavelengths of the visible spectrum. Organic dyes contain chromophoric and auxochromic groups. Chromophoric groups are responsible for colouring properties, and the salt-forming auxochromes for the dying properties. The chromophores are usually composed of double-bonds (carbon-carbon, nitrogen-nitrogen, carbon-nitrogen), aromatic and heteroaromatic rings (containing oxygen, nitrogen or sulphur).

4.1 Classification

Textiles are dyed using many different colorants, that may be classified in several ways. With regard to the methods and domains during usage, the dyes are classified into acid, reactive, direct, basic, disperse, metal complex, vat, mordant, and sulphur dyes. Reactive and direct dyes are commonly in use for cotton and viscose-rayon dyeing, whilst disperse dyes are used for dyeing polyester. Dyes can be classified with regard to: their chemical structures, the methods and domains of usage, and chromogen (a dye or pigment precursor, containing chromophores), as presented in Table 6.

Azo dyes contain one or more azo bonds, and can be used for dyeing natural fibres (cotton, silk, and wool) as well as synthetic fibres (polyesters, polyacrylic, rayon, etc.). Azo dyes are the most-widely used synthetic dyes and are present in 60-70 % of all textile dyestuffs produced (ETAD, 2003). They are mainly used for yellow, orange, and red colours. The biodegradation of more than 100 azo dyes were tested and only a few of them were degraded aerobically. The degree of stability of azo dyes under aerobic conditions depends on the structure of the molecule. C. I. Acid Orange 7 is one of the rare aerobically biodegradable azo dyes (Vandervivere, 1998). Under anaerobic conditions, azo dyes are cleaved into aromatic amines, which are not further metabolized under anaerobic conditions but readily biodegraded within an aerobic environment (Fig. 2).

	Classification	
	Subclass	Characteristic
With regard to chemical structure (C.I.)	E.g. azo, anthraquinone, triphenylmethane, indigo,…	The classification of a dye by chemical structure into a specific group is determined by the chromophore.
With regard to method and domain of usage (C.I.)	E. g. direct, acid, basic, reactive, reductive, sulphuric, chromic, metal complex, disperse, pigments,…	Dyes used in the same technological process of dyeing and with similar fastness are classified into the same group.
With regard to chromogen n→π*	E. g. absorptive, fluorescent and dyes with energy transfer,…	This classification is based on the type of excitation of electrons, which takes place during light adsorption.
With regard to the nature of donor – acceptor couple	E. g. 1-aminoanthraquinone, p-nitroaniline,…	These chromogenes contain a donor of electrons (non-bound electron couple), which directly bonds to the system of conjugated π electrons.
With regard to the nature of polyenes: a) Acyclic and cyclic	E. g. polyolefins, annulenes, carotenoids, rhodopsin,…	Polyene chromogen contains sp^2 (or sp) hybridised atoms. The molecules enclose single and double-bonds that form open chains, circles, or a combination of both.
b) Cyanine	E. g. cyanines, amino substituted di- and tri-arylmethane, oxonols, hydroxyarylmethanes,…	Cyanine chromogens have a system of conjugated π electrons, in which the number of electrons matches the number of p orbitals.

Table 6. Classification of dyes

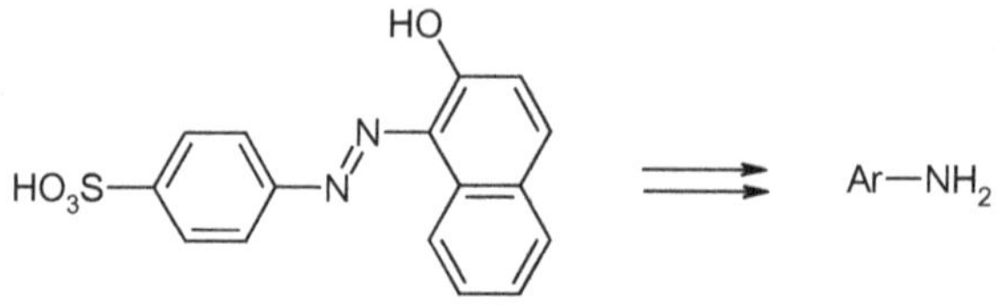

Fig. 2. Degradation of the azo dye C. I. Acid Orange 7 under anaerobic conditions

Carcinogenic amines that can be formed by the cleavage of certain azo dyes are: 4-aminodiphenyl, benzidine, 4-chloro-o-toluidine, 2-naphthylamine, o-aminoazotoluene, 2-amino-4-nitrotoluene, p-chloroaniline, 2,4-diaminoanisol, 4,4'-diaminodiphenylmethane, 3,3'-dichlorobenzidine, 3,3'-dimethoxybenzidine, 3,3'-dimethylbenzidine, 3,3'-dimethyl-4-4'-diaminodiphenylmethane, p-cresidine, 4,4'-methylene-bis-(2-chloraniline), 4,4'-oxydianiline, 4,4'-thiodianiline, o-toluidine, 2,4-diaminotoluene, 2,4,5-trimethylaniline, 4-aminobenzene, and o-anisidine (BAT, 2003).

More than 100 azo dyes with the potential to form carcinogenic amines are still available on the market (Euratex, 2000). The usage of these azo dyes that may cleave into one of the potentially carcinogenic aromatic amines from Table 7, is banned according to the 19th amendment of Directive 76/769/EWG on dangerous substances.

Fifty years ago, large amounts and numbers of azo colorants based on benzidine, 3,3-dichlorobenzidine, 3,3-dimethylbenzidine (o-toluidine), and 3,3-dimethoxybenzidine (o-dianisidine), have been synthesized, especially within the German chemical industry. 447 of the azo colorants from a list of 2000 in The Colour Index (1987), were based on 2-naphthylamine, benzidine or benzidine derivatives. The manufacturing of these kinds of azo dyes was stopped in 1971, with the exception of the dye Direct Black 4, and the manufacturing of this dye was continued until 1973 (Golka, 2004). The problem of carcinogenicity regarding azo dyes was first officially addressed by *the German Commission for Investigating Health Hazards of Chemical Compounds within the Work Area* ("MAK-Commission") (DFG, 1988). According to current EU regulations, azo dyes based on benzidine, 3,3-dimethoxybenzidine, and 3,3-dimethylbenzidine, are classified as carcinogens of category 2 as 'substances which should be regarded as if they are carcinogenic to man'. This is not the case for 3,3'-dichlorobenzidine-based azo pigments (BIA-Report, 2/2003).

The second more important class of textile dyes is anthraquinone dyes. They have a wide-range of colours over almost the whole visible spectrum, but they are more-commonly used for violet, blue, and green colours. Anthraquinone dyes are more resistant to biodegradation due to their fused aromatic structures, and thus remain coloured for a long time in wastewater.

Acid dyes are water-soluble compounds applied to wool, nylon, silk, and some modified acrylic textiles. Acid dyes have one or more sulphonic or carboxylic acid groups in their molecular structure. The dye-fibre affinity is the result of ionic bonds between the sulphonic acid part of the dye and the basic amino groups in wool, silk, and nylon fibres.

Reactive dyes are water soluble and are mainly used for dyeing cellulosic fibres such as cotton and rayon, and also for wool, silk, nylon, and leather. Reactive dyes form covalent chemical bonds with the fibre and become part of it. They are used extensively within textile industries in regard to their favourable characteristics of bright colour, water-fastness, and simple application techniques with low-energy consumption.

Metal-complex dyes show great affinity towards protein (wool) and polyamide fibres. Generally they are chromium or cobalt complexes (Zollinger, 2003). Chromium complex dyes are formed through chemical reactions between Cr_2O_3 and a variety of azo organic compounds (Zhao et al., 2005). Chromium occurs primarily in the trivalent state (III) and in the hexavalent state (VI). Hexavalent chromium is known to be toxic, can irritate the nose, throat, and lungs, provoke permanent eye damage, cause dermatitis and skin ulcers, and exhibit carcinogenic effects (Baral & Engelken, 2002). When reduced to chromium(III), it may be significantly less-harmful. As chromium compounds were used in dyes and paints and the tanning of leather, these compounds are often found in soil and groundwater at abandoned industrial sites, now needing environmental clean-ups and remediation regarding the treatment of brownfill land.

Disperse dyes have a very low water solubility, so they are applied as a dispersion of fine grounded powders in the dye-bath (Banat et al., 1996). Disperse dyes are used for oleophyllic fibres (polyester and other synthetics) that reject water-soluble dyes. Disperse dye-inks are used in the ink-jet textile printing of polyester fabrics. They have good fastness to light, perspiration, laundering, and dry cleaning. Some disperse dyes also have a tendency to bioaccumulate. Skin sensitisation risks are likely to be within acceptable limits.

Various intermediates used in the manufacture of disperse dyes are given below (Science tech entrepreneur, 2003): p-amino acetanilide, 1-amino-4-bromo-2-anthraquinone-sulphonic acid, p-amino phenol, 4-amino xanthopurpurin, aniline, anilino methane sulphonic acid, 1-benzamido-4-chloro-anthraquinone, 3-bromo benzanthrone, 1-bromo-4-methyl amino anthraquinone, 1-chloro-2, 4-dinitrobenzene, 2-chloro-4-nitroaniline, 4-chloro-3-nitrobenzene sulfonyl chloride, 2, 2-(m-chlorophenylamino)-diethanol, cresidune, o-cresol, p-cresol, 1,4-diamino, anthraquinone, 1,5-diamino anthraquinone, 2,6-dichloro-4-nitroanioline, 1,5-dihydroxy-4, 8-dinitro anthraquinone, N, N-dimethyl aniline, 1,5-dinitro anthraquinone, 1,8-dinitro anthraquinone, diphenylamine, ethanolamine, 2-(N-ethykamino)ethanol, 1-hydroxy anthraquinone, 1-hydroxy-4-nitro anthraquinone, leuco-1, 4, 5, 8-tetrahydroxy anthraquinone, methylamine, 1-methoxy anthraquinone, 3-methyl-1-phenyl-5-pyrazolone, 1-naphthylamine, p-nitroaniline, phenol, p-phenyl azoaniline, quinizarin, p-toluene sulfoanamide, 2, 2-(m-tolyamino) diethanol.

Basic dyes have high brilliance and intensity of colour, and are highly visible even in a very low concentration (Chu & Chen, 2002). Basic dyes are the most acutely toxic dyes for fish, especially those with tri-aryl-methane structures. Occasionally, some people develop a contact allergy to dyes in clothing. This is ample evidence that there is reason to be cautious in the use of some basic dyes. The absence of a known hazard does not prove that the substance is non-carcinogenic or mutagenic.

5. Possible sources of persistent and hazardous organic pollutants in textile chain

Persistent organic pollutants are a wide group of compounds with specific properties: persistency, high bioaccumulative coefficient, ability for transmission to large distances. POPs emission inventory has a great significance as a starting-point for pollutant flows' modelling, impact and risk assessment. The most toxic among POPs are polychlorinated dibenzo-p-dioxins and furans (PCDD/Fs), often simply termed as dioxins. There are 75 isomers of dibenzo-p-dioxins (PCDDs) and 135 isomers of polychlorinated dibenzofurans differentiated from each other by the number and location of the chlorine atom addition. Chemical and biological properties (including the toxicological) depend on the positions of the chlorine atoms. In this group, isomers with chlorine atoms at positions 2,3,7,8 are especially toxic. The so-called 'dirty group' comprises 17 isomers of PCDD/Fs, among which 2,3,7,8-tetrachlorodibenzodioxin (2,3,7,8-TCDD) is the most toxic (Fig. 3). There are also brominated dioxins (PBDD/Fs), fluoro dioxins and mixed dioxins PXDD/Fs (X = Cl, Br, F). PBDD/Fs are contaminants with properties similar to PCDD/Fs, together with their persistence and toxicity. Fluorinated congeners of dioxins are taken to be less dangerous to humans and the environment, due to their short life and low toxicity.

$X + Y = 1\text{-}8$ (75 congeners) $X + Y = 1\text{-}8$ (135 congeners)

Polychlorinated dibenzo-p-dioxins   Polychlorinated dibenzofurans

Fig. 3. Molecular structures of polychlorinated dibenzo-p-dioxins and dibenzofurans

Possible sources of dioxins among the textile chains will be presented in the following chapters.

5.1 Natural fibre

As already-mentioned in chapter 2.2.1, pentachlorophenol and other organochlorine pesticides are used during the production and transportation of natural textile fibres. The production and use of PCP is a significant source of PCDD/Fs. During the synthesis of pentachlorophenol from chlorinated phenols, PCDD/Fs are formed as by-products. Furthermore, PCDD/Fs are formed from chlorinated phenols during natural UV radiation conditions.

5.2 Textile dyes

The main source of dioxins in the textile industry are dioxazine and antraquinone dyes and pigments. These classes of dyes are produced from chloranil. PCDD/Fs are formed during the synthesis of chloranil from chlorinated phenols. Dioxins are also found in other classes of dyes. Considerable levels of PCDD/Fs have also been determined in some phtalocyanine dyes, and in printing inks (Križanec, 2006). Recently some dispersive aromatic azo dyes were found as sources of dioxins. In dual-black disperse dyes (mixtures of antraquinone and azo substances); concentrations of 50 and 170 ng PCDD/F TEQ/kg have been determined. Octachlorodibenzodioxin (OCDD) was the dominant compound (Križanec, 2005). Disperse azo dyes produced from chlorinated anilines or chlorinated nitro anilines contain dioxins as by-products during synthesis. Dioxins are formed after the dediazonation of aromatic diazonium salts via chlorinated phenols and chlorinated nitrophenols. The hydrolysis of diazonium salts leads to phenols (Križanec, 2007). Chlorinated phenols and chlorinated nitrophenols are present in dyes' formulations as impurities, and dioxins may form further during dyeing processes (Fig.4) (Križanec, 2007).

5.3 Halogenated organic compounds

Halogenated organic compounds, especially aromatic halogenated compounds, are also precursor compounds for the synthesis of dioxins. Halogenated organic compounds and metals such as copper are involved in the thermal formation processes of dioxins. During the combustion processes of waste textile materials, the formation of dioxins is possible via the 'The Precursor Concept' or 'De novo Mechanism' (Križanec, 2006).

All the halogenated flame-retardant agents presented in chapter 3.2.4 are involved in the formation of dioxins and furans, when submitted to high temperature treatments. Dioxins and furans can be formed in small amounts during the syntheses of these compounds, and as a side-reaction when they are subject to combustion/burning for disposal.

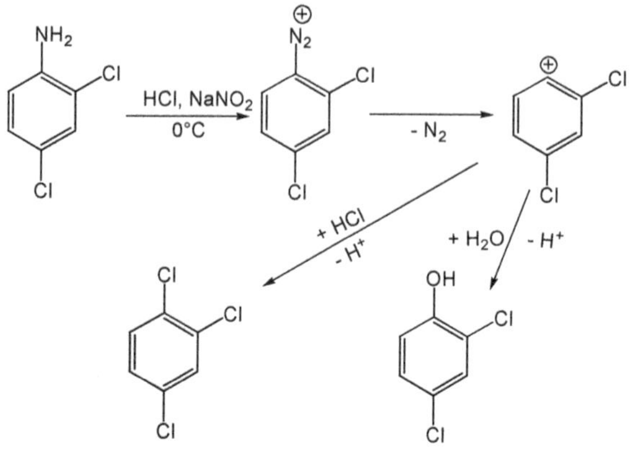

Fig. 4. Heterolytic pathway during the dediazonation of chlorinated aromatic diazonium salts, and the formation of chlorinated benzenes and chlorinated phenols

5.4 Formation of dioxins during finishing processes

The generation of persistent organic pollutants in the textile industry during finishing processes is caused by (Križanec, 2006):

- high temperatures (>150 °C);
- alkaline conditions;
- UV radiations or other radical starters;
- the presence of chlorinated organic compounds;
- the presence of metal catalysts.

The distribution of dioxins and their fate during textile dyeing processes was investigated by Križanec (Križanec, 2005). Two dyeing experiments were conducted at laboratory scale using disperse dye contaminated with PCDD/Fs. After the dyeing and finishing process, the PCDD/Fs, and especially OCDD, increased compared to the input. The authors concluded that PCDD/Fs are formed during the textile process, most probably from precursors present in the dyestuffs (Križanec, 2005).

5.5 Incineration of textile materials

Incineration processes are a known source of dioxins. Several textile products are also potential sources for the formation of dioxins during waste incineration. As described in previous sections, dioxins, and especially dioxin precursor compounds, may be present in textile products. According to the dioxin mass-balance present in a Spanish municipal waste incinerator, the incineration of textiles results in the highest dioxin emission levels (Abad et al., 2000).

6. Textile wastewater

6.1 Characteristics of wastewater after finishing processes

The diversity of textile finishing processes results in variable compositions of textile wastewater. In general, the wastewater is loaded with complex mixtures of organic and inorganic chemicals (Volmajer Valh et al., 2011). Moreover, textile finishing processes are, despite the gradual reduction of fresh water consumption by introducing, for example, jet dyeing machines and short liquor-dyeing systems, still very water intensive, leading to large volumes of produced wastewater that have to be treated before being discharged. For illustration purposes, 70 L of freshly-softened water is used per kg of cotton material during dark-colour reactive dyeing processes (data from Slovene textile company). With further optimization, it is possible to reach values between 25 to 40 L of fresh water per kilogram of cotton for a dark reactive dyeing, in contrast to the 100 to 150 L water consumption for this type of dyeing, as used in the mid-nineteen seventies.

When considering both the volume generated and the effluent composition, textile industry wastewater is considered to be the most polluted of all the industrial sectors.

The pollutants of major concern are recalcitrant or hazardous organics, such as dyes or some surfactants, metals, and salts. During the textile dyeing process, dyes are always used in combination with other chemicals such as acids, alkali, salts, fixing agents, carriers, dispersing agents, and surfactants that are partly or almost completely discharged in the wastewater. Dye-fixation rates vary considerably among the different classes of dyes and may be especially low for reactive dyes (in the case of cotton 20-50 % residual dyestuff) and for sulphur dyes (30-40 % residual). Moreover, large variations are round even within a given class of colorants. This is particularly significant in the case of reactive dyes.

As already-mentioned, from among all the dye classes, azo dyes are the group of colorants most used. Whilst most azo dyes themselves are non-toxic, a significantly larger portion of their metabolites are (Isik & Sponza, 2007; Van der Zee & Villaverde 2005). Brown and DeVito (Brown & DeVito, 1993) postulated that azo dyes may be toxic only after the reduction and cleavage of the azo linkage, producing aromatic amines. Azo dyes with structures containing free aromatic amine groups that can be metabolically oxidized without azo reduction, may cause toxicity. Azo dye toxic activation may occur following direct oxidation of the azo linkage, producing highly reactive electrophilic diazonium salts.

Most dyes shown to be carcinogenic are no longer used; however, a complete investigation of all dyestuffs available on the market is impossible.

Substituted benzene and naphthalene rings are common constituents of azo dyes, and have been identified as potentially carcinogenic agents (IARC, 1982).

Other concerns are the impurities within commercial dye products and the additives used during the dyeing process. Understanding the dye structures and how they are degraded is crucial to understanding how toxic by-products are created. The colour of wastewater is one of the major problems facing industries involved in dyeing processes. Wastewaters from dye-houses often carry high concentrations of excess, unfixed dye.

A study conducted on 45 combined effluents from textile finishing plants showed that 27 percent of the wastewater samples were mutagenic during the Ames test (McCarthy, 1997). The potential for toxic effects to humans, resulting from exposure to dyes and dye metabolites, is not a new concern.

6.2 Wastewater treatment processes

Textile wastewater is the main source of organic contamination regarding pollution within the textile industry. Several cleaning processes may be used to remove organic pollutants from textile wastewater. In general, we distinguish between physical methods (adsorption, filtration methods, coagulation and flocculation processes), chemical methods (oxidation, advanced oxidation, Fenton´s reagent) and, more recently, more and more attractive biological treatment (anaerobic, aerobic) as an effective option for relatively inexpensive effluent decolouration. Non-destructive physical techniques just transfer the pollutants to other mediums (sludge, concentrate in filtration techniques), and cause secondary pollution. From this point of view, the chemical destruction of pollutants is more desirable, but could have some drawbacks such as the formation of aromatic amines, when the degradation/mineralisation is incomplete. An overview of the treatment methods for textile wastewater treatment, as well as their advantages and disadvantages, are gathered in 'Decolouration of textile wastewaters' (Volmajer Valh & Majcen Le Marechal, 2009). The following paragraph focuses on chemical methods, especially on advanced oxidation processes (AOPs).

According to their definition, AOPs combine ozone (O_3), ultraviolet (UV) irradiation, hydrogen peroxide H_2O_2 and/or a catalyst in order to offer a powerful water treatment solution for the reduction (removal) of residual organic compounds, as measured by COD, BOD or TOC, without producing additional hazardous by-products or sludge, which requires further handling (Arslan-Alaton, 2004). All AOPs are designed to produce hydroxyl radicals that act with high-efficiency to destroy organic compounds. The most widely-applied advanced oxidation processes (AOP) are: H_2O_2/UV, O_3/UV, H_2O_2/O_3, H_2O_2/O_3/UV (Kurbus et al. 2003), and ultrasound (US) (Vajnhandl & Majcen Le Marechal, 2007) have several advantages such as: rapid reaction rates, small foot-print, reduction of toxicity, and complete mineralization of treated organics, no concentration of waste for further treatment (as membranes), no production of materials that require further treatment, such as 'spent carbon' from activated carbon absorption, no creation of sludge as with physical-chemical processes or biological processes (wasted biological sludge), and a non-selective pathway allows for the treatment of multiple organics, at once. On the contrary, these processes are capital-intensive and, in the case of complex chemistry, must be tailored to specific applications (Slokar & Majcen Le Marechal, 1998).

AOPs are marked as treatment methods for the effective removal of organic pollutants in terms of total organic carbon (TOC), COD, and BOD reduction, but less information is available regarding the chemical structures of formed degradation products and their toxicity aspect. Moreover, with modern advanced oxidation processes, caution is necessary when dioxins and other halogenated persistent organic-pollutants could be present. During UV or US irradiation, for example, high chlorinated dioxins and other halogenated pollutants can be de-chlorinated. Lower chlorinated dioxins are known as more toxic compounds compared to high chlorinated pollutants. The very efficient processes for removing persistent organic pollutants from wastewater are those coagulation and adsorption processes used in municipal-waste plants (Križanec, 2007).

6.3 Toxicity of organic pollutants within textile wastewater

The toxicity of textile wastewater varies depending on the different processes applied within the textile industry. The wastewaters of some processes have high aquatic toxicity, whilst others show little or no toxicity. An identification of all toxic compounds used within the textile industry is impossible due to the huge variety of used chemicals and a lack of data about their toxicities. Usually overall toxicity is determined by toxicity-testing the whole effluent stream of aquatic organisms, which is a cost-effective method.

The sources of organic pollutants that can cause aquatic toxicity can be dyes, surfactants, toxic organic chemicals, or biocides. Examples of compounds in each of these classes and their sources, are shown in Table 7.

Textile wastewaters contain different polar and non-polar compounds, but the polar ones are predominant. Polar organic pollutants are non-biodegradable and their elimination is often incomplete.

Agent	Chemical example	Source
Surfactants	Ethoxylated phenols	Multiple processes
Organics	Chlorinated solvents	Scour, machine cleaning
Biocides	Pentachlorophenol	Wool fibres contaminant

Table 7. Typical causes of aquatic toxicity

Non-polar organic pollutants such as dioxins, may be present in textile wastewater in trace amounts. Due to their high toxicity and persistent organic pollutant properties, the restricted limits on wastewater are relatively low. According to EU regulations there are no restrictive limits for dioxins in textile wastewater. For wastewater from incineration plants the restrictive limit is 0.3 ng TEQ-ITF/L. According to our study the concentrations of dioxins in textile wastewater with disperse dyes are considerable (Križanec, 2007). The concentration of PCDD/Fs found in the wastewater sample polluted with disperse dyes was 0.44 ng TEQ-ITF/L. The dominant PCDD/F congener in the disperse dyes' wastewater sample was 2,3,4,6,7,8-HxCDF, the contribution of which to the TEQ was more than 85 % (Fig. 5). The concentration of PCDD/F in this selected wastewater polluted with disperse dyes exceeded the limit of 0.3 ng TEQ-ITF/L, as determined by the European regulation for wastewaters from incineration plants.

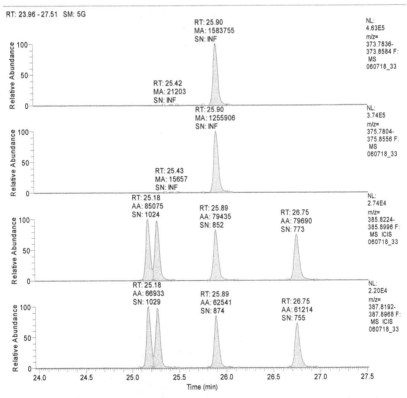

Fig. 5. Dioxin homologue profile found in textile wastewater. There is evident that 2,3,4,6,7,8-HxCDF congener is dominant

7. Tracking persistent organic pollutants in textile finishing and AOP processes

Information about the emissions of persistent organic pollutants from the textile industry is limited due to a lack of complete and accurate emission data. Indirect measurements are necessary regarding the reduction of persistent organic pollution from the textile industry. The presence and concentrations of pollutants should be controlled throughout the textile industry regarding incoming materials and also for textile products and wastewater.

Several instrumental and analytical approaches are in use for determining persistent organic pollutants. For the determination of semi-volatile organic pollutants, the most appropriate are gas chromatographic methods connected with appropriate detectors. For tracking low-level halogenated organic pollutants, the use of gas chromatography with an electron capture detector (ECD), is a useful and relatively cost-effective method.

In order to determine the trace-levels of organic pollutants (dioxins), gas chromatography with mass spectrometry or high-resolution mass spectrometry is required (UEPA method 1613, 1994). Liquid chromatography with mass spectrometry is used to determine water solubility and highly-volatile organic pollutants(Dolman & Pelzing, 2011).

In our pilot studies, two samples of textile wastewater (a mix of disperse dyes' wastewater and mix of metal-complex dyes' wastewater) were analyzed for persistent organo-halogen compounds (POPs). Polychlorinated dibenzo-p-dioxins (PCDDs), polychlorinated dibenzofurans PCDFs), dioxin-like PCBs, and polybrominated diphenyl ethers (PBDEs), were determined using the HRGC/HRMS technique. Further decolouration experiments of these wastewaters were performed using the advanced oxidation process (AOP). The POPs' analyses were performed on the samples after the AOP process, and the results were then compared with the initial values.

Wastewater samples were collected from a local textile-dyeing facility, the main activity of which is the dyeing (wet processes) of polyester yarns with disperse dyes. Samples of wastewater were collected at the out-flow of the facility, and stored in pre-cleaned 2.5 L borosilicate amber bottles at 5 °C until analysis or experiment. The disperse dyes' and metal complex dyes' wastewaters were a mix of at least four different dispersals, and at least four different metal complex dyes, respectively. The exact composition of the wastewater was unknown. Wastewater samples (0.5 L) were transferred in separate funnels and spiked with an internal standard mixture containing ^{13}C-labelled isomers of analytes. Hexane (100 mL) was used as the liquid-liquid extraction solvent. The sample extract clean-up was performed according to the USEPA method 1613 (USEPA method 1613, 1994). By using adsorption chromatography on a graphitized carbon column (Carbopack C), the PCBs and PBDEs (60 mL dichloromethane-direct flow),were separated from the PCDD/Fs (60 mL of toluene-opposite flow). The obtained fractions were concentrated to a final volume of 20 µL, and analyzed with GC/HRMS.

Extracts were analyzed on a HP 6890 gas chromatograph GC (Hewlett-Packard, Palo Alto, CA, USA) coupled to a Finnigan MAT 95 XP high resolution mass spectrometer. Aliquot (1.5 µL) of PCDD/F toluene sample extracts were injected into the GC system, equipped with a Restek Rtx Dioxin2 capillary column (60 m x 0.25 mm i.d., film thickness 0.25 µm) in splitless mode. The mass spectrometer operated within the electron impact ionization mode using selected ion-monitoring (SIM), at a minimum resolution of 10,000. The samples were analyzed for PCDD/Fs concentrations, using the isotope dilution method based on USEPA 1613 protocol (USEPA method 1613, 1994). A similar determination/quantification approach was also used for the determining of PCBs and PBDEs in the dichloromethane extracts. In each set of experiments the blank samples were analyzed, and in cases of some PCBs and PBDEs congeners, the blank values were subtracted from the sample values.

H_2O_2/UV experiments were carried out in a batch photo-reactor using a low-pressure mercury UV lamp emitting at 254 nm. The high-frequency plate type URS 1000 L-3 communications system ELAC Nautik was used for ultrasonic irradiation. A system consisting of an AG 1006 LF generator/amplifier (200 W maximum output) and USW 51 ultrasonic transducer (817 kHz) with a working volume of 500 mL. The experimental parameters for the power of the UV lamp was 1600 W, and the reaction time was 60 minutes. Ultrasonic irradiation was performed at 820 kHz.

The concentrations of PCDD/Fs in the wastewater samples at different stages of the experiment, are presented in Table 8. The concentrations of PCDD/Fs in the metal complex dyes wastewater samples were low, and close to both the quantification limits and the concentrations of the background levels. As expected, the concentrations of PCDD/Fs in the samples of disperse dyes' wastewater, were relatively high (Fig. 5).

After AOP treatment the concentrations of PCDD/Fs were similar to those of the original sample, for both types of wastewater. This indicates that PCDD/Fs are stable enough in these types of wastewater to resist those conditions used during our AOP experiments. Also, no new PCDD/F congener was observed.

The concentrations of dioxin-like PCBs in the wastewater samples at different stages of the experiment, are presented in Table 9. The concentrations of dioxine-like PCBs and also PBDEs after AOPs experiments are incomparable with the PCDD/F concentrations of the original wastewater samples. This is because PCBs and PBDEs are not as resistant as PCDD/Fs when some transformations were observed

The concentrations in the samples of H_2O_2/UV-treated wastewater were, for most of the dioxine-like PCBs and the selected PBDEs, higher in comparison with the original samples. We suggest that this is due to the dechlorination or debromination of the higher halogenated congeners and/or the cleavages of the polychlorinated terphenyls. After the sonification AOP method, the concentrations of dioxine-like PCBs and selected PBDEs were relative low. These results suggest that local rigorous conditions (high-pressure, high- temperature) cause the destruction of PCBs and PBDEs via a thermal and/or radical mechanism.

The HGC/HRMS analytical method was used for tracking POP during this pilot-study. This analytical method is expensive and time-consuming but gives us very important information regarding POP's congener distribution at trace levels. In addition, other analytical methods should be tested on samples from the textile industry. When tracking POPs in the textile industry, appropriate screening methods (for example GC/ECD) should be tested and used during routine controls, in order to prevent POPs' pollution from this sector.

With the aim of reducing organic pollution within the textile industry, all the important steps towards sustainable thinking and the introduction of green chemistry are necessary at all stages of the textile-chain, in-line with the recommendations of industrial platforms (WSSTP, Textile Platform). The development of new wet-processing equipment is recommended in order to minimize the amount of fresh water and, consequently, to reduce the volume of generated wastewater and production costs. New wastewater treatment approaches should be implemented through pollution prevention in order to prevent environmental problems, such as waste minimization and reuse of treated water during the production processes. Finally, regulation of pollution by developing strategies for characterization and monitoring (IPPC) the most dangerous pollutants, should certainly also be expanded to cover the textile sector.

Congener/Group	Concentration (ng/L)					
	1	2	3	4	5	6
TCDD	< 0.01	0.04	< 0.01	< 0.01	0.01	< 0.01
PeCDD	0.02	0.03	0.03	< 0.01	< 0.01	< 0.01
HxCDD	0.08	0.19	0.10	0.01	0.03	0.01
HpCDD	0.20	< 0.01	0.23	0.01	0.05	0.02
OCDD	0.25	0.51	0.23	0.06	0.05	0.02
TCDF	0.15	1.02	0.50	< 0.01	0.06	< 0.01
PeCDF	0.22	1.54	0.24	< 0.01	0.03	< 0.01
HxCDF	3.77	3.05	6.44	< 0.01	0.08	0.30
HpCDF	0.12	0.10	0.09	< 0.01	0.01	< 0.01
OCDF	0.05	0.03	0.02	< 0.01	0.01	< 0.01
2.3.7.8-TCDD	< 0.01	0.01	< 0.01	< 0.01	< 0.01	< 0.01
1.2.3.7.8-PeCDD	< 0.01	0.01	0.01	< 0.01	< 0.01	< 0.01
1.2.3.4.7.8-HxCDD	< 0.01	0.12	< 0.01	< 0.01	< 0.01	< 0.01
1.2.3.6.7.8-HxCDD	0.01	0.06	0.01	< 0.01	< 0.01	< 0.01
1.2.3.7.8.9-HxCDD	0.01	0.03	0.01	< 0.01	< 0.01	< 0.01
1.2.3.4.6.7.8-HpCDD	0.14	0.58	0.17	0.01	0.02	0.01
1.2.3.4.6.7.8.9-OCDD	0.25	0.51	0.23	0.06	0.05	0.02
2.3.7.8-TCDF	< 0.01	0.07	< 0.01	< 0.01	0.01	< 0.01
1.2.3.7.8-PeCDF	0.01	0.04	< 0.01	< 0.01	0.01	< 0.01
2.3.4.7.8-PeCDF	0.01	0.04	0.01	< 0.01	0.01	< 0.01
1.2.3.4.7.8-HxCDF	0.01	0.02	< 0.01	< 0.01	0.01	< 0.01
1.2.3.6.7.8-HxCDF	0.04	0.01	< 0.01	< 0.01	0.01	< 0.01
2.3.4.6.7.8-HxCDF	4.23	4.00	6.73	0.01	0.07	0.30
1.2.3.7.8.9-HxCDF	< 0.01	< 0.01	< 0.01	< 0.01	< 0.01	< 0.01
1.2.3.4.6.7.8.-HpCDF	0.09	0.08	0.07	< 0.01	0.01	< 0.01
1.2.3.4.7.8.9-HpCDF	< 0.01	0.01	0.01	< 0.01	< 0.01	< 0.01
1.2.3.4.6.7.8.9-OCDF	0.05	0.03	0.02	< 0.01	0.01	< 0.01
Sum TEQ-ITF	0.44 ± 0.09	0.47 ± 0.09	0.69 ± 0.14	< 0.01	0.02 ± 0.01	0.03 ± 0.01

Legend: 1- Wastewater with disperse dyes. 2- Wastewater with disperse dyes treated with H_2O_2/UV. 3- Wastewater with disperse dyes treated with US. 4- Wastewater with metal complex dyes. 5- Wastewater with metal complex dyes treated with H_2O_2/UV. 6- Wastewater with metal complex dyes treated with US

Table 8. Mass-balance of PCDD/Fs in wastewater polluted with disperse and metal complex dyes treated with H_2O_2/UV and US processes

	Concentration (ng/L)					
PCB congener	1	2	3	4	5	6
PCB 81	< 0.05	< 0.05	< 0.05	< 0.05	< 0.05	< 0.05
PCB 77	0.35	< 0.05	0.37	0.15	0.41	< 0.05
PCB 126	< 0.05	< 0.05	< 0.05	< 0.05	< 0.05	< 0.05
PCB 169	< 0.05	< 0.05	< 0.05	< 0.05	< 0.05	< 0.05
PCB 105	0.36	< 0.05	2.81	0.38	< 0.05	< 0.05
PCB 114	0.05	< 0.05	0.17	< 0.05	< 0.05	< 0.05
PCB 118	0.88	< 0.05	5.52	1.10	< 0.05	< 0.05
PCB 123	0.11	< 0.05	0.45	0.08	< 0.05	< 0.05
PCB 156	0.08	< 0.05	1.20	0.16	< 0.05	< 0.05
PCB 157	< 0.05	< 0.05	0.29	< 0.05	0.28	< 0.05
PCB 167	< 0.05	< 0.05	0.46	0.07	< 0.05	< 0.05
PCB 189	< 0.05	< 0.05	< 0.05	< 0.05	< 0.05	< 0.05

Legend: 1- Wastewater with disperse dyes. 2- Wastewater with disperse dyes treated with H_2O_2/UV. 3- Wastewater with disperse dyes treated with US. 4- Wastewater with metal complex dyes. 5- Wastewater with metal complex dyes treated with H_2O_2/UV. 6- Wastewater with metal complex dyes treated with US

Table 9. Mass-balance of PCB congener in wastewater polluted with disperse and metal complex dyes treated with H_2O_2/UV and US processes.

8. References

Abad. E.; Adrados M. A.; Caixach J.; Fabrellas B. & Rivera J. (2000). Dioxin mass balance in a municipal waste incinerator. *Chemosphere*. Vol. 40. pp. 1143-1147. ISSN 0045-6535

Arslan-Alaton; I. (2004); Advanced oxidation of textile industry dyes. In: Advanced Oxidation Processes for Water and Wastewater Treatment. Simon Pearsons. 302-323. IWA Publishing. ISBN 1-84339-017-5. Tunbridge Wells.

Banat. I. M.; Nigam. P.; Singh. D. & Marchant. R. (1996). Microbial decolourization of textile-dye containing effluents: a review. *Bioresour Technol.* Vol. 58. pp. 217-227. ISSN 0960-8524

Baral. A. & Engelken. R.D. (2002). Chromium-based regulations and greening in metal finishing industries in the USA. *Environmental Science & Policy.* Vol. 5. pp. 121-133. ISSN 1462-9011

BASF. (2000) "Technical Information about BASF Products for Resin Finishing". TI/T 344.

BAT (2003) Integrated Pollution Prevention and Control (IPPC) Reference Document on Best Available Techniques for the Textiles Industry

BIA-Report (2003) 2/2003. Grenzwerteliste 2003. HVBG. Hauptverband der gewerblichen Berufsgenossenschaften. Sankt Augustin.

Brookes. G. & Barfoot P. (2008). Global impact of biotec crops: socio-economic and environmental effects 1996-2007. *AgBioForum.* Vol.11. pp. 21-38. ISSN 1522-936X

Brown. M.A.; DeVito. S.C. (1993). Predicting azo dye toxicity. *Critical Reviews in Environmental Science and Technology.* Vol. 23. No. 3. pp. 249-324. ISSN 1064-3389.

Chemical Inspection & Regulation Service. (www.cirs-reach.com/textile/)

Chu. H. C. & Chen. K. M. (2002). Reuse of activated sludge biomass: I. Removal of basic dyes from wastewater by biomass. *Process Biochem.* Vol. 37. pp.595-600. ISSN 0032-9592

DFG (Deutsche Forschungsgemeinschaft). (1988) List of MAK and BAT Values. VCH Publishers. Weinheim.

Dolman. S. & Pelzing. M. (2011) An optimized method for the determination of perfluorooctanoic acid perfluorooctane sulfonate and other perfluoro-chemicals in different matrices using liquid chromatography/ion-trap mass spectrometry. *Journal of Chromatography B.* Vol. 879. pp. 2043-2050. ISSN 0378-4347.

ETAD (2003). ETAD information on the 19th amendment of the restriction on the marking and use of certain azocolourants. ETAD-Ecology and Toxicology Association of Dyes and Organic Pigments Manufacturers.

EURATEX. E.-D. (2000) "Textile Industry BREF document (Chapter 2-3-4-5-6)"

Golka. K.; Kopps. S. & Myslak. Z. W. (2004). Carcinogenicity of azo colorants: influence of solubility and bioavailability. *Toxicology Letters.* Vol. 151. pp.203-210. ISSN 0378-4274

Gupta. G. S.; Prasad G. & Singh V. H. (1990). Removal of chrome dye from aqueous solutions by mixed adsorbents: fly ash and coal. *Water Research.* Vol. 24. pp. 45-50. ISSN 0043-1354

Humans. "Some Industrial Chemicals and Dyestuffs." Vol. 29. Lyon. France.

IARC (1982). World Health Organization International Agency for Research on Cancer. Monographs on the Evaluation of the Carcinogenic Risk of Chemicals to

Isik. M.; Sponza. T.S. (2007). Fate and toxicity of azo dye metabolites under batch long-term anaerobic incubations. *Enzyme and Microbial Technology.* 40. pp. 934-939. ISSN 0141-0229.

Križanec. B. (2007) *Obstojne organske halogenirane spojine v tekstilni industriji.* PhD Thesis.

Križanec. B.; Majcen Le Marechal. A. (2006). Dioxins and Dioxin-like Persistent Organic Pollutants in Textiles and Chemicals in the Textile Sector. *Croatica Chemica Acta.* Vol. 79. No. 2. pp. 177-186. ISSN 0011-1643

Križanec. B.; Majcen Le Marechal. A.; Vončina. E. & Brodnjak Vončina. D. (2005). Presence of dioxins in textile dyes and their fate during the dyeing processes. *Acta Chimica Slovenica.* Vol.52. (January 2005). pp. 111-118. ISSN 1318-0207

Kurbus. T.; Majcen Le Marechal. A. & Brodnjak Vončina. D. (2003). Comparison of H_2O_2/UV. H_2O_2/O_3 and H_2O_2/Fe^{2+} processes for the decolorisation of vinylsulphone reactive dyes. *Dyes and Pigments.* Vol. 58. pp. 245-252. ISSN 0143-7208.

Lacasse, K. & Baumann, W. (2004). *Textile Chemicals Environmental Data and Facts,* Springer, ISBN 3-540-40815-0, Berlin.

Mattioli. D.; Malpei, F.; Borone, G.; Rozzi, A. (2002). Water minimisation and reuse in the textile industry. In: Water Recycling and Resource Recovery in Industry. Piet Lens, Look Hulshoff Pol, Peter Wilderer and Takashi Asano. IWA Publishing. ISBN 1-84339-005-1. Cornwall.

McCarthy. B.J. (1997). Biotechnology and Coloration. *Coloration Technology.* Vol. 27. No.1. pp. 26-31. ISSN 1472-3581.

Naranjo. E. S. (2009). Impact of Bt crops on non-target invertebrates and insecticide use patterns. CAB Reviews: *Perspectives in Agriculture. Veterinary Science. Nutrition and Natural Resources.* Vol.4. No.11. (January 2009). pp. 1-23. ISSN 1749-8848

Science tech entrepreneur (May 2003) http://www.techno-preneur.net/information-desk/sciencetech-magazine/2006/may06/Disperse_dyes.pdf

Shaw. T. (1989). Environmental issues in wool processing. International Wool Secretariat (IWS) Development center monograph. IWS. West Yorkshire. England

Slokar. Y. M. & Majcen Le Marechal. A. (1998). Methods of decoloration of textile wastewater. *Dyes and Pigments.* Vol. 37. No. 4. pp.335-356. ISSN 0143-7208.

UK. (2001) "Comments made by UK to the First Draft of the BREF Textiles"

US EPA. (1996) "Manual – Best Management Practices for Pollution Prevention in the Textile Industry"

USEPA method 1613 (1994). Tetra-trough Octa-Chlorinated Dioxins and Furans by isotopic Dilution HRGC-HRMS. USEPA. Washington.

Vajnhandl. S. & Majcen Le Marechal. A. (2007). Case study of the sonochemical decolouration of thextile azo dye Reactive Black 5. *Journal of hazardous materials.* Vol. 141. pp. 329-335. ISSN 0304-3894.

Van der Zee. F.; Villaverde. S. (2005). Combined anaerobic-aerobic treatment of azo dyes-Ashort review of bioreactor studies. *Water Research.* 39. pp.1425-1440. ISSN 0043-1354.

Vandervivere. P. C.; Bianchi. R. & Verstraete. W. (1998). Treatment and reuse of wastewater from the textile wet-processing industry: Review of emerging technologies. *J. Chem. Technol. Biotechnol.* Vol. 72. pp.289-302. ISSN 0268-2575

Volmajer Valh. J.; Majcen Le Marechal A. (2009). Decolouration of Textile Wastewaters. In: *Dyes and Pigments New Research.* Arnold R. Lang. 175-199. Nova Science Publishers. Inc.. ISBN 978-1-60692-027-5. New York.

Volmajer Valh. J.; Majcen Le Marechal. A.; Vajnhandl. S.; Jerič. T.; Šimon E. (2011) Water in the Textile Industry. In: Peter Wilderer (ed.) *Treatise on Water Science*. vol. 1. pp. 685–706 Oxford: Academic Press.

Zollinger. H. (2003). *Color chemistry syntheses, properties, and applications of organic dyes and pigments* (3th edition). Wiley-VCH. ISBN 3-906390-23-3. Zürich

2

Textile Organic Dyes – Characteristics, Polluting Effects and Separation/Elimination Procedures from Industrial Effluents – A Critical Overview

Zaharia Carmen and Suteu Daniela
'Gheorghe Asachi' Technical University of Iasi,
Faculty of Chemical Engineering and Environmental Protection,
Romania

1. Introduction

The residual dyes from different sources (e.g., textile industries, paper and pulp industries, dye and dye intermediates industries, pharmaceutical industries, tannery, and Kraft bleaching industries, etc.) are considered a wide variety of organic pollutants introduced into the natural water resources or wastewater treatment systems.

One of the main sources with severe pollution problems worldwide is the textile industry and its dye-containing wastewaters (i.e. 10,000 different textile dyes with an estimated annual production of $7 \cdot 10^5$ metric tonnes are commercially available worldwide; 30% of these dyes are used in excess of 1,000 tonnes per annum, and 90% of the textile products are used at the level of 100 tonnes per annum or less) (Baban et al., 2010; Robinson et al., 2001; Soloman et al., 2009). 10-25% of textile dyes are lost during the dyeing process, and 2-20% are directly discharged as aqueous effluents in different environmental components.

In particular, the discharge of dye-containing effluents into the water environment is undesirable, not only because of their colour, but also because many of dyes released and their breakdown products are toxic, carcinogenic or mutagenic to life forms mainly because of carcinogens, such as benzidine, naphthalene and other aromatic compounds (Suteu et al., 2009; Zaharia et al., 2009). Without adequate treatment these dyes can remain in the environment for a long period of time. For instance, the half-life of hydrolysed Reactive Blue 19 is about 46 years at pH 7 and 25°C (Hao et al., 2000).

In addition to the aforementioned problems, the textile industry consumes large amounts of potable and industrial water (Tables 1, 2 and Fig. 1) as processing water (90-94%) and a relatively low percentage as cooling water (6-10%) (in comparison with the chemical industry where only 20% is used as process water and the rest for cooling). The recycling of treated wastewater has been recommended due to the high levels of contamination in dyeing and finishing processes (i.e. dyes and their breakdown products, pigments, dye intermediates, auxiliary chemicals and heavy metals, etc.) (Tables 3, 4 and 5) (adapted from Bertea A. and Bertea A.P., 2008; Bisschops and Spanjers, 2003; Correia et al., 1994; Orhon et al., 2001).

Type of finishing process	Water consumption, 10^{-3} m^3/kg textile product		
	Minimum	Medium	Maximum
Raw wool washing	4.2	11.7	77.6
Wool finishing	110.9	283.6	657.2
Fabric finishing			
• Short process	12.5	78.4	275.2
• Complex processing	10.8	86.7	276.9
Cloth finishing			
• Simplified processing	8.3	135.9	392.8
• Complex process	20	83.4	377.8
• Panty processing	5.8	69.2	289.4
Carpet finishing	8.3	46.7	162.6
Fibre finishing	3.3	100.1	557.1
Non-fabrics finishing	2.5	40	82.6
Yarn finishing	33.4	212.7	930.7

Table 1. Specific water consumption in textile finishing processes (adapted from Bertea A. & Bertea A.P., 2008)

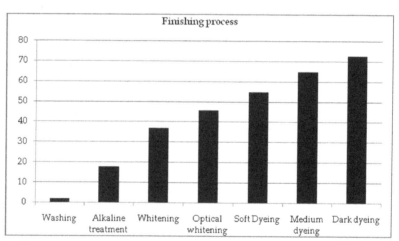

Fig. 1. Specific water consumption in different operations of textile finishing (EPA, 1997)

Operation/Process	Water consumption (% from total consumption of the textile plant)			Organic load (% from total organic load of the textile plant)		
	Minimum	Medium	Maximum	Minimum	Medium	Maximum
General facilities	6	14	33	0.1	2	8
Preparation	16	36	54	45	61	77
Dyeing	4	29	53	4	23	47
Printing	42	55	38	42	59	75
Wetting	0.3	0.4	0.6	0	0.1	0.1
Fabrics washing	3	28	52	1	13	25
Finishing	0.3	2	4	0.1	3	7

Table 2. Water consumption and organic load in different textile finishing steps (EWA, 2005)

Textile Organic Dyes – Characteristics, Polluting Effects and Separation/Elimination Procedures from Industrial
Effluents – A Critical Overview

31

The most common textile processing technology consists of desizing, scouring, bleaching, mercerizing and dyeing processes (EPA, 1997):

- *Sizing* is the first preparation step, in which sizing agents such as starch, polyvinyl alcohol (PVA) and carboxymethyl cellulose are added to provide strength to the fibres and minimize breakage.
- *Desizing* is used to remove sizing materials prior to weaving.
- *Scouring* removes impurities from the fibres by using alkali solution (commonly sodium hydroxide) to breakdown natural oils, fats, waxes and surfactants, as well as to emulsify and suspend impurities in the scouring bath.
- *Bleaching* is the step used to remove unwanted colour from the fibers by using chemicals such as sodium hypochlorite and hydrogen peroxide.
- *Mercerising* is a continuous chemical process used to increase dye-ability, lustre and fibre appearance. In this step a concentrated alkaline solution is applied and an acid solution washes the fibres before the dyeing step.
- *Dyeing* is the process of adding colour to the fibres, which normally requires large volumes of water not only in the dye bath, but also during the *rinsing step*. Depending on the dyeing process, many chemicals like metals, salts, surfactants, organic processing aids, sulphide and formaldehyde, may be added to improve dye adsorption onto the fibres.

In general, the textile industry uses a large quantity of chemicals such as:

- *Detergents and caustic*, which are used to remove dirt, grit, oils, and waxes. Bleaching is used to improve whiteness and brightness.
- *Sizing agents*, which are added to improve weaving.
- *Oils*, which are added to improve spinning and knitting.
- *Latex and glues*, which are used as binders.
- *Dyes, fixing agents, and many in-organics*, which are used to provide the brilliant array of colours the market demands.
- *A wide variety of special chemicals*, which are used such as softeners, stain release agents, and wetting agents.

Many of these chemicals become part of the final product whereas the rest are removed from the fabric, and are purged in the textile effluent.

Type of finished textile product	Dyes, g/kg textile product	Auxiliaries, g/kg textile product	Basic chemical compounds, g/kg textile product
Polyester fibres	18	129	126
Fabrics from synthetic fibres	52	113	280
Fabrics from cotton	18	100	570
Dyed fabrics from cellulose fibres	11	183	200
Printed fabrics from cellulose fibres	88	180	807

Table 3. Principal pollutants of textile wastewaters (EWA, 2005)

The annual estimated load with pollutants of a textile wastewater is of: 200,000-250,000 t salts; 50,000-100,000 t impurities of natural fibres (including biocids) and associated materials (lignin, sericine, etc.); 80,000-100,000 t blinding agents (especially starch and its derivatives, but also polyacrylates, polyvinyl alcohol, carboxymethyl cellulose); 25,000-

30,000 t preparation agents (in principal, mineral oils); 20,000-25,000 t tensides (dispersing agents, emulsifiers, detergents and wetting agents); 15,000-20,000 tonnes carboxylic acids (especially acetic acid); 10,000-15,000 t binders; 5,000-10,000 t urea; 5,000-10,000 t ligands, and < 5,000 t auxiliaries (EWA, 2005). The environmental authorities have begun to target the textile industry to clean up the wastewater that is discharged. The principal quality indicators that regulators are looking for polluting effect or toxicity are the high salt content, high Total Solids (TS), high Total Dissolved Solids (TDS), high Total Suspended Solids (TSS), Biological Oxygen Demand (BOD) and Chemical Oxygen Demand (COD), heavy metals, colour of the textile effluent (ADMI color value - American Dye Manufacturer Institute color value), and other potential hazardous or dangerous organic compounds included into each textile processing technological steps (Tables 4 and 5).

Process	Textile effluent
Singering, Desizing	High BOD, high TS, neutral pH
Scouring	High BOD, high TS, high alkalinity, high temperature
Bleaching, Mercerizing	High BOD, high TS, alkaline wastewater
Heat-setting	Low BOD, low solids, alkaline wastewater
Dyeing, Printing & Finishing	Wasted dyes, high BOD, COD, solids, neutral to alkaline wastewater

Table 4. Wet processes producing textile wastewater (adapted from Naveed S. et al., 2006)

Process	COD, g O$_2$/L	BOD, g O$_2$/L	TS, g/L	TDS, g/L	pH	Colour (ADMI)	Water usage, L/kg product
Desizing	4.6-5.9	1.7-5.2	16.0-32.0	-	-	-	3-9
Scouring	8.0	0.1-2.9	7.6-17.4	-	10-13	694	26-43
Bleaching	6.7-13.5	0.1-1.7	2.3-14.4	4.8-19.5	8.5-9.6	153	3-124
Mercerising	1.6	0.05-0.1	0.6-1.9	4.3-4.6	5.5-9.5	-	232-308
Dyeing	1.1-4.6	0.01-1.8	0.5-14.1	0.05	5-10	1450-4750	8-300

Table 5. Principal characteristics of a cotton wet processing wastewater (Cooper, 1995)

The wastewater composition is depending on the different organic-based compounds, inorganic chemicals and dyes used in the industrial dry and wet-processing steps. Textile effluents from the dyeing and rinsing steps represent the most coloured fraction of textile wastewaters, and are characterized by extreme fluctuations in many quality indicators such as COD, BOD, pH, colour, salinity and temperature.

The colour of textile wastewater is mainly due to the presence of textile dyes, pigments and other coloured compounds. A single dyeing operation can use a number of dyes from different chemical classes resulting in a complex wastewater (Correia et al., 1994). Moreover, the textile dyes have complex structures, synthetic origin and recalcitrant nature, which makes them obligatory to remove from industrial effluents before being disposed into hydrological systems (Anjaneyulu et al, 2005).

The dye removal from textile effluent is always connected with the decolourization treatment applied for textile wastewater in terms of respectation the local environmental quality requirements and standards (Table 6) (i.e. removal values of COD, BOD, TS, TSS, TDS, colour, total nitrogen, and total phosphorus from textile wastewater higher than 70-

85% or concentration values of the specific quality indicators under the imposed or standard limits) (Zaharia, 2008).
The decolorization treatments applied for different textile effluents include current and also advanced non-biological (i.e. specific mechano-physical, chemical, electrochemical processes, etc.) and also biological processes (Suteu et al., 2009; Zaharia, 2006; Zaharia et al, 2011).

Quality indicator	M.A.C.*, mg/L		
	Discharge directly in water bodies	Discharge in urban WW sewerage network	Water bodies quality, class I -natural non-polluted state
pH	6.5-8.5	6.5-8.5	6.5-8.5
BOD$_5$	25	300	3
COD	125	500	10
TSS	35	350	-
TDS	2000	250	< 500
Total N	10	15	1.5
Total P	1	5	0.1
Sulphates, SO$_4$$^{2-}$	600	600	80
Chlorides, Cl-	500	500	<100
Sulphides (S^{2-}) + H$_2$S	0.5	1.0	<0.5
Synthetic detergents	0.5	25	<0.5
Others (Oil & grease)	20	30	<0.1
*M.A.C. – Maximum Admissible Concentration			

Table 6. Romanian national wastewater and water quality standard adapted to European and international standards (adapted from Zaharia, 2008)

Studies on the behavior of textile organic dyes in water and wastewater treatment processes refer predominantly to laboratory tests or investigations of semi-technical plants, sometimes under conditions related to waterworks practice. In addition, textile operators, water supply companies, local environmental authorities have collected a lot of data on the behavior of textile dyes during textile wastewater treatment, but have seldom published their results.
However, information on the behavior of textile organic dyes is needed, because the limited number of reports available that are based on realistic operating conditions or which reproduce practical conditions that are already several years old.

2. Textile organic dyes – Classification and characteristics

The dyes are natural and synthetic compounds that make the world more beautiful through coloured products. The textile dyes represent a category of organic compounds, generally considered as pollutants, presented into wastewaters resulting mainly from processes of chemical textile finishing (Suteu et al., 2011a; Zaharia et al., 2009).
The textile coloration industry is characterised by a very large number of dispersed dyehouses of small and medium size that use a very wide range of textile dyes.

2.1 Textile organic dye classification

The nature and origin are firstly considered as criteria for the general classification in natural and synthetic textile dyes.

The natural textile dyes were mainly used in textile processing until 1856, beginning in 2600 BC when was mentioned the use of dyestuff in China, based on dyes extracted from vegetable and animal resources. It is also known that Phoenicians were used Tyrian purple produced from certain species of crushed sea snails in the 15th century BC, and indigo dye produced from the well-known indigo plant since 3000 BC. The dyes from madder plants were used for wrapping and dyeing of Egyptian mummies clothes and also of Incas fine textures in South America.

The synthetic dyes were firstly discovered in 1856, beginning with ,mauve' dye (aniline), a brilliant fuchsia colour synthesed by W.H. Perkin (UK), and some azo dyes synthesed by diazotisation reaction discovered in 1958 by P. Gries (Germany) (Welham, 2000). These dyes are aromatic compounds produced by chemical synthesis, and having into their structure aromatic rings that contain delocated electrons and also different functional groups. Their color is due to the chromogene-chromophore structure (acceptor of electrons), and the dyeing capacity is due to auxochrome groups (donor of electrons). The chromogene is constituted from an aromatic structure normally based on rings of benzene, naphthaline or antracene, from which are binding chromofores that contain double conjugated links with delocated electrons. The chromofore configurations are represented by the azo group ($-N=N-$), ethylene group ($=C=C=$), methine group ($-CH=$), carbonyl group ($=C=O$), carbon-nitrogen ($=C=NH$; $-CH=N-$), carbon-sulphur ($=C=S$; $\equiv C-S-S-C\equiv$), nitro ($-NO_2$; $-NO-OH$), nitrozo ($-N=O$; $=N-OH$) or chinoid groups. The auxochrome groups are ionizable groups, that confer to the dyes the binding capacity onto the textile material. The usual auxochrome groups are: $-NH_2$ (amino), $-COOH$ (carboxyl), $-SO_3H$ (sulphonate) and $-OH$ (hydroxyl) (Suteu et al, 2011; Welham, 2000). Five examples of textile dyes are presented in Fig. 2.

The textile dyes are mainly classified in two different ways: (1) based on its application characteristics (i.e. CI Generic Name such as acid, basic, direct, disperse, mordant, reactive, sulphur dye, pigment, vat, azo insoluble), and (2) based on its chemical structure respectively (i.e. CI Constitution Number such as nitro, azo, carotenoid, diphenylmethane, xanthene, acridine, quinoline, indamine, sulphur, amino- and hydroxy ketone, anthraquinone, indigoid, phthalocyanine, inorganic pigment, etc.) (Tables 7 and 8).

Excepting the colorant precursors such as azoic component, oxidation bases and sulphur dyes, almost two-third of all organic dyes are azo dyes ($R_1-N=N-R_2$) used in a number of different industrial processes such as textile dyeing and printing, colour photography, finishing processing of leather, pharmaceutical, cosmetics, etc. The starting material or intermediates for dye production are aniline, chloroanilines, naphthylamines, methylanilines, benzidines, phenylenediamines, and others.

Considering only the general structure, the textile dyes are also classified in anionic, nonionic and cationic dyes. The major anionic dyes are the direct, acid and reactive dyes (Robinson et al., 2001), and the most problematic ones are the brightly coloured, water soluble reactive and acid dyes (they can not be removed through conventional treatment systems).

The major nonionic dyes are disperse dyes that does not ionised in the aqueous environment, and the major cationic dyes are the azo basic, anthraquinone disperse and

Textile Organic Dyes – Characteristics, Polluting Effects and Separation/Elimination Procedures from Industrial
Effluents – A Critical Overview

35

reactive dyes, etc. The most problematic dyes are those which are made from known carcinogens such as benzidine and other aromatic compounds (i.e. anthroquinone-based dyes are resistant to degradation due to their fused aromatic ring structure). Some disperse dyes have good ability to bioaccumulation, and the azo and nitro compounds are reduced in sediments, other dye-accumulating substrates to toxic amines (e.g. R_1-N=N-R_2 + $4H^+$ + $4e^- \rightarrow$ R_1-NH_2 + R_2-NH_2).

The organic dyes used in the textile dyeing process must have a high chemical and photolytic stability, and the conventional textile effluent treatment in aerobic conditions does not degrade these textile dyes, and are presented in high quantities into the natural water resources in absence of some tertiary treatments.

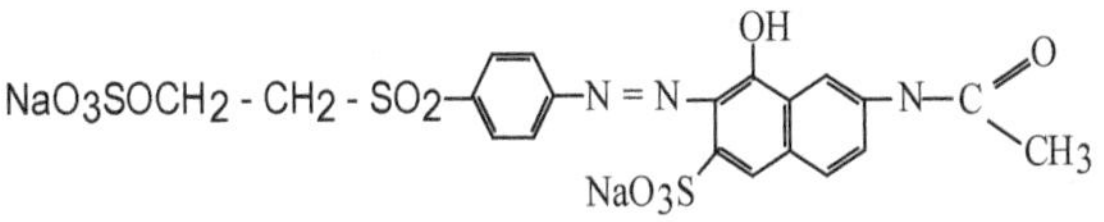

Reactive Orange 16
C.I. 18097
Anionic monoazo reactive dye;
MW = 617.54 g/mol;
λ_{max} = 495 nm

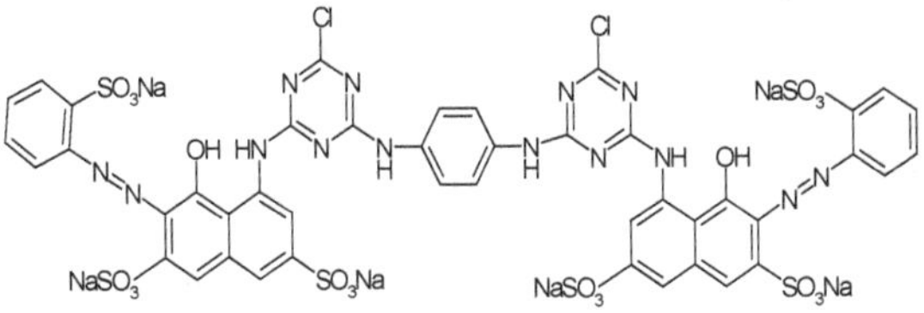

Brilliant Red HE-3B
(Reactive Red 120)
C.I. 25810
Anionic, bifunctional
azo reactive dye;
MW = 1463 g/mol;
λ_{max} = 530 nm

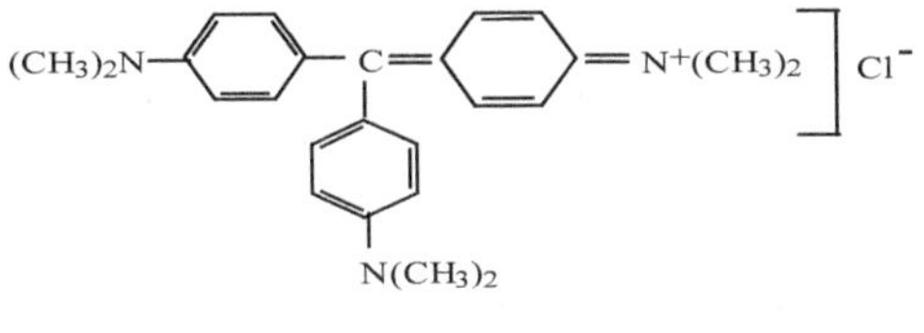

Crystal Violet
(Basic Violet 3)
C.I. 42555
Cationic triphenylmethane dye;
MW = 407.99 g/mol;
λ_{max} = 590 nm

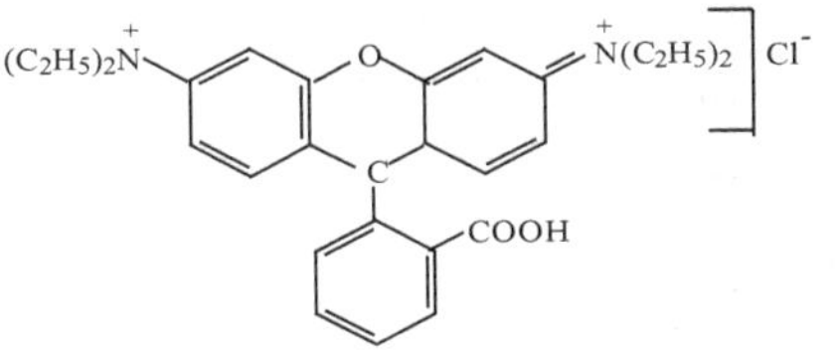

Rhodamine B
(Basic Violet 10)
C.I. 45170
Cationic, Xantenic dye;
MW =479.2 g/mol;
λ_{max} = 550 nm

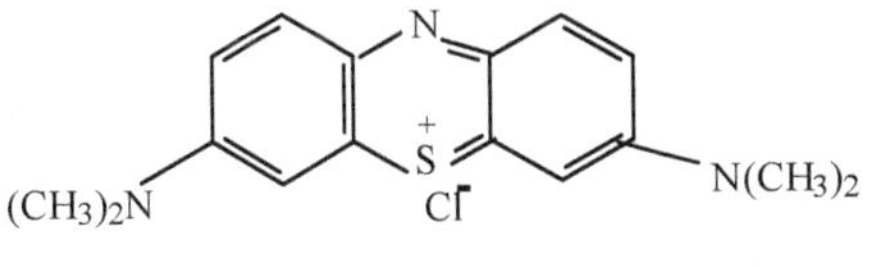

Methylene Blue
(Basic Blue 9)
C.I. 52015
Cationic, phenothiazine dye;
MW =319.85 g/mol;
λ_{max}= 660 nm

Fig. 2. Chemical structure and principal characteristics of different textile dyes
(Suteu et al., 2011a)

Chemical class	C.I. Constitution numbers	Chemical class	C.I. Constitution numbers
Nitroso	10000-10299	Indamine	49400-49699
Nitro	10300-10099	Indophenol	49700-49999
Monoazo	11000-19999	Azine	50000-50999
Disazo	20000-29999	Oxazine	51000-51999
Triazo	30000-34999	Thiazine	52000-52999
Polyazo	35000-36999	Sulphur	53000-54999
Azoic	37000-39999	Lactone	55000-56999
Stilbene	40000-40799	Aminoketone	56000-56999
Carotenoid	40800-40999	Hydroxyketone	57000-57999
Diphethylmethane	41000-41999	Anthraquinone	58000-72999
Triarylmethane	42000-44999	Indigoid	73000-73999
Xanthene	45000-45999	Phthalocyanine	74000-74999
Acridine	46000-46999	Natural	75000-75999
Quinoline	47000-47999	Oxidation base	76000-76999
Methine	48000-48999	Inorganic pigment	77000-77999
Thiazole	49000-49399		

Table 7. Colour index classification of dye chemical constituents (Cooper, 1995)

Chemical class	Distribution between application ranges, %								
	Acid	Basic	Direct	Disperse	Mordant	Pigment	Reactive	Solvent	Vat
Unmetallised azo	20	5	30	12	12	6	10	5	-
Metal complex	65	-	10	-	-	-	12	13	-
Thiazole	-	5	95	-	-	-	-	-	-
Stilbene	-	2	98	-	-	-	-	-	-
Anthraquinone	15	2	-	25	3	4	6	9	36
Indigoid	2	-	-	-	-	17	-	-	81
Quinophthalene	30	20	-	40	-	-	10	-	-
Aminoketone	11	-	-	40	8	-	3	8	20
Phtalocyanine	14	4	8	-	4	9	43	15	3
Formazan	70	-	-	-	-	-	30	-	-
Methine	-	71	-	23	-	1	-	5	-
Nitro, nitroso	31	2	-	48	2	5	-	12	-
Triarylmethane	35	22	1	1	24	5	-	12	-
Xanthene	33	16	-	-	9	2	2	38	-
Acridine	-	92	-	4	-	-	-	4	-
Azine	39	39	-	-	-	3	-	19	-
Oxazine	-	22	17	2	40	9	10	-	-
Thiazine	-	55	-	-	10	-	-	10	25

Table 8. Distribution of each chemical class between major application ranges (adapted from Cooper, 1995)

The major textile dyes can be included in the two high classes: azo or anthraquinone (65-75% from total textile dyes). The azo dyes are characterised by reactive groups that form covalent bonds with HO-, HN-, or HS- groups in fibres (cotton, wool, silk, nylon). Azo dyes are mostly used for yellow, orange and red colours. Anthraquinone dyes constitute the second most important class of textile dyes, after azo dyes, and have a wide range of colours in almost the whole visible spectrum, but they are most commonly used for violet, blue and green colours (Fontenot et al., 2003). Considering the nature of textile fibres that are dyeing, the textile dyes can be classified as into Table 9.

Class	Subclass	PES	CA	PAN	PA	Silk	Wool	Cotton
Disperse		+++	+++	++	++	-	-	-
Cationic		-	~	+++	++	-	-	-
Acid	Standard	-	-	~	+++	+++	+++	-
	1 : 1	-	-	-	P	+	+++	-
	1 : 2	-	-	-	++	+	+++	-
Reactive		-	-	-	~	++	++	+++
Direct		-	-	-	++	++	P	+++
Stuff		~	-	~	~	~	–	+++
Indigoid		-	-	–	~	~	~	P
Sulphur		-	+	–	–	–	–	+++
Insoluble azo		+++	-	~	~	~	–	+++
Legend: +++ very frequent; ++ frequent; + sometimes; ~ possible; P especially printing								

Table 9. Dye classes and dyeing textile substrates (adapted from Bertea A. & Bertea A.P., 2008)

The textile dyeing process is due to physico-chemical interactions developed at contacting of textile material with dye solution or dispersion, which contains a large variety of chemicals (salts, acids) and dyeing auxiliaries (tensides, dispersing agents, etc.).

2.2 Textile dye characterisation

The identification of individual unknown dyes in a coloured effluent or watercourse is difficult to be done and implies advanced analytical methods (i.e. individual and/or coupled spectrophotometry, G/L chromatography and mass spectrometry procedures), and also the colour determination and appreciation in different operating situations.

The characterisation and identification data of the textile dyes as main chemicals in dyeing process must consist of:

- dye identity data (i.e. name, C.I. or CAS number, molecular and structural formula; composition, degree of purity, spectral data; methods of detection and determination) (e.g., some examples illustrated in Fig. 2),

- dye production information (i.e. production process, proposed uses, form, concentration in commercially available preparations, estimated production, recommended methods and precautions concerning handling, storage, transport, fire and other dangers, emergency measures, etc.) (e.g. some indications in Table 3),

- dye physico-chemical properties (i.e. boiling point (b.p.), relative density, water solubility, partition coefficient, vapour pressure, self-ignition, oxidising properties, granulometry, particle size distribution, etc.) (e.g., for the first synthetic discovered dye: Aniline – 184 (b.p.), k_H= 2.05E-01, $C_{sat}{}^w$= 3.6E+04, pK_a= 4.6, log K_{ow}= 0.90; for 4,4'-

Methylenedianiline - 398 (b.p.), k_H= 5.67E-06, $C_{sat}{}^w$= 1000, pK_a= n.s., log K_{ow}= 1.59; for 4-Aminodiphenylamine - 354 (b.p.), k_H= 3.76E-01, $C_{sat}{}^w$= 1450, pK_a= 5.2, log K_{ow}= 1.82; where, k_H - the Henry's constant at 1013 hPa and 25°C (Pa·m^3/mol), $C_{sat}{}^w$ - the water solubility (mg/L) at 25°C, pK_a – dissociation constant of the protonated azo dye at 25°C, K_{ow} – n-octanol/water partition coefficient, n.s. – not specified)

- toxicological studies (i.e. acute toxicity-oral, inhalation, dermal, skin or eyes irritation, skin sensitisation, repeated dose toxicity-28 days, mutagenicity, toxicity to reproduction, toxicokinetic behaviour),
- ecotoxicological studies (i.e. acute toxicity to fish, daphnia, growth inhibition on algae, bacteriological inhibition, degradation: ready biodegradability, abiotic degradation-hydrolysis as a function of pH, BOD, COD, BOD/COD ratio).

These characteristics and indentification data are given obligatory by the dye producers or distributors of textile products on the free market of homologated colorants, and also exist in the library data of some operating programs of advanced analysis apparatus.

The textile azo dyes are characterized by relatively high polarity (log K_{ow} up to 3) and high recalcitrance. Recalcitrance is difficult to evaluate because of the dependence of degradation on highly variable boundary conditions (e.g., redox milieu or pH). For example, aniline (the first synthetic discovered dye) is known to be easily degradable, but under specific anoxic conditions it has been proven to be easily stable (Börnick and Schmidt, 2006). Furthermore, the azo dyes are relevant in terms of eco- and human toxicity, industrially produced in high quantities, and known to occur in hydrosphere.

The azo dyes can accept protons because of the free electron pair of the nitrogen, and the free electron pair of nitrogen interacts with the delocalized π-orbital system.

Acceptor substituients at the aromatic ring such as –Cl or –NO$_2$ cause an additional decrease in the basic character of aminic groups. Donor groups such as –CH$_3$ or –OR (in metha and para position) lead to an increase in the basicity of aromatic aminic groups. However, donor substituients in the ortho position can sterically impede the protonation and consequently decrease the basicity of aminic groups. The azo dyes are characterized by amphoteric properties when molecules contain additional acidic groups such as hydroxyl, carboxyl or sulfoxyl substituents.

Depending on pH value, the azo dyes can be anionic (deprotonation at the acidic group), cationic (protonated at the amino group) or non-ionic. Accordingly, knowledge of the acidity constants is indispensable for the characterization of the behavior of azo dyes. Environmental partitioning is influenced by substituents as well as the number of carbon atoms and aromatic structure of the carbon skeleton. The presence of an amino group causes a higher boiling point, a higher water solubility, a lower Henry's law constant, and a higher mobility in comparison with hydrocarbons (the amino group can also reduce the mobility by specific interactions with solids via covalent bonding to carbonyl moieties or cation exchange) (Börnick and Schmidt, 2006). The volatility of azo dyes in aqueous solution is in most cases very low. Some colour characteristics from different studies, tests and literature data are presented in Table 10, especially for reactive dyes.

The chromophore distribution in reactive dyes indicated that the great majority of unmetallised azo dyes are yellow, orange and red. Contrarily, the blue, green, black and brown contain a much more proportion of metal-complex azo, anthraquinone, triphenyldioxazine or copper phthalocyanine chromophores (Table 10).

Chemical class	Yellow	Orange	Red	Violet	Blue	Green	Brown	Black	Proportion of all reactive dyes
Unmetallised azo	97	90	90	63	20	16	57	42	66
Metal complex azo	2	10	9	32	17	5	43	55	15
Anthraquinone	-	-	-	5	34	37	-	3	10
Phthalocyanine	-	-	-	-	27	42	-	-	8
Miscellanous	1	-	1	-	2	-	-	-	1

Table 10. Distribution of chemical classes in reactive dye range (adapted from Cooper, 1995)

In general, colour in wastewater is classified in terms of true/real colour (i.e. colour of turbidity-free water sample), or apparent colour (i.e. colour of non-treated water sample). The most common methods to measure the colour of dye solution or dispersion, and/or wastewater are visual comparison and spectrophotometry, although there is still a lack of an universal method to classify coloured wastewater discharges. By *visual comparison*, colour is quantified by comparing the colour of sample with either known concentrations of coloured standards (normally a platinum-cobalt solution), or properly calibrated colour disks, and is less applicable for highly coloured industrial wastewaters. In the *spectrophotometric method*, colour-measuring protocols differ between the methodologies, of which the most commonly used are Tristimulus Filter Method, American Dye Manufacturer Institute (ADMI) Tristimulus Filter Method, and Spectra record (Table 11).

Spectrophotometric method	Description
Tristimulus	Three tristimulus light filters combined with a specific light source (i.e. tungsten lamp) and a photoelectric cell inside a filter photometer. The output transmittance is converted to trichromatic coefficient and colour characteristic value.
ADMI Tristimulus	The ADMI colour value provides a true watercolour measure, which can be differentiated in 3 (WL) ADMI (i.e. the transmittance is recorded at 590, 540 and 438 nm) or 31 (WL) ADMI (i.e. the transmittance is determined each 10 nm in the range of 400-700 nm).
Spectra record	The complete spectrum is recorded, and the entire spectrum, or a part of it, is used for comparison. A modified method has been suggested in which areas beneath an extinction curve represent the colour intensity, being expressed as space units.

Table 11. Spectrophotometric methods for colour determination in dye solution or dispersion, water and wastewater (adapted from Dos Santos et al., 2004).

2.3 Dye fixation on textile fibres

In general, textile fibres can catched dyes in their structures as a result of van der Waals forces, hydrogen bonds and hydrophobic interactions (physical adsorption). The uptake of the dye in fibres depends on the dye nature and its chemical constituents. But the strongest dye-fibre attachment is a result of a covalent bond with an additional electrostatic interaction where the dye ion and fibre have opposite charges (chemisorption).

In alkaline conditions (i.e. pH 9-12), at high temperatures (30-70°C), and salt concentration from 40-100 g/L, reactive dyes form a reactive vinyl sulfone ($-SO_3-CH=CH_2$) group, which creates a bond with the fibres. However, the vinyl sulfone group undergoes hydrolysis (i.e. a spontaneous reaction that occurs in the presence of water), and because the products do not have any affinity with the fibres, they do not form a covalent bond (Dos Santos et al., 2004). Therefore, a high amount of dye constituents are discharged in the wastewater.

The fixation efficiency varies with the class of azo dye used, which is around 98% for basic dyes and 50% for reactive dyes (Table 12) (Bertea A. & Bertea A.P., 2008; O'Neill et al., 1999). Large amounts of salts such as sodium nitrate, sodium sulphate and sodium chloride are used in the dyebath, as well as sodium hydroxide that is widely applied to increase the pH to the alkaline range. It is estimated that during the mercerising process the weight of these salts can make up 20% of the fibre weight (EPA, 1997).

Dye class	Fibre type	Fixation degree, %	Loss in effluent, %
Acid	Polyamide	80-95	5-20
Basic	Acrilic	95-100	0-5
Direct	Cellulose	70-95	5-30
Disperse	Polyester	90-100	0-10
Metal complex	Wool	90-98	2-10
Reactive	Cellulose	50-90	10-50
Sulphur	Cellulose	60-90	10-40
Dye-stuff	Cellulose	80-95	5-20

Table 12. Fixation degree of different dye classes on textile support (EWA, 2005).

The problem of high coloured effluent or dye-containing effluent has become identified particularly with the dyeing of cellulose fibres (cotton – 50% of the total consumed fibres in the textile industry worldwide), and in particular with the use of reactive dyes (10-50% loss in effluent), direct dyes (5-30% loss in effluent), vat dyes (5-20% loss in effluent), and sulphur dyes (10-40% loss in effluent).

The research for dynamic response and improved dyeing productivity has served to focus the attention of the textile coloration industry on right-first-time production techniques that minimise wastes, make important contribution to reduce colour loads in the effluent by optimisation of processes, minimising of dye wastage, and control automatically the dyeing and printing operation.

After the textile dyeing and finishing processes, a *predicted environmental concentration* of a dye in the receiving water can be estimated based on the following factors: (*i*) daily dye usage; (*ii*) dye fixation degree on the substrate (i.e. textile fibres or fabrics); (*iii*) dye removal degree into the effluent treatment process, and (*iv*) dilution factor in the receiving water.

Some scenario analyses were mentioned the values of dye concentration in some receiving rivers of 5-10 mg/L (average value, 50 days each year) or 1300-1555 mg/L (the worst case, 2 days each year) for batchwise dyeing of cotton with reactive dyes, and of 1.2-3 mg/L (average value, 25 days each year) or 300-364 mg/L (the worst case, 2 days each year) for batchwise dyeing of wool yarn with acid dyes (adapted from Cooper Ed., 1995).

Limits on dye-containing organic loads will become more restrictive in the future, which makes cleaning exhausts an environmental necessity.

3. Textile organic dyes – Environmental problems and polluting effects

The environmental issues associated with residual dye content or residual colour in treated textile effluents are always a concern for each textile operator that directly discharges, both sewage treatment works and commercial textile operations, in terms of respecting the colour and residual dye requirements placed on treated effluent discharge (Zaharia et al., 2011).

Dye concentrations in watercourses higher of 1 mg/L caused by the direct discharges of textile effluents, treated or not, can give rise to public compliant. High concentrations of textile dyes in water bodies stop the reoxygenation capacity of the receiving water and cut-off sunlight, thereby upsetting biological activity in aquatic life and also the photosynthesis process of aquatic plants or algae (Zaharia et al., 2009).

The colour in watercourses is accepted as an aesthetic problem rather than an eco-toxic hazard. Therefore, the public seems to accept blue, green or brown colour of rivers but the 'non-natural' colour as red and purple usually cause most concern.

The polluting effects of dyes against aquatic environment can be also the result of toxic effects due to their long time presence in environment (i.e. half-life time of several years), accumulation in sediments but especially in fishes or other aquatic life forms, decomposition of pollutants in carcinogenic or mutagenic compounds but also low aerobic biodegradability. Due to their synthetic nature and structure mainly aromatic, the most of dyes are non-biodegradable, having carcinogenic action or causing allergies, dermatitis, skin irritation or different tissular changes. Moreover, various azo dyes, mainly aromatic compounds, show both acute and chronic toxicity. High potential health risk is caused by adsorption of azo dyes and their breakdown products (toxic amines) through the gastrointestinal tract, skin, lungs, and also formation of hemoglobin adducts and disturbance of blood formation. LD_{50} values reported for aromatic azo dyes range between 100 and 2000 mg/kg body weight (Börnick & Schmidt, 2006).

Several azo dyes cause damage of DNA that can lead to the genesis of malignant tumors. Electron-donating substituents in ortho and para position can increase the carcinogenic potential. The toxicity diminished essentially with the protonation of aminic groups. Some of the best known azo dyes (e.g. Direct Black 38 azo dye, precursor for benzidine; azodisalicylate, precursor for 4-phenylenediamine) and their breakdown derivatives inducing cancer in humans and animals are benzidine and its derivatives, and also a large number of anilines (e.g. 2-nitroaniline, 4-chloroaniline, 4,4'-dimethylendianiline, 4-phenylenediamine, etc.), nitrosamines, dimethylamines, etc.

The main pollution characteristics and category (pollution risk) of the principal products used in processing of textile materials are summarized in Table 13 (EWA, 2005).

In different toxicological studies are indicated that 98% of dyes has a lethal concentration value (LC_{50}) for fishes higher than 1 mg/L, and 59% have an LC_{50} value higher than 100 mg/L (i.e. 31% of 100-500 mg/L and 28% higher than 500 mg/L).

Other ecotoxicological studies indicated that over 18% of 200 dyes tested in England showed significant inhibition of the respiration rate of the biomass (i.e. wastewater bacteria) from sewage, and these were all basic dyes (adapted from Cooper, 1995).

The bioaccumulation potential of dyes in fish was also an important measure estimating the bioconcentration factor (dye concentration in fish/dye concentration in water). No bioaccumulation is expected for dyes with solubility in water higher than 2000 mg/L.

Dyes are not biodegradable in aerobic wastewater treatment processes and some of them may be intactly adsorbed by the sludge at wastewater biological treatment (i.e. bioelimination by adsorptive removal of dyes).

Products used in textile industry	Pollution characteristics	Pollution category
Alkali, mineral acids, salts, oxidants	Inorganic pollutants, relatively inofensive	1
Sizing agents based on starch, natural oils, fats, waxes, biodegradable surfactants, organic acids, reducing agents	Easy biodegradables; with a moderate - high BOD_5	2
Colorants and optic whitening agents, fibres and impurities of polymeric nature, synthetic polymeric resins, silicones	Difficult to be biodegraded	3
Polyvinyl alcohols, mineral oils, tensides resistant to biodegradation, anionic or non-ionic emolients	Difficult to be biodegraded; moderate BOD_5	4
Formaldehyde or N-methylolic reagents, coloured compounds or accelerators, retarders and cationic emolients, complexants, salts of heavy metals	Can not be removed by conventional biological treatment, low BOD_5	5

Table 13. Characterization of products used in textile industry vs. their polluting effect

Some investigations of adsorption degree onto sludge of some azo dyes indicated typically high levels of adsorption for basic or direct dyes, and high to medium range for disperse dyes, all the others having very low adsorption, which appears to depend on the sulphonation degree or ease of hydrolysis. These tested azo dyes are presented in Table 14.

Group 1		Group 2	Group 3
Dyes unaffected by biological treatment		*Dyes eliminated by adsorption on sludge*	*Dyes with high biodegradability*
CI Acid Yellow 17	CI Acid Red	CI Acid Red 151	CI Acid Orange 7
CI Acid Yellow 23	CI Acid Red 14	CI Acid Blue 113	CI Acid Orange 8
CI Acid Yellow 49	CI Acid Red 18	CI Direct Yellow 28	CI Acid Red 88
CI Acid Yellow 151	CI Acid Red 337	CI Direct Violet 9	
CI Acid Orange 10	CI Acid Black 1		
CI Direct Yellow 4			

Table 14. Fate of water-soluble azo dyes in the activated sludge treatment (Cooper, 1995)

The high degree of sulphonation of azo dyes in Group 1 enhanced their water solubility and limited their ability to be adsorbed on the biomass. The dyes in Group 2 were also highly sulphonated but permitted a relatively good adsorption performance on sludge.

Other information in bioelimination of different reactive dyes mentioned that monoazo dyes are particularly poorly adsorbed, and disazos, anthraquinones, triphendioxazines and phthalocyanines are generally much better adsorbed than monoazos.

It is important to underline that toxic compounds (e.g. toxic aromatic amines, benzidine and its derivatives) can be formed in the environment via transformation of textile dye-precursors (e.g., reduction or hydrolysis of textile azo dyes). The textile dye-precursors are

Textile Organic Dyes – Characteristics, Polluting Effects and Separation/Elimination Procedures from Industrial
Effluents – A Critical Overview

43

introduced in water environment due to industrial production of dyes and industrial production of textile fibres, fabrics and clothes via wastewater, sludge, or solid deposits. The quality problem of dye content and/or colour in the dyehouse effluent discharged in watercourses can be solved by using of a range of advanced decolourisation technologies investigated by the major dye suppliers, textile operators and customers who are under pressure to reduce colour and residual dye levels in their effluents.

4. Textile organic dyes – Separation and elimination procedures from water environment (especially industrial wastewater)

The textile organic dyes must be separated and eliminated (if necessary) from water but especially from industrial wastewaters by effective and viable treatments at sewage treatment works or on site following two different treatment concepts as: (1) separation of organic pollutants from water environment, or (2) the partial or complete mineralization or decomposition of organic pollutants. Separation processes are based on fluid mechanics (sedimentation, centrifugation, filtration and flotation) or on synthetic membranes (micro-, ultra- and nanofiltration, reverse osmosis). Additionally, physico-chemical processes (i.e. adsorption, chemical precipitation, coagulation-flocculation, and ionic exchange) can be used to separate dissolved, emulsified and solid-separating compounds from water environment (Anjaneyulu et al., 2005; Babu et al., 2007; Robinson et al., 2001; Suteu et al., 2009a; Suteu et al., 2011a; Zaharia, 2006; Zaharia et al., 2009; Zaharia et al., 2011).

The partial and complete mineralization or decomposition of pollutants can be achieved by biological and chemical processes (biological processes in connection with the activated sludge processes and membrane bioreactors, advanced oxidation with ozone, H_2O_2, UV) (Dos Santos et al., 2004 ; Oztekin et al., 2010 ; Wiesmann et al., 2007 ; Zaharia et al., 2009).

A textile operator will decide on options available to plan forward strategy that will ensure compliance with the environmental regulators' requirements on a progressive basis focused on some options and applied solutions of different separation processes (sedimentation, filtration, membrane separation), and some physico-chemical treatment steps (i.e. adsorption; coagulation-flocculation with inorganic coagulants and organic polymers; chemical oxidation; ozonation; electrochemical process, etc.) integrated into a specific order in the technological process of wastewater treatment for decolourization or large-scale colour and dye removal processes of textile effluents.

To introduce a logical order in the description of treatment methods for textile dye and colour removal, the relationship between pollutant and respective typical treatment technology is taken as reference. The first treatment step for textile wastewater and also rainwater is the separation of suspended solids and immiscible liquids from the main textile effluents by gravity separation (e.g., grit separation, sedimentation including coagulation/flocculation), filtration, membrane filtration (MF, UF), air flotation, and/or other oil/water separation operations.

The following treatment steps are applied to soluble pollutants, when these are transferred into solids (e.g., chemical precipitation, coagulation/flocculation, etc.) or gaseous and soluble compounds with low or high dangerous/toxic effect (e.g., chemical oxidation, ozonation, wet air oxidation, adsorption, ion exchange, stripping, nanofiltration/reverse osmosis). Solid-free wastewater can either be segregated into a biodegradable and a non-biodegradable part, or the contaminants responsible for the non-biodegradable wastewater part that can be decomposed based on physical and/or chemical processes. After an adequate treatment, the treated wastewater (WW) can either be discharged into a receiving water body, into a subsequent central biological wastewater treatment plant (BWWTP) or a municipal wastewater treatment plant (MWWTP).

Some selected treatment processes for dyes and colour removal of industrial wastewater applied over the time into different textile units are summarized in Table 15. Some of these methods will be further detail and some of authors' results summarized.

Treatment methodology	Treatment stage	Advantages	Limitations
Physico-chemical treatments			
Precipitation, coagulation-flocculation	Pre/main treatment	Short detention time and low capital costs. Relatively good removal efficiencies.	Agglomerates separation and treatment. Selected operating condition.
Electrokinetic coagulation	Pre/main treatment	Economically feasible	High sludge production
Fenton process	Pre/main treatment	Effective for both soluble and insoluble coloured contaminants. No alternation in volume.	Sludge generation; problem with sludge disposal. Prohibitively expensive.
Ozonation	Main treatment	Effective for azo dye removal. Applied in gaseous state: no alteration of volume	Not suitable for dispersed dyes. Releases aromatic dyes. Short half-life of ozone (20 min)
Oxidation with NaOCl	Post treatment	Low temperature requirement. Initiates and accelerates azo-bond cleavage	Cost intensive process. Release of aromatic amines
Adsorption with solid adsorbents such as:			
Activated carbon	Pre/post treatment	Economically attractive. Good removal efficiency of wide variety of dyes.	Very expensive; cost intensive regeneration process
Peat	Pre treatment	Effective adsorbent due to cellular structure. No activation required.	Surface area is lower than activated carbon
Coal ashes	Pre treatment	Economically attractive. Good removal efficiency.	Larger contact times and huge quantities are required. Specific surface area for adsorption are lower than activated carbon
Wood chips/ Wood sawdust	Pre treatment	Effective adsorbent due to cellular structure. Economically attractive. Good adsorption capacity for acid dyes	Long retention times and huge quantities are required.

Textile Organic Dyes – Characteristics, Polluting Effects and Separation/Elimination Procedures from Industrial Effluents – A Critical Overview

45

Silica gels	Pre treatment	Effective for basic dyes	Side reactions prevent commercial application
Irradiation	Post treatment	Effective oxidation at lab scale	Requires a lot of dissolved oxygen (O_2)
Photochemical process	Post treatment	No sludge production	Formation of by-products
Electrochemical oxidation	Pre treatment	No additional chemicals required and the end products are non-dangerous/hazardous.	Cost intensive process; mainly high cost of electricity
Ion exchange	Main treatment	Regeneration with low loss of adsorbents	Specific application; not effective for all dyes
Biological treatments			
Aerobic process	Post treatment	Partial or complete decolourization for all classes of dyes	Expensive treatment
Anaerobic process	Main treatment	Resistant to wide variety of complex coloured compounds. Bio gas produced is used for stream generation.	Longer acclimatization phase
Single cell (Fungal, Algal & Bacterial)	Post treatment	Good removal efficiency for low volumes and concentrations. Very effective for specific colour removal.	Culture maintenance is cost intensive. Cannot cope up with large volumes of WW.
Emerging treatments			
Other advanced oxidation process	Main treatment	Complete mineralization ensured. Growing number of commercial applications. Effective pre-treatment methodology in integrated systems and enhances biodegradability.	Cost intensive process
Membrane filtration	Main treatment	Removes all dye types; recovery and reuse of chemicals and water.	High running cost. Concentrated sludge production. Dissolved solids are not separated in this process
Photocatalysis	Post treatment	Process carried out at ambient conditions. Inputs are no toxic and inexpensive. Complete mineralization with shorter detention times.	Effective for small amount of coloured compounds. Expensive process.
Sonication	Pre treatment	Simplicity in use. Very effective in integrated systems.	Relatively new method and awaiting full scale application.
Enzymatic treatment	Post treatment	Effective for specifically selected compounds.	Enzyme isolation and purification is tedious.

		Unaffected by shock loadings and shorter contact times required.	Efficiency curtailed due to the presence of interferences.
Redox mediators	Pre/ supportive treatment	Easily available and enhances the process by increasing electron transfer efficiency	Concentration of redox mediator may give antagonistic effect. Also depends on biological activity of the system.
Engineered wetland systems	Pre/ post treatment	Cost effective technology and can be operated with huge volumes of wastewater	High initial installation cost. Requires expertise and managing during monsoon becomes difficult

Table 15. Various current and emerging dye separation and elimination treatments applied for textile effluents with their principal advantages and limitations (adapted from Anjaneyulu et al., 2005; Babu et al., 2007; Robinson et al., 2001)

4.1 Physical treatments
4.1.1 Adsorption

One of the most effective and proven treatment with potential application in textile wastewater treatment is adsorption. This process consists in the transfer of soluble organic dyes (solutes) from wastewater to the surface of solid, highly porous, particles (the adsorbent). The adsorbent has a finite capacity for each compound to be removed, and when is 'spent' must be replaced by fresh material (the 'spent' adsorbent must be either regenerated or incinerated).

Adsorption is an economically feasible process for dyes removal and/or decolourization of textile effluents being the result of two mechanisms: adsorption and ion exchange. The principal influencing factors in dye adsorption are: dye/adsorbent interaction, adsorbent surface area, particle size, temperature, pH, and contact time. Adsorbents which contain amino nitrogen tend to have a significantly larger adsorption capacity in acid dyes.

The most used adsorbent is activated carbon, and also other commercial inorganic adsorbents. Some 'low cost' adsorbents of industrial or agricultural wastes (i.e. peat, coal ashes, refused derived coal fuel, clay, bentonite and modified bentonite, red soil, bauxite, ebark, rice husk, tree barks, neem leaf powder, wood chips, ground nut shell powder, rice hulls, bagasse pith, wood sawdust, grounded sunflower seed shells, other ligno-cellulosic wastes, etc.) are also used for removal of dye and organic coloured matter from textile effluents (i.e. a removal of 40-90% basic dyes and 40% direct dyes, with maximum adsorption capacities for basic dyes of 338 mg/g) (Anjaneyulu et al., 2005; Bhattacharyya & Sarma, 2003; Gupta et al., 1992; Nigam et al., 2000; Ozcan et al., 2004; Robinson et al., 2001; Suteu & Zaharia, 2008; Suteu et al., 2009b; Suteu et al., 2011a,b; Zaharia et al., 2011). The use of these materials is advantageous mainly due to their widespread availability and cheapness. Sometimes the regeneration is not necessary and the 'spent' material is conventionally burnt although there is potential for solid state fermentation (SSF) for protein enrichment. The use of 'low cost' adsorbents for textile dye removal is profitable but requires huge quantity of adsorbents, being lower efficient than activated carbon. Some authors' results for dye adsorption are summarized in Table 16.

Textile Organic Dyes – Characteristics, Polluting Effects and Separation/Elimination Procedures from Industrial Effluents – A Critical Overview

47

Adsorptive material	Operating conditions / Optimal doses (Adsorption + Sedimentation/Filtration)	Adsorption efficiency for some tested textile dyes, %
Peat	pH= 2 (BRed); 5.7 (MB, RhB); C_{dye}= (20-300) mg/L BRed; (19-134) mg/L MB; (28-155) mg/L RhB; $C_{adsorbant}$= 12 g/L	(67.50 – 85.60) BRed (65.70– 89.30) MB (98.30– 99.00 RhB
Wood sawdust	pH= 2 (BRed); 5,7 (MB; CV; RhB); 1 (RO); C_{dye}= (20-150) mg/L BRed; (6-40) mg/L MB; (8-50) mg/L CV; (9-58) mg/L RhB; (24-160) mg/L (RO); $C_{adsorbant}$= 20 g/L (BRed); 4 g/L (MB, CV, RhB); 8 g/L (RO)	(63.90– 83.00) BRed (69.40 – 91.80) MB (66.00 – 80.00) CV (52.20 – 72.00) RhB (38.30 – 50.80) RO
Sunflower seed shell	pH= 1 (RO); 6 (MB); C_{dye}= (25-160) mg/L RO; (25–280) mg/L MB; $C_{adsorbant}$= 8 g/L (RO), 4 g/L (MB)	(80.60 – 82.60) RO (84.40 – 93.00) MB
Corn cobs	pH= 1 (RO) or 6 (MB); C_{dye}= (24-160) mg/L RO; (25-280) mg/L MB; $C_{adsorbant}$= 8 g/L (RO), 4 g/L (MB)	(83.00 – 86.20) RO (87.20 – 98.10) MB
Lignine	pH= 1.5 (BRed); 1 (RO); 6 (MB); C_{dye}= (50-300) mg/L BRed; (25.60-280) mg/L MB; (30-150) mg/L RO; $C_{adsorbant}$= 14 g/L (BRed); 4 g/L (MB); 12 g/L (RO)	(49.50-59.50) BRed (43.50-60.70) RO (55.30-64.30) MB
Cellolignine	pH=6; C_{dye}= (25.6-281.6) mg/L MB; $C_{adsorbant}$= 4 g/L	(96.60 – 98.40) MB
Abbreviations: BRed - Brilliant Red HE-3B (Reactive Red 120)/CI 25810; RO - Reactive Orange 16/CI 17757; MB - Methylene Blue (Basic Blue 9)/CI 52015; RhB - Rhodamine B (Basic Violet 10)/CI 45170; CV - Crystal Violet (Basic Violet 3)/CI 42555		

Table 16. Dye adsorption performance onto some natural adsorptive materials (adapted from Suteu et al., 2009b; Suteu et al., 2011a,b).

Adsorption with activated carbon. Activated carbon have been engineered for optimal adsorption of the contaminants found in dyehouse effluents: large, negatively charged or polar molecules of dyes. Powdered or granular activated carbon (specific surface area of 500-1500 m^2/g; pore volume of 0.3-1 cm^3/g; bulk density of 300-550 g/L) has a reasonably good colour removal capacity when is introduced in a separate filtration step. The activated carbon is used as granulate (GAC) in columns or as powder (PAC) in batchwise treatment into a specific treatment tank or basin. High removal rates are obtained for cationic mordant and acid dyes (Anjaneyulu et al., 2005), and a slightly lesser extent (moderate) for dispersed, direct, vat, pigment and reactive dyes (Cooper, 1995; Nigam et al., 2000) with consumable doses of 0.5-1.0 kg adsorbent/m^3 wastewater (i.e. dye removal of 60-90%).

Most recent studies mentioned that an effective irreversible adsorption of dye molecules onto the adsorbent particles takes place via a combination of physical adsorption of dye onto adsorbent surfaces within the microporous structure of the particles, enhanced by an ion-exchange process wherein the interlayer anions of the adsorbent are displaced by the dye molecules, and also inter-particle diffusion processes. Removal of pollutants can take place at any pH between 2 and 11, and at any temperature between 0 and 100°C (effluent temperature as received, generally 30-40°C).

The adsorption on activated carbon without pretreatment is imposible because the suspended solids rapidly clog the filter, and may be feasible in combination with flocculation-sedimentation treatment or a biological treatment (Masui et al., 2005; Ramesh Babu et al., 2007). The main important disadvantage of this process is attributed to the high cost of activated carbon. Performance is dependent on the type of activated carbon used and wastewater characteristics, and can be well suited for one particular wastewater system and ineffective in another. The activated carbon has to be reactivated otherwise disposal of the concentrates has to be considered (reactivation results in 10-15% loss of the adsorbent).

4.1.2 Irradiation
The irradiation treatment is a simple and efficient procedure for eliminating a wide variety of organic contaminants, and as well disinfecting harmful microorganism using gamma rays or electron beams (e.g., source for irradiation can be a monochromatic UV lamps working under 253.7 nm). A high quantity of dissolved oxygen is required for an organic dye to be effectively broken down by irradiation. The dissolved oxygen is consumed very rapidly and so a constant and adequate supply is required. Irradiation treatment of a secondary effluent from sewage treatment plant reduced COD, TOC and colour up to 64%, 34% and 88% respectively, at a dose of 15 K Gy gamma-rays (Borrely et al., 1998). The efficiency of irradiation treatment increases when is used catalyst as titanium dioxide (Krapfenbauer et al., 1999). A lot of data are reported with the practical results obtained at the simple exposure of different dye solutions or dispersions and dye-containing textile wastewaters to sunlight for a period of a half, one or two months (direct photolysis with natural sunlight into open basins). All these reports indicated high removals of colour (>84%), dye destruction by photooxidation following first order kinetics at treatment of some vat dye effluents. But the direct photolysis of textile organic dye in the natural aquatic environment has proven difficult due to strong dependence of the decay rates on dye reactivity and photosensitivity. Most of all commercial dyes are usually designed to be light resistant. Therefore, the recent researches have been directed towards investigation of organic dye photodegradation by sensitizers or catalysts in aqueous/dispersion systems by UV irradiation. Moreover, there are reported high removal of indigo-colour when is initiated a laser fading process for indigo coloured denim textile mainly based on basic interaction of laser beam with indigo-coloured textile (Dascalu et al., 2000).

4.1.3 Membrane processes
The increasing of water cost and necessity of reduction of water consumption implies treatment process which is integrated with in-plant water circuits rather than subsequent treatment (Baban et al., 2010; Machenbach, 1998). From this point of view, membrane filtration offers potential application in combination with other textile effluent treatments.
Membrane processes for wastewater treatment are pressure-driven processes, capable to clarify, concentrate, and most important, separate dye discontinuously from effluent (Xu & Lebran, 1991). These are new technologies, which can restrict organic contaminants and microorganisms presented in wastewater (i.e. color removal, BOD reduction, salt reduction, Polyvinyl Acetate (PVA) recovery, and latex recovery). The common membrane filtration types are: Micro-Filtration (MF), Ultra-Filtration (UF), Nano-Filtration (NF), and Reverse Osmosis (RO). The choice of the membrane process must be guided by the required quality of the final effluent.

Textile Organic Dyes – Characteristics, Polluting Effects and Separation/Elimination Procedures from Industrial
Effluents – A Critical Overview

49

Micro-filtration is mainly used for treatment of dye baths containing pigment dyes as well as for subsequent rinsing baths (Ramesh Babu et al., 2007). Chemicals that can not be filtrated by microfiltration will remain in the dye bath. Microfiltration can be used as a pretreatment for nanofiltration or reverse osmosis (Ghayeni et al., 1998), and also to separate suspended solids, colloids from effluents or macromolecules with pores of 0.1 to 1 micron. MF performance is typically of >90% for turbidity or silt density index. Microfiltration membranes are made of specific polymers such as Poly (Ether Sulfone), Poly (Vinylidiene Fluoride), Poly (Sulfone), Poly (Vinylidene Difluoride), Polycarbonate, Polypropylene, Poly Tetrafluoroethylene (PTFE), etc. Ceramic, glass, carbon, zirconia coated carbon, alumina and sintered metal membranes have been employed where extraordinary chemical resistance or where high temperature operation is necessary. MF and UF operate at 20 to 100 psi transmembrane pressures (Ptm) (low pressure membrane procees) and velocities of 20 to 100 cm/s (Naveed et al., 2006).

Ultra-filtration is used to separate macromolecules and particles, but the elimination of polluting substances such as dyes is never complete (only 31-76% dye removal). The quality of treated wastewater does not permit its reuse for sensitive processes, such as textile dyeing (Ramesh Babu et al., 2007) but permit recycling of 40% treated wastewater in stages in which salinity is not a problem, such as rinsing, washing, etc. Ultrafiltration can only be used as a pretreatment for reverse osmosis (Ciardelli & Ranieri, 2001) or in combination with a biological reactor (Mignani et al., 1999) or to remove metal hydroxides (reducing the heavy metal content to 1 ppm or less) (Naveed et al., 2006). UF membranes are made of polymeric materials (i.e. polysulfone, polypropylene, nylon-6, polytetrafluoroethylene (PTFE), polyvinyl chlorides (PVC), acrylic copolymer etc.

Nano-Filtration was used for the treatment of coloured effluents from the textile industry, mainly in a combination of adsorption (for decreasing of concentration polarization during the filtration process) and nanofiltration (NF modules are extremely sensitive to fouling by colloidal material and macromolecules). NF membranes are generally made of cellulose acetate and aromatic polyamides, and retain low-molecular weight organic compounds, divalent ions, large monovalent ions, hydrolized reactive dyes, and dyeing auxiliaries. Inorganic materials, such as ceramics, carbon based membranes, zirconia, are also used in manufacturing NF and RO membranes. Typical NF flux rates are 5 to 30 GFD (Gross Flow per Day) (Naveed et al., 2006). A performance of above 70% colour removal for a NF plant was reported working at 8 bar/18°C, with four polyethersulphonate membranes with molecular weight cut offs of 40, 10, 5 and 3 kda for three different effluents coming from dyeing cycle of textile industry (Alves & Pinho, 2000). Values of colour removal higher than 90% were reported for single NF process, and also combination MF and NF, in the case of different effluents from textile fabrics processing. Harmful effects of high concentrations of dye and salts in the dye house effluents were frequently reported (i.e. concentration of dye > 1.5 g/L, and of mineral salts >20 g/L) (Tang & Chen, 2002). An important problem is the acculumation of dissolved solids, which makes discharge of treated effluents in watercourses almost impossible. NF treatment can be an alternative fairly satisfactory for textile effluent decolourization.

Reverse Osmosis is used to remove in a single step most types of ionic compounds, hydrolized reactive dyes, chemical auxiliaries, and produce a high quality of permeate (Ramesh Babu et al., 2007). Like NF, RO is very sensitive to fouling and the influent must be carefully pretreated. RO membranes are generally made of cellulose acetate and aromatic polyamides but also of inorganic materials. The Ptm in RO is typically 500 to 1000 psi, with cross flows of 20 to 100 cm/s. The range of typical RO fluxes is 5 to 15 GFD (Naveed et al., 2006). In combination with physio-chemical treatment, the membrane processes has advantages over the other conventional treatments, such as the ability to recover materials with valuable recyclable water, reducing fresh water consumption and wastewater treatment costs, small disposal volumes which minimizes waste disposal costs, reduction of regulatory pressure and fine improved heat recovery systems. Membrane processes have many cost-effective applications in textile industry.

4.2 Chemical treatment
4.2.1 Oxidative processes
Chemical oxidation represents the conversion or transformation of pollutants by chemical oxidation agents other than oxygen/air or bacteria to similar but less harmful or hazardous compounds and/or to short-chained and easily biodegradable organic components (aromatic rings cleavage of dye molecules).

The modern textile dyes are resistant to mild oxidation conditions such those existing in biological treatment systems. Therefore, efficient dye and colour removal must be accomplished by more powerful oxidising agents such as chlorines, ozone, Fenton reagents, UV/peroxide, UV/ozone, or other oxidising procedures or combinations.

Oxidative processes with hydrogen peroxide. The oxidation processes with hydrogen peroxide (H_2O_2) (oxidation potential, $E° = 1.80$ V at pH 0, and $E° = 0.87$ V at pH 14) can be explored as wastewater treatment alternatives in two systems: (1) homogenous systems based on the use of visible or ultraviolet light, soluble catalysts (Fenton reagents) and other chemical activators (e.g. ozone, peroxidase etc.) and (2) heterogenous systems based on the use of semiconductors, zeolites, clays with or without ultraviolet light, such as TiO_2, stable modified zeolites with iron and aluminium (i.e. FeY_5, $FeY_{11.5}$ etc.) (difficulty encountered in the separation of the solid photocatalysts at the end of the process) (Neamtu et al., 2004; Zaharia et al., 2009).

Fenton reagent is usually hydrogen peroxide (H_2O_2) that is activated by some iron salts (i.e. Fe^{2+} salts) (without UV irradiation) to form hydroxyl radicals (HO·) which are strong oxidants (oxidation potential, $E° = 3.06$ V) than H_2O_2 and ozone. The Fenton oxidation reactions are detailed in other chapters of this book, and the treatment efficiency depends mainly of effluent characteristics, and operating parameters (e.g., colour removal of 31.10 or 56.20%, at a pH of 4.00, for Fenton oxidation of textile Remazol Arancio 3R, Remazol Rose RB dye-containing effluents working with 0.18-0.35 M H_2O_2 and 1.45 mM Fe^{2+}, after 30 or 120 min) (Zaharia et al., 2011).

Heterogenous catalytic oxidation with 20 mM H_2O_2 and $FeY_{11.5}$ (1 g/L) of Procion Marine H-EXL dye-containing effluents lead to colour removal of 53-83%, COD removal of 68-76% and TOC removal of 32-37% at pH=3-5, after 10 min of oxidation (Neamtu et al., 2004). Working with FeY_5 (1 g/L) and 20 mM H_2O_2, at pH=3 and 5, for the same textile effluent the treatment *efficiency*, was of 95 and 35% for colour and COD removal after 10 min of oxidation, and 97% for colour after 60 min of oxidation; COD removal (60 min) of 64.20% (Zaharia, 2006).

When small quantities of wastewater are involved or when there is no biotreatment available at the textile site, chemical oxidation might be recommendable treatment option instead of installing a central biological WWT plant. Advantages of this oxidative treatment include reduction of effluent COD, colour and toxicity, and also the possibility to be used to remove both soluble and insoluble dyes (i.e. disperse dyes). Complete decolourization was obtained after the complete Fenton reagent stage (generally 24 hours).

Ozonation process. Ozone is a powerful oxidising agent (oxidation potential, $E° = 2.07$ V) capable of cleavage the aromatic rings of some textile dyes and descomposition of other organic pollutants from industrial effluents. The ozone decomposes the organic dyes with conjugated double bonds forming smaller molecules with increased carcinogenic or toxic properties, and so ozonation may be used alongside a physical method to prevent this (i.e. irradiation, membrane separation, adsorption, etc). Ozone can react directly or indirectly with dye molecules. In the direct pathway, the ozone molecule is itself the electron acceptor, and hydroxide ions (i.e. pH > 7-8) catalyze the auto decomposition of ozone to hydroxyl radicals ($\cdot OH$) in aqueous effluents (very strong and non-selective oxidants) which react with organic and inorganic chemicals. At low pH ozone efficiently reacts with unsaturated chromophoric bonds of a dye molecule via direct reactions (Adams & Gorg, 2002). The main advantage is that ozone can be applied in its gaseous state and therefore does not increase the volume of wastewater and sludge. A disadvantage of ozonation is its short half-life, tipically being 20 min, the destabilisation by the presence of salts, pH, and temperature, and the additional costs for the installation of ozonation plant. The improvement of ozonation preformance is obtained in combination with irradiation (Surpateanu & Zaharia, 2004a; Zaharia et al., 2009) or with a membrane filtration technique (Lopez et al., 1999). Treatment of dye-containing wastewater with ozone followed by chemical coagulation using $Ca(OH)_2$ indicated 62% colour removal after ozonation (Sarasa et al., 1998).

Oxidation process with sodium hypochlorite. This treatment implies the attack at the amino group of the dye molecule by Cl^+, initiating and accelerating azo-bond cleavage. The increasing of chlorine concentration favors the dye removal and decolourization process, and also the decreasing of pH. The dye containing amino or substituted amino groups on the naphthalene ring (i.e. dyes derived from amino-naphtol- and naphtylamino-sulphonic acids) are most susceptible for chlorine decolourization (Omura, 1994). This treatment is unsuitable for disperse dyes, and is becoming less frequent due to the negative effects at releasing into watercourses of aromatic amines or otherwise toxic molecules. Moreover, although about 40% of the pigments used worldwise contain chlorine this corresponds to only less than 0.02% of the total chlorine production (Slokar & Le Marechal, 1997).

Photochemical oxidation process. The UV treatment in the presence of H_2O_2 can descomposed dye molecules to low weight organic molecules, or even to CO_2, H_2O, other inorganic oxides, hydrides, etc. There can be also produced additional by-products such as halides, metals, inorganic acids, organic aldehydes and organic acids depending on initial materials and the extent of decolourisation treatment (Yang et al., 1998). The dye decomposition is initiated by the generated hydroxyl radicals ($H_2O_2 + h\upsilon \rightarrow 2HO\cdot$) and hydroperoxide radicals ($H_2O_2 + HO\cdot \rightarrow HO_2\cdot + H_2O$).

The treatment may be set-up in a batch or continuous column unit, and is influenced by the intensity of the UV radiation, pH, dye structure and the dye bath composition (Slokar & Le Marechal, 1997). The performance of photooxidation treatment in the presence of hydrogen peroxide are high (i.e. >60-90% for colour removal, working with 400-500 mg/L H_2O_2 at pH 3-7, for Red M5B, H-acid and Blue MR dye-containing effluents) (Anjaneyulu et al., 2005) or 81-94% dye removal after 60 min, working with 88 mM H_2O_2 at pH of 4-6, for Acid Red G dye-containing effluent (Surpateanu & Zaharia, 2004b; Zaharia et al., 2009).

Electrochemical oxidation process. As an advanced process, the electrochemical treatment of dye-containing effluents is a potentially powerful method of pollution control, offering high removal efficiencies (Anjaneyulu et al., 2005) especially for acid dyes as well as disperse and metal complex dyes. The main advantages of this treatment are considered the requirement of simple equipment and operation, low temperature in comparison with other non-electrochemical treatments, no requirement of any additional chemicals, easy control but crucial for pH, the electrochemical reactors (with electrolytic cells) are compact, and prevent the production of unwated by-products. The principal oxidising agent in electrochemical process is hypochlorite ion or hypochlorous acid produced from naturally occuring chloride ions. Hydroxyl radical and other reactive species also participate in electrochemical oxidation of organics (Kim et al., 2002) that can be achieved directly or indirectly at the anode. The breakdown compounds are generally not hazardous being discharged into watercourses without important environmental and health risks.

The electrochemical oxidation is considered an efficient and economic treatment of recycling textile wastewater for the dyeing stage. The environmental advantage mainly achieved is the minimization of all emissions: emission of gases, solid waste, and liquid effluent. Other important advantage is its capacity of adaptation to different volumes and pollution loads. The main disadvantage is the generation of metallic hydroxide sludge (from the metallic electrodes in the cell), that limits its use (Ramesh Babu et al., 2007).

Some studies reported colour removals of about 100% for dyeing wastewater within only 6 min of electrolysis (Vlyssides et al., 2000) (e.g., complete decolourization of textile effluent containing blue-26 anthraquinone dye by electrochemical oxidation with lead dioxide coated anode - Titanium Substract Insoluble Anode (TSIA), at neutral pH, in the presence of sodium chloride, current density of 4.5 A/dm^2, electrolysis time of 220 min, or maximum 95.2% colour and 72.5% COD removal of textile azo dye-containing effluent in a flow reactor working at rate of 5 mL/min and current density of 29.9 mA/cm^2) (Anjaneyulu et al., 2005).

4.2.2 Coagulation-flocculation and precipitation

It is clearly known that the coloured colloid particles from textile effluents cannot be separated by simple gravitational means, and some chemicals (e.g., ferrous sulphate, ferric sulphate, ferric chloride, lime, polyaluminium chloride, polyaluminium sulphate, cationic organic polymers, etc.) are added to cause the solids to settle. These chemicals cause destabilisation of colloidal and small suspended particles (e.g. dyes, clay, heavy metals, organic solids, oil in wastewater) and emulsions entrapping solids (coagulation) and/or the agglomeration of these particles to flocs large enough to settle (flocculation) or highly improve further filtration (Zaharia et al., 2006; Zaharia et al., 2007). In the case of flocculation, anionic and non-ionic polymers are also used.

The mechanism by which synthetic organic polymer removes dissolved residual dyes from effluents is best described in terms of the electrostatic attraction between the oppositely charged soluble dye and polymer molecules. Many of the most problematic dye types, such as reactive dyes, carry a residual negative charge in their hydrolysed dissolved form, and so positively charged groups on the polymers provide the neccesary counter for the interaction and subsequent precipitation to occur. The immediate result of this coprecipitation is the almost instantaneous production of very small coloured particles, having little strength and breaking down at any significant disturbances. The agglomeration of the coloured precipitates by using appropriate high polyelectrolyte flocculants produces stable flocs (Zaharia et al., 2007, 2011). The main disadvantages of this treatment are the process control that is a little difficult, the potential affection of precipitation rate and floc size by impurities such as non-ionic detergents remaining in the effluent, and the sludge production which has to be settled, dewatered and pressed into a cake for subsequent landfilling tipping.

Textile Organic Dyes – Characteristics, Polluting Effects and Separation/Elimination Procedures from Industrial
Effluents – A Critical Overview

53

There are reported very effective chemical coagulation-flocculation (C-F) and precipitation of textile wastewater which reduced the load on the biological treatment, working with polyaluminium chloride along with an organic polymer (Lin & Chen, 1997) or ferrous/ferric chloride and a commercial organic coagulant aid at pH of 6.7-8.3 (colour removal > 80%) (Venkat Mohan et al., 1999) or alum at pH=8.2 (54-81% colour removal) with addition of bentonite (3 g/L) for Remazol Violet dye-containing effluent (Sanghi et al., 2001). Other efficient textile treatments mentioned by different textile operators consist in coagulation-flocculation followed by membrane technology (especially for recycling textile effluents).

Some of authors' results in different (C-F) treatments are summarized in Table 17.

Process	Coagulant/ Flocculant	Wastewater characteristics or dye type / Results
Coagulation-flocculation/ sedimentation/ filtration	Ferrous sulphate (5 mg/L) + Ponilit GT-2 anionic polyelectrolyte (15 mg/L) + bentonite (3 g/L)	*Wastewater characteristics*: pH=6.5-7; TSS=250-1000 mg/L; colour=650 UH; COD$_{Cr}$=152.7-272 mg O$_2$/L *Process efficiency*: turbidity removal=70.31-91.34%; colour removal=70.20-90.50% and COD$_{Cr}$ removal=34.90-45.20%
Precipitation/ flocculation/ flotation with dissolved air	NaOH + Na$_2$CO$_3$ + Ca(OH)$_2$, Ponilit GT-2 anionic polyelectrolyte	*Wastewater characteristics*: pH=7.0-9.5; total metallic ions= 48.30 mg/L; extractable substances in organic solvents= 980 mg/L *Process efficiency*: Metallic ions removal=78.67-92.33%
Coagulation-flocculation/ sedimentation/ filtration	Ferric sulfate (2-5 mg/L) +Prodefloc CRC 301 (0.25-1.5 mg/L) cationic polyelectrolyte	*Wastewater characteristics*: pH=6.98; T=20°C, Turbidity=556 FTU; COD$_{Cr}$=152.60 mg O$_2$/L; colour=1320 UH *Process efficiency*: maximal turbidity removal=95.87% and colour removal= 93.90-98.10%

Table 17. Some applications of coagulation- flocculation in wastewater treatment (adapted from Zaharia, 2006; Zaharia et al., 2006, 2007).

4.2.3 Electrocoagulation

An advanced electrochemical treatment for dye and colour removal is electrocoagulation (EC) that has as main goal to form flocs of metal hydroxides within the effluent to be cleaned by electro-dissolution of soluble anodes. EC involves important processes as electrolytic reactions at electrodes, formation of coagulants in aqueous effluent and adsorption of soluble or colloidal pollutants on coagulants, and removal by sedimentation and flotation. This treatment is efficient even at high pH for colour and COD removals being strongly influenced by the current density and duration of reaction. The EC treatment was applied with high efficiency for textile Orange II and Acid red 14 dye-containing effluents (i.e. > 98% colour removal) (Daneshvar et al., 2003) or Yellow 86 dye-containing wastewater (i.e. turbidity, COD, extractible substances, and dye removal of 87.20%, 49.89%, 94.67%, and 74.20%, after 30 min of operation, current intensity of 1 A, with monopolar electrodes) (Zaharia et al, 2005) when iron is used as sacrificial anode. In general, decolorization performance in EC treatment is between 90-95%, and COD removal between 30-36% under optimal conditions (Ramesh Babu et al., 2007).

4.2.4 Ionic exchange

The ion exchange process has not be widely used for treatment of dye-containing effluents, mainly because of the general opinion that ion exchangers cannot accomodate a wide range of dyes (Slokar & Le Marechal, 1997).

The ionic exchange occurs mainly based on the interaction of ionic species from wastewater with an adsorptive solid material, being distinguished from the conventional adsorption by nature and morphology of adsorptive material or the inorganic structure containing functional groups capable of ionic exchange (Macoveanu et al., 2002). The mechanisms of ionic exchange process are well known, and two principal aspects must be mentioned: (1) ionic exchange can be modeling as well as adsorption onto activated coal; (2) the ion exchangers can be regenerated without modifying the equilibrium condition (e.g., by passing of a salt solution containing original active groups under ion exchanger layer). In the case of wastewater treatment, the effluent is passed over the ion exchanger resin until the available exchange sites are saturated (both cationic and anionic dyes are removed).

The ionic exchange is a reversible process, and the regenerated ion exchanger can be reused. The essential characteristic of ionic exchange that makes distinction of adsorption is the fact that the replace of ions takes place in stoechiometric proportion (Macoveanu et al., 2002).

The effluent treatment by ionic exchange process contributes to the diminishing of energetic consumption and recovery of valuable components under diverse forms, simultaneously with the wastewater treatment. In practice, the ion exchangers are used in combination with other wastewater treatments. The main advantages of ion exchange are removal of soluble dyes, no loss of adsorbent at regeneration, and reclamation of solvent after use. The important disadvantages of this process is the cost, organic solvents are expensive, and ion exchange treatment is not efficient for disperse dyes (Robinson et al., 2001). Our results in batchwise treatment of Brilliant Red HE-3B dye-containing effluents (0.05 – 0.3 mg/mL) using anionic Purolite A-400 and Purolite A-500 indicated dye removal of 48-89% working in the optimal conditions, or dye removal of 56-78% for Crystal Violet (Basic Violet 3) using Purolite C-100 or dye removal between 78-89% for Reactive Blue M-EB using ion exchange celluloses (Suteu et al., 2002).

4.3 Biological treatments

Biological treatments are considered reproduction, artificially or otherwise, of self-purification phenomena existing in natural environment. There are different biological treatments, performed in aerobic or anaerobic or combined anaerobic/aerobic conditions. The processing, quality, adaptability of microorganisms, and the reactor type are decisive parameters for removal efficiency (Börnick & Schmidt, 2006).

Biological treatment process for decolorization of industrial effluents is ambiguous, different and divergent (Anjaneyulu et al., 2005). Previous subchapters indicate that dyes themselves are not biologically degradable since microorganisms do not use the coloured constituents as a source of food. The most currently used biodegradation involve aerobic micro-organisms, which utilize molecular oxygen as reducing equivalent acceptor during the respiration process. But biodegradation in anaerobic environment conditions (anoxic and hypoxic environments) also occurs, and survival of microorganisms is possible by using sulphates, nitrates and carbon dioxide as electron acceptors (Birch et al., 1989).

Research data indicates that certain dyes are susceptible to anoxic/anaerobic decolourization, and also that an anaerobic step followed by an aerobic step may represent a significant advancement in biological decolourization treatment in future (Ong et al., 2005). The treatment plant that receives dye-containing effluents has high potential to form toxic biodegradation products such as toxic amines, benzidine and its derivates, etc. To avoid that risk, anaerobic/aerobic sequential reactor systems seem to be an efficient procedure (i.e. efficient colour removal takes place during the anaerobic treatment, and high reduction of aromatic amines and other organic compounds occurs during the subsequent aerobic treatment). Under aerobic conditions, most of the azo dye metabolites are quickly degraded by oxidation of the substituents or of the side branches. However, some of them are still rather recalcitrant. Successful removal of poorly degradable amines was often achieved by adaptation of microorganisms (i.e. acclimatization of biological sludge to nitroaniline-containing wastewaters; after gradual adaptation, the microorganisms are able to eliminate 3- and 4-nitroaniline simultaneously) (Börnick & Schmidt, 2006). Difficult to be biodegraded are also the aromatic amines containing sulfo substituents in metha position (e.g., 3-aminobenzenesulfonate) that can be treated with good efficiencies with flow-through bioreactors within 28 days. Contrarily, some experimental results found anaerobic mineralization to be efficient for aromatic amines.

The main advantage of biological treatment in comparison with certain physico-chemical treatments is that over 70% of organic matter expressed by COD_{Cr} may be converted to biosolids (Anjaneyulu et al., 2005).

4.3.1 Aerobic biological treatment

Biological treatment with 'activated sludge' was the most used in large scale textile effluent treatment, and the trickling filter or biological aerated filter (BAF) is an alternative, permitting a 34-44% dye-colour removal for different high dyeing loads of industrial effluents. The main microorganisms contributing to biodegradation of organic compounds are bacteria (e.g. *Bacillus subtilis, Aeromonas Hydrophilia, Bacillus cetreus, Klebsiella pneunomoniae, Acetobacter liquefaciens, Pseudomonas* species, *Pagmentiphaga kullae, Sphingomonas*, etc.), fungi (e.g., white-rot fungi: *Phanerochaete chrysosporium, Hirschioporus larincinus, Inonotus hispidus, Phlebia tremellosa, Coriolus versicolor*, etc.), algae (e.g. *Chlorella* and *oscillotoria* species) etc. Moreover, some bacteria, white-rot fungi, mixed microbial cultures from a wide variety of habitats are found to be able to degrade dyes using enzymes, such as lignin peroxidases (LiP), manganese dependent peroxidases (MnP), H_2O_2-producing enzyme such as glucose-1-oxidase and glucose-2-oxidase, along with laccase, and a phenoloxidase. Biological aerated filters involve the growth of an organism on inert media that are held stationary during normal operation and exposed to aeration. In aerobic conditions, the mono- and dioxygenase enzymes catalyse the incorporation of dissolved oxygen into the aromatic ring of organic compounds prior to ring fission. Although azo dyes are aromatic compounds, their nitro and sulfonic groups are quite recalcitrant to aerobic bacterial degradation (Dos Santos et al., 2004). However, in the presence of specific oxygen-catalysed enzymes called azo reductases, some aerobic bacteria are able to reduce azo compounds and produce aromatic amines (Stolz, 2001). The batch experiments with aerobic activated sludge confirmed the biodegradability of sulphonated azo dyes. Only aerobic degradation of the azo dyes is possible by azo reduction (i.e. high colour removal (>90%) of Red RBN azo dye-containing effluents (3000 mg/L) working with *Aeromonas hydrophilla* in the specific optimal conditions of pH (5.5-10), temperature (20°-38°C), and time (8 days)), and mineralization does not occur. The subsequential anaerobic and aerobic bioreactor was able to completely remove the sulphonated azo dye (i.e. MY10) at a maximum loading rate of 210 mg/L per day (Tan et al., 2000). The degradation of azo dyes (i.e. Acid Red 151; Basic Blue 41; Basic Red 46, 16; Basic Yellow 28, 19) in an aerobic biofilm system indicated 80% colour removal (Anjaneyulu et al., 2005). The improvement of dye biodegradation performance (i.e. >90% colour removal) are made by adding activated carbon (PAC) or bentonite in aeration tank.

4.3.2 Anaerobic biological treatment

Anaerobic biodegradation of azo and other water-soluble dyes is mainly reported as an oxidation-reduction reaction with hydrogen, and formation of methane, hydrogen sulphide, carbon dioxide, other gaseous compounds, and releasing electrons. The electrons react with the dye reducing the azo bonds, causing the effluent decolourization. Azo dye is considered an oxidising agent for the reduced flavin nucleotides of the microbial electron chain, and is reduced and decolourized concurrently with reoxidation of the reduced flavin nucleotides (Robinson et al., 2001). An additional carbon organic source is necessary, such as glucose which is a limiting factor in scale set-up technology application. The azo and nitro-components are reduced in anoxic sediments and in the intestinal environment, with regeneration of toxic amines (Banat et al., 1996). A major advantage of anaerobic system along with effluent decolourization is the production of biogas, reusable for heat and power generation that will reduce energy costs. Since textile industry wastewaters are generally discharged at high temperatures (40–70°C), thermophilic anaerobic treatment could serve as an interesting option, especially when closing process water cycles is considered. Anaerobic decolourization of textile effluents (e.g., colour removal of >99% for a Orange II, Black 2HN under anaerobic condition, more than 72 h) is not yet well established although successful pilot-scale and full-scale plants are very well operating (Tan et al., 2000). Among the different studied reactors, anaerobic filter and UASB thermophilic anaerobic reactor gave good colour removals, using or not redox mediator (e.g., antraquinone-2,6, disulphonic acid) as catalyst capable of acceleration the colour removal of azo dye-containing wastewaters.

5. Conclusions

The dyes are natural and synthetic compounds that make the world more beautiful through coloured products but are also considered as pollutants of some water resources.

The textile sector will continue to be vitally important in the area of water conservation due to its high consumption of water resources, and its individual or combined effluents' treatments for no environmental pollution generation (i.e. polluting colourants).

The satisfaction of both discharge criteria for sewerage systems, watercourses and textile reuse standards within economically viable limits implies critical analyses of industrial effluents (total wastewater and raw reusable stream characterisation) and removal of all pollutants from final effluents. The special category of organic pollutants - textile organic dyes - must respect the strict limits in final effluents discharged or not in natural water resources. This fact imposes the colour and/or dye removal from final effluents (especially industrial effluents).

Dye removal from textile effluents in controlled conditions and strict reproductibility is an environmental issue achievable by application of adequate mechano-physico-chemical and also biological treatment procedures.

Textile Organic Dyes – Characteristics, Polluting Effects and Separation/Elimination Procedures from Industrial
Effluents – A Critical Overview

57

6. References

Adams, C.D. & Gorg, S. (2002). Effect of pH and gas-phase ozone concentration on the decolourization of common textile dyes. *J.Environ.Eng.*, Vol.128, No.3, pp. 293-298

Alves, B.M.A. & Pinho, D.N.M. (2000). Ultrafiltration for colour removal of tannery dyeing wastewaters., *Desalination*, Vol.1303, pp. 147-154

Anjaneyulu, Y.; Sreedhara Chary, N. & Suman Raj, D.S. (2005). Decolourization of industrial effluents – available methods and emerging technologies – a review. *Reviews in Environmental Science and Bio/Technology*, Vol.4, pp. 245-273, DOI 10.1007/s11157-005-1246-z

Baban, A.; Yediler, A. & Ciliz, N.K. (2010). Integrated water management and CP implementation for wool and textile blend processes. *Clean*, Vol.38, No.1, pp. 84-90

Balchioglu, I.A.; Aslan, I. & Sacan, M.T. (2001). Homogeneous and heterogeneous advanced oxidation of two commercial reactive dyes. *Environ.Technol.*, Vol.22, pp.813-822

Banat, M.E.; Nigam, P.; Singh, D. & Marchant, R. (1996) Microbial decolourization of textile dye containing effluents, a review. *Biores. Technol.*, Vol.58, pp. 217-227

Bhattacharyya, K.G. & Sarma, A. (2003). Adsorption characteristics of the dye, Brilliant Green, on Neem leaf powder. *Dyes Pigments*, Vol.57, pp. 211-222

Birch, R.R.; Biver, C. Campagna, R.; Gledhill, W.E.; Pagga, U.; Steber, J.; Reust, H. & Bontinck, W.J. (1989). Screening of chemicals for anaerobic biodegradability. *Chemosphere*, Vol.19, No.(10-11), pp. 1527-1550

Bisschops, I.A.E. & Spanjers, H. (2003). Literature review on textile wastewater characterisation. *Environmental Technology*, Vol.24, pp. 1399-1411

Bertea, A. & Bertea, A.P. (2008). *Decolorisation and recycling of textile wastewater* (in Romanian), Performantica Ed, ISBN 978-973-730-465-0, Iasi, Romania

Börnick, H. & Schmidt, T.C. (2006). Amines, In: *Organic pollutants in the water cycle. Properties, occurrence, analysis and environmental relevance of polar compounds*, T. Reemtsma & M. Jekel, (Eds.), pp. 181-208, Wiley-VCH Verlag GmbH & Co. KGaA, ISBN 978-3-527-31297-9, Weinheim, Germany

Borrely, S.I.; Cruz, A.C.; Del Mastro, N.L.; Sampa, M.H.O. & Somssari, E.S. (1998). Radiation process of sewage and sludge – A review. *Prog.Nucl.Energy*, Vol.33, No.1/2, pp. 3-21

Carliell, C.M.; Barclay, S.J.; Naidoo, N.; Buckley, C.A.; Mulholland, D.A. & and Senior, E. (1994). Anaerobic decolorisation of reactive dyes in conventional sewage treatment processes. *Water SA*, Vol.20, pp. 341-344

Ciardelli, G. & & Ranieri, N. (2001). The treatment and reuse of wastewater in the textile industry by means of ozonation and electroflocculation. *Water Res.*, Vol.35, pp. 567-572

Cooper P. (1995). *Color in dyehouse effluent*, Society of Dyers and Colourists, ISBN 0 901956 694, West Yorkshire BDI 2JB, England

Correia, V.M.; Stephenson, T. & Judd, S.J. (1994). Characterization of textile wastewaters – a review. *Environmental Technology*, Vol.15, pp.917-929

Daneshvar, N.; Sorkhabi, H.A. & Tizpar, A. (2003). Decolorization of orange II by electrocoagulation method. *Sep.Purifi.Technol.*, Vol.31, pp.153-162

Dascalu T.; Acosta-Ortiz, E.S.; Morales, O.M. & Compean, I. (2000). Removal of the indigo colour by laser beam-denim interanction. *Opt. Lasers Eng.*, Vol.34, pp.179-189

Dos Santos, A.B.; Cervantes, F.J. & Van Lier, J.B. (2004). Azo dye reduction by thermophilic anaerobic granular sludge, and the impact of the redox mediator AQDS on the reductive biochemical transformation. *Applied Microbiology and Biotechnology*, Vol.64, pp.62-69

EPA. (1997). *Profile of the textile industry*. Environmental Protection Agency, Washington, USA

EWA. (2005). *Efficient use of water in the textile finishing industry*, Official Publication of the European Water Association (EWA), Brussels, Belgium

Fontenot, E.J.; Lee, Y.H.; Matthews, R.D.; Zhu, G. & Pavlostathis, S.G. (2003). Reductive decolorization of a textile reactive dyebath under methanogenic conditions. *Applied Biochemistry and Biotechnology*, Vol.109, pp. 207-225

Ghayeni, S.B.; Beatson, P.J.; Schneider, R.P. & Fane, A.G. (1998). Water reclamation from municipal wastewater using combined microfiltration-reverse osmosis (ME-RO): Preliminary performance data and microbiological aspects of system operation. *Desalination*, Vol.116, pp. 65-80

Gupta, G.S.; Singh, A.K.; Tayagi, B.S.; Prasad, G. & Singh, V.N. (1992). Treatment of carpet and metallic effluent by China clay. *J.Chem.Technol.Biotechnol.*, Vol.55, pp. 227-283

Hao, O.J.; Kim, H. & Chang, P.C. (2000). Decolorization of wastewater. *Critical Reviews in Environmental Science and Technology*, Vol.30, pp. 449-505

Kim, T.H.; Park, C.; Lee, J.; Shin, E.B. & Kim, S. (2002). Pilot scale treatment of textile wastewater by combined processes (fluidized biofilm process– chemical coagulation – electrochemical oxidation). *J.Hazar.Mat. B*, Vol.112, pp.95-103

Krapfenbauer, K.F.; Robinson, M.R. & Getoff, N. (1999). Development of and testing of TiO_2-Catalysts for EDTA-radiolysis using γ-rays (1st part). *J.Adv.Oxid.Technol.*, Vol.4, No.2, pp.213-217

Lin, S.H. & Chen, L.M. (1997). Treatment of textile waste waters by chemical methods for reuse. *Water Res.*, Vol.31, No.4, pp.868-876

Lopez, A.; Ricco, G.; Ciannarella, R.; Rozzi, A., Di Pinto, A.C. & Possino, R. (1999). Textile wastewater reuse: ozonation of membrane concentrated secondary effluent. *Water Sci. Technol.*, Vol.40, pp. 99-105

Machenbach, I. (1998). Membrane technology for dyehouse effluent treatment. *Membrane Technol.*, Vol.58, pp.7-10

Macoveanu, M.; Bilba, D.; Bilba, N.; Gavrilescu, M. & Soreanu, G. (2002). *Ionic exchange processes in environmental protection* (in Romanian), MatrixRom Ed., Bucuresti, Romania

Matsui, Y.; Murase, R.; Sanogawa, T.; Aoki, N.; Mima, S.; Inoue, T. & Matsushita, T. (2005). Rapid adsorption pretreatment with submicrometre powdered activated carbon particles before microfiltration. *Water Science and Technology*, Vol.51, pp.249-256

Mignani, M.G.; Nosenzo, G. & Gualdi, A. (1999). Innovative ultrafiltration for wastewater reuse. *Desalination*, Vol.124, pp.287-292

Naveed, S.; Bhatti, I. & Ali, K. (2006). Membrane technology and its suitability for treatment of textile waste water in Pakistan. *Journal of Research (Science)*, Bahauddin Zakariya University, Multan, Pakistan, Vol. 17, No. 3, pp.155-164

Neamtu, M.; Zaharia, C.; Catrinescu, C.; Yediler, A.; Kettrup, A. & Macoveanu, M. (2004). Fe-exchanged Y zeolite as catalyst for wet peroxide oxidation of reactive azo dye Procion Marine H-EXL. *Applied Catalysis B: Environmental*, Vol.78, No.2, pp.287-294

Nigam, P.; Armou, G.; Banat, I.M.; Singh, D. & Marchant, R. (2000). Physical removal of textile dyes and solid-state fermentation of dye-adsorbed agricultural residues. *Biores. Technol.*, Vol.72, pp.219-226

Omura, T. (1994). Design of chlorine – fast reactive dyes – part 4; degradation of amino containing azo dyes by sodium hydrochlorite. *Dyes Pigments*, Vol.26, pp.33-38

Ong, S.A.; Toorisaka, E.; Hirata, M. & Hano, T. (2005). Treatment of azo dye Orange II in aerobic and anaerobic-SBR systems. *Proc.Biochem.*, Vol.40, pp.2907-2914

Orhon, D.; Babuna, F.G. & Insel, G. (2001). Characterization and modelling of denim-processing wastewaters for activated sludge. *Journal of Chemical Technology and Biotechnology*, Vol.76, pp.919-931

Ozcan, A.S.; Erdem, B. & Ozcan, A. (2004). Adsorption of Acid Blue 193 from aqueous solutions onto Na-bentonite and DTMA-bentonite. *J.Coll.Interf.Sci.*, Vol.280, No.1, pp.44-54

Oztekin, Y; Yazicigil, Z.; Ata, N. & Karadayl, N. (2010). The comparison of two different electro-membrane processes' performance for industrial application. *Clean-Soil, Air, Water*, Vol.38, No.5-6, pp.478-484

Ramesh Babu, B.; Parande, A.K.; Raghu, S. & Prem Kumar, T. (2007). Textile technology. Cotton Textile Processing: Waste Generation and Effluent Treatment. *The Journal of Cotton Science*, Vol.11, pp.141-153

Robinson, T.; McMullan, G.; Marchant, R. & Nigam, P. (2001). Remediation of dyes in textile effluent: a critical review on current treatment technologies with a proposed alternative. *Bioresource Technology*, vol.77, pp.247-255

Slokar, Y.M. & Le Marechal, A.M. (1997). Methods of decoloration of textile wastewaters. *Dyes Pigments*, Vol.37, pp.335-356

Sarasa, J.; Roche, M.P.; Ormad, M.P.; Gimeno, E.; Puig, A. & Ovelleiro, J.L. (1998). Treatment of wastewater resulting from dye manufacturing with ozone and chemical coagulation. *Water Res.*, Vol.32, No.9, pp.2721-2727

Soloman, P.A.; Basha, C.A.; Ramamurthi, V.; Koteeswaran, K. & Balasubramanian, N. (2009). Electrochemical degradation of Remazol Black B dye effluent. *Clean*, Vol.37, No.11, pp.889-900

Stolz, A. (2001). Basic and applied aspects in the microbial degradation of azo dyes. *Appl.Microbiol.Biotechnol.*, Vol.56, pp.69-80

Surpăţeanu, M. & Zaharia, C. (2004a). Advanced oxidation processes. Decolorization of some organic dyes with hydrogen peroxide. *Environmental Engineering and Management Journal*, Vol.3, No.4, pp.629-640

Surpateanu, M. & Zaharia C. (2004b). Advanced oxidation processes for decolorization of aqueous solution containing Acid Red G azo dye. *Central European Journal of Chemistry*, Vol.2, No.4, pp.573-588

Suteu, D.; Bîlbă, D. & Zaharia, C. (2002). Kinetic study of Blue M-EB dye sorption on ion exchange resins. *Hungarian Journal of Industrial Chemistry*, Vol.30, pp.7-11

Suteu, D.; Zaharia, C.; Bilba, D.; Muresan, A.; Muresan, R. & Popescu, A. (2009a). Decolorization wastewaters from the textile industry – physical methods, chemical methods. *Industria Textila*, Vol.60, No.5, pp.254-263

Suteu, D.; Zaharia, C.; Muresan, A.; Muresan, R. & Popescu, A. (2009b). Using of industrial waste materials for textile wastewater treatment. *Environmental Engineering and Management Journal*, Vol.8, No.5, pp.1097-1102

Suteu, D. & Zaharia, C. (2008). Removal of textile reactive dye Brilliant Red HE-3B onto materials based on lime and coal ash, *ITC&DC, Book of Proceedings of 4th International Textile, Clothing & Design Conference – Magic World of Textiles*, pp.1118-1123, ISBN 978-953-7105-26-6, Dubrovnik, Croatia, October 5-8, 2008

Suteu, D.; Zaharia, C. & Malutan, T. (2011a). Biosorbents Based On Lignin Used In Biosorption Processes From Wastewater Treatment (chapter 7). In: *Lignin: Properties and Applications in Biotechnology and Bioenergy*, Ryan J. Paterson (Ed.), Nova Science Publishers, 27 pp., ISBN 978-1-61122-907-3, New York, U.S.A.

Suteu, D.; Zaharia, C. & Malutan, T. (2011b). Removal of Orange 16 reactive dye from aqueous solution by wasted sunflower seed shells. *Journal of the Serbian Chemical Society*, Vol.178, No.3, pp.907-924

Tan, N.C.G.; Borger, A.; Slenders, P., Svitelskaya, A.; Lettinnga, G. & Field, J.A. (2000). Degradation of azo dye mordent yellow 10 in a subsequential anaerobic and bioaugmented aerobic bioreactor. *Water Sci.Technol.*, Vol.42, No.(5-6), pp.337-344

Tang, C. & Chen, V. (2002). Nanofiltration of textile wastewater for water reuse. *Desalination*, Vol.143, pp.11-20

Vandevivere, p.C.; Bianchi, R. & Verstraete, W. (1998). Treatment and reuse of wastewater from the textile wet-processing industry review of emerging technologies. *J.Chem.Technol.Biotechnol.*, Vol.72, No.4, pp.289-302

Venkat Mohan, s.; Srimurli, M.; Sailaja, P. & Karthikeyan, J. (1999). A study of acid dye colour removal using adsorption and coagulation. *Environ.Eng.Poly.*, vol.1, pp.149-154

Vlyssides, A.G.; Papaioannou, D.; Loizidoy, M.; Karlis, P.K. & Zorpas, A.A. (2000). Testing an electrochemical method fortreatment of textile dye wastewater. *Waste Manag.*, Vol.20, pp.569-574

Xu, Y. & Lebrun, R.E. (1991). Treatment of textile dye plant effluent by nanofiltration membrane. *Separ.Sci.Technol.*, Vol.34, pp.2501-2519

Welham, A. (2000). The theory of dyeing (and the secret of life). *Journal of the Society of Dyers and Colourists*, Vol.116, pp.140-143

Wiesmann, U.; Choi, I.S. & Dombrowski, E.M. (2007). *Fundamentals of Biological Wastewater Treatment*. Wiley-VCH Verlag GmbH&Co. KgaA, Weinheim, Germany

Yang, Y.; Wyatt II, D.T & Bahorsky, M. (1998). Decolorisation of dyes using UV/H_2O_2 photochemical oxidation. *Text.Chem.Color.*, Vol.30, pp.27-35

Zaharia, C. (2006). *Chemical wastewater treatment* (in Romanian), Performantica Ed., ISBN 978-973-730-222-9, Iasi, Romania

Zaharia, C. (2008). *Legislation for environmental protection* (in Romanian), Politehnium Ed., ISBN 978-973-621-219-2, Iasi, Romania

Zaharia, C.; Diaconescu, R. & Surpăţeanu, M. (2006). Optimization study of a wastewater chemical treatment with PONILIT GT-2 anionic polyelectrolyte. *Environmental Engineering and Management Journal*, Vol.5, No.5, pp.1141-1152

Zaharia, C.; Diaconescu, R. & Surpăţeanu, M. (2007). Study of flocculation with Ponilit GT-2 anionic polyelectrolyte applied into a chemical wastewater treatment. *Central European Journal of Chemistry*, Vol.5, No.1, pp.239-256

Zaharia, C.; Surpăţeanu, M.; Creţescu, I.; Macoveanu, M. & Braunstein, H. (2005). Electrocoagulation/electroflotation – methods applied for wastewater treatment. *Environmental Engineering and Management Journal*, Vol.4, No.4, pp.463-472

Zaharia, C.; Suteu, C. & Muresan, A. (2011). Options and solutions of textile effluent decolourization using some specific physico-chemical treatment steps. *Proceedings of 6th International Conference on Environmental Engineering and Management ICEEM'06*, pp. 121-122, Balaton Lake, Hungary, September 1-4, 2011

Zaharia, C.; Suteu, D.; Muresan, A.; Muresan, R. & Popescu, A. (2009). Textile wastewater treatment by homogenous oxidation with hydrogen peroxide. *Environmental Engineering and Management Journal*, Vol.8, No.6, pp.1359-1369

3

The Inputs of POPs into Soils by Sewage Sludge and Dredged Sediments Application

Radim Vácha
Research Institute for Soil and Water Conservation, Prague
Czech Republic

1. Introduction

The use of sewage sludge and dredged sediments in agriculture belong to the most important ways of possible pollutants inputs into agricultural soils in many countries including Czech Republic.

The application of sewage sludge on agricultural soils is connected with following facts:

- increasing amounts of sewage sludge thanks to intensive waste water treatment
- the characteristic of sludge as the material with increased content of organic matter and nutrients

The application of sludge into soil could lead to an increase of the contents of organic matter or macro elements, but the contamination by potentially risky elements and persistent organic pollutants could be relevant also. The problems connecting with increased persistent organic pollutants (POPs) contents in sewage sludge were confirmed by many authors (Markard, 1988; Melcer et al., 1988; Starke, 1992; Oleszczuk, 2007; Clarke et al., 2008; Natal-da-Luz et al., 2009). The load of soil by POPs after sludge application can influence their transfer into food chains (Passuello et al., 2010). Increased contents of polycyclic aromatic hydrocarbons (PAHs) limit not only direct application of sewage sludge on the soil but also the use of sludge in composting processes for example (Rosik-Dulewska et al., 2009). The inputs of POPs into agricultural soils by biosolids use in agriculture plays an important role. This problematic is documented on the example of following study realised in the Czech Republic where the contents of POPs in the soil and plants after sewage sludge and sediments application were observed.

The number of waste water factories increased after implementation of Czech Republic into European Union when the obligation of waste water factory existence in every settlement over 10 000 inhabitants till 2010 year had to be fulfilled. The necessity of legislative regulation existence controlling this process was obvious since the beginning of ninetieth years and the Directive No. 382/2001 was the first version of legislative adaptation. The Directive was modified under the No. 504/2004 Sb. in 2004 year.

The directive of Czech Ministry of Environment No. 504/2004 Sb. regulates the application of the sludge on agricultural soils. The directive determines the conditions of sludge application on agricultural soils, including limit values of potentially risk elements and some persistent organic pollutants (sum of halogenated organically bound substances - AOX, sum of six congeners of polychlorinated biphenyls - PCB6) in sludge. The directive 86/278/EEC regulates the sludge application in EU legislation. Only the contents of 6 potentially risk elements in sludge (Cd, Cu, Hg, Pb, Ni and Zn) are limited in the directive. The proposal of limit values of potentially risk elements and persistent organic pollutants was presented in Working Document on Sludge that was available for professional community, too. This proposal altered existing criteria and installed new criteria for persistent organic pollutants especially. The contents of seven POPs groups were regulated, the sum of halogenated organic compounds (AOX), linear alkylbenzene sulphonates (LAS), di(2-ethylhexyl)phtalate (DEHP), nonylphenol and nonylphenoletoxylates substances with 1 or 2 ethoxy groups (NPE), sum of polycyclic aromatic hydrocarbons (PAHs), the sum of seven congeners of PCB (28+52+101+118+138+153+180) and polychlorinated dibenzo-p-dioxins and dibenzofuranes (PCDD/F). The acceptance of Working Document on Sludge for the legislation was complicated by the lobbies and by economical needs for the determination of the pollutants. The proposal was refused and the directive 86/278/EEC is valid in original form.

The second group of problematic materials including into our research are the sediments dredged from river or pond bottoms. The volumes of dredged river and pond sediments reach huge amounts because of the necessity of periodical maintenance of river channels and water reservoirs. The existence of 97 millions m3 of ponds sediments and 5 millions m3 of river and irrigation channel sediments was reported in Czech Republic (Gergel, 1995). The problem of the liquidation or suitable use of extracted sediments of these amounts is evident. In spit of the traditional use of the sediments as the fertilizers on agricultural soils till to first halve of 20th century is not current approach unified, especially thanks to misgivings of their hygienic standards and environmental merits.

The elaboration of complex methodological approach including the assessment and testing of sediment conditions, the contamination and possible negative effects and the evaluation of positives and negatives of their application is highly needed. This approach must follow current EU politics of soil protection, sewage sludge application and the use of the other wastes (European Parliament, 2003; ISO 15799, 2003; EN 14735, 2006). The complex system should use chemical and biological methods concluded by risk assessment where contact ecotoxicity tests cannot be missing (Domene et al., 2007; Pandard et al., 2006).

The sedimentation of soil particles originated from agricultural soil erosion seems to be the most important way of sediments inputs into water systems. This process is described in Czech Republic also where about 50% of soil fund is endangered by water erosion (Janeček et al., 2005). The accumulation of nutrients and organic matter especially in pond and downstream sediments belongs to the positives of sediment application. The sediment could be valuable substrate useful in soil and landscape reclamation for example (Santin et al., 2009).

The other hand must be accepted that eroded soil particles are under the influence of many factors in water environment resulting to the changeover of their quality especially from the viewpoint of elements and substances sorption. The sediment characteristics are changing by particles sedimentation process in different parts of the stream and this process influences sorption of risky substances (Tripathy & Praharaj, 2006; Fuentes et al., 2008). It could lead to the problems of water eutrofization or sediment contamination. The sediments are known as the "chemical time bomb" thanks to their function of final deposits of pollutants in the river basins (Hilscherová et al., 2007; Holoubek et al., 1998). The sediment load by risky substances is connected with the presence of pollution sources like industrial or urban zones or wastes outputs from mining activities. The negative impact of these sources can be confirmed by chemical methods (Gomez-Alvarez et al., 2007) or by toxicity tests (Riba et al., 2006). The inputs of risky elements into sediments from geochemical anomalous substrates or from the other natural sources respectively can play an important role (Liu et al., 2008). The increased loads of risky substances lead to the complications of sediment use in the same way as the use of sewage sludge and the other organically reach materials (Vácha et al., 2005a).

The potential contamination of the sediments by wide spectrum of hazardous substances could not be eliminated. We accept the fact that fluvisols developed on alluvial sediments in river fluvial zones belong to the most loaded soils in our conditions by risky elements Cd>Hg>Zn>Cu>Pb and Cr (Podlešáková et al., 1994) and by persistent organic pollutants (POPs). Increased contents of polycyclic aromatic hydrocarbons (PAHs), chlorinated pesticides (sum of DDT), petroleum hydrocarbons and polychlorinated biphenyls (PCBs) on some localities were observed (Podlešáková et al., 1994; Vácha et al., 2003). The monitoring of fluvisols load by polychlorinated dibenzo-p-dioxins and dibenzofurans (PCDDs/Fs) resulted into similar trends (Podlešáková et al., 2000; Vácha et al., 2005b). The contamination of sediments from water reservoirs by PCDDs/Fs was confirmed (Urbaniak et al., 2009). At the same time, POPs degradation is strongly influenced by sediment conditions, oxygenation conditions belong to the most important (Devault et al., 2009). The sediment quality was monitored in Labe river basin by germen researchers. They observed increasing water quality in Labe River after collapsing of communist regime in central Europe thanks to increasing number of wastewater factories and the other modern pollution-controlling technologies (Netzband et al., 2002). In spit of this fact the concentrations of several contaminants are still remaining in sediments of Labe River and their use for agriculture is questionable (Heininger et al., 2004). The most problematic are the contents of Cd, Hg, As, Zn, HCB, PCBs and PCDDs/Fs in the sediments (Heise et al., 2005).

The other hand, realised monitoring of pond sediments load by risky elements in the Czech Republic confirmed relatively low contamination (Benešová & Gergel, 2003). The authors did not find the exceeding of risky elements limit values in the Czech Direction for soil protection No. 13/1994 Sb. The database of sediment load by risky elements and some POPs separated into groups following sediment origin (field ponds, village ponds, forest ponds and rivers) is available in the Central Institute for Supervising of Testing of Czech Republic (Čermák et al., 2009). The results of this monitoring show only sporadically increased values of risky elements (Cd and Zn usually) in the sediments but these load can reach extremely increased contents namely in village ponds (1660 mg/kg for Cd or 1630 mg/kg for Zn) in some cases. The contents of risky elements and observed POPs (AOX, PCB_7) were under background values of agricultural soils in the most observed sediment samples.

Long-term prepared legislative regulation (Direction No. 257/2009 Sb.) for sediment application on agricultural soils is valid in the Czech Republic since 2009 year. The Direction regulates selected characteristics and conditions for the application of extracted sediments. The limits of potentially risky elements (As, Be, Cd, Co, Cr, Cu, Hg, Ni, Pb, V and Zn) and persistent organic pollutants (BTEX, sum of PAHs, PCB_7, sum of DDT and $C_{10} - C_{40}$ hydrocarbons) in the sediment and soil of the locality for the application are defined. The limits of risky elements and substances in the soil were derived from the background values of Czech agricultural soils proposed originally (Podlešáková et al., 1996; Němeček et al., 1996). The limits in the Direction use total contents of risky elements only.

The paper shows the results of the research of risky substances contents in the set of sediment samples collected in 2008 year. These contents are compared with sediment characteristics depending on sediment origin and the way of sediment processing. The experiences following from real use of Czech legislative on the field of sediments use in agriculture can contribute to the process of European legislative formulation.

2. Materials and methods

2.1 Sewage sludge analyses

The research focused on the contents of POPs in sewage sludge resulting in the proposal of their recommended maximum contents in the sludge for application on agricultural soils was based on:

The POPs monitoring in 45 wastewater factories in Czech Republic,

the realisation of pot and micro field trial,

the synthesis of the results and their comparison with the proposal of EU directive amendment (EU 2000, Working Document on Sludge), table 1.

The monitoring of POPs in sewage sludge covered the area of the Czech Republic. The waste-water factories were separated into following groups:

- Areas of regional and district towns (including capital city of Prague),
- areas of towns with the presence of industrial activities,
- areas of settlements under 15 000 inhabitants.

The waste-water factories with comparable technologies of wastewater treatment were collected. The contents of polychlorinated dibenzo-p-dioxins and dibenzofurans (PCDDs/Fs) were analysed in the samples from 16 wastewater factories. The example of wastewater characteristics for sludge sampling show table 2.

The list of POPs analyses realised in sludge samples shows table 3.

Two sludge samples form Nord-Moravian region with increased contents of PAHs and PCB6 (table 4) were used in pot and field trials. The application of sludge followed the criteria of Czech directive 382/2001 Sb. and the dose of sludge in trials was derived from the dose of 5 t/ha of dry matter.

2.2 Sewage sludge experiments

Three soil types (typic Chernozem, typic Cambisol and arenic Cambisol) were used in the pot trial (6 kg of soil in Mitscherlich pots). The pot trial was run in three replications.

The field trial was set up on typic Cambisol in the area of Bohemian and Moravian highlands. The field trial was realised in four variants (ploughed and not ploughed, two sludge samples) each in three replications. Ploughed and not ploughed variants were focused on the influence of soil treatment on the decomposition of POPs in the soil (photo degradation, increased input of the air, stimulation of microbial activity). The ploughed variant was treated every two weeks in the layer of humic horizon (cca 20 cm). The characteristics of all used soils are presented in table 5.

The mustard (*Brassica alba*) was used in both (pot and field) trials in first year. The pot trial was sowed by radish (*Raphanus sativus*) and the field trial by parsnip (*Pastinaca sativa*) in the second year. The samples of soil and plants were taken after the harvest, the yield was measured and the contents of POPs in soil and plant samples were analysed. The list of POPs substances and analytical methods for POPs determination in sludge and soil is identical with table 2, except of PCDDs/Fs. The identical analytical methods were used for POPs determination in digested plant samples. The standard elementary statistic methods (file characteristics) were used for the evaluation of the results.

2.3 Sediment sampling

The pond sediment samples from 29 locations were collected in 2008. The samples from pond bottoms and from sediment heaps were used. Field ponds, village ponds and forest ponds were observed. Probe poles with a length of 50 cm for the sampling of bottom sediments and 100 cm for the sampling of heap sediments were used. The individual samples consist of 10 partial samples. The samples were stored in plastic bags and closed jars (for POPs analyse). Closed jars were stored in a deep-freeze condition before chemical analysis. The summary of collected samples is presented (Table 6).

2.4 Sediment analysis

The following characteristics were analysed in sediment samples by the Research Institute for Soil and Water Conservation (RISWC):

- Dry matter content (%)
- Organic matter content (%) – 550°, (CSN EN 12879, 2001)
- pH (H2O), pH (KCl) (CSN ISO 10390, 1996)
- Indicators of the cation exchange capacity CEC (CSN ISO, 13536), BS – the rate of complex saturation adsorption (%)
- Al-exchangeable – titration method (Hraško et al., 1962)

The content and quality of primary organic matter and humus substances were analysed in RISWC using the following approach:

- C_{ox} – organic carbon indicative of the carbon content in primary soil organic matter (SOM). The determination procedure is based on the chromic acid oxidation of organic carbon under the abundance of sulphuric acid and at elevated temperature. Unexpended chromic acid is determined by the iodometric method. This method is a modification of CSN ISO, 14235. The assay of loosely and tightly bound humus materials includes the determination of the humic acid carbon (C-HA), fulvic acid carbon (C-FA), humus matter carbon (C-FA+C-HA) and the assessment of the colour coefficient (Q4/6) indicating the humus quality. The determination procedure is based on the sample extraction method using a mixed solution of sodium diphosphate and sodium hydroxide (Zbíral et al. 2004). Carbon contents (C-FA, C-HA) are determined by titration and the coefficient Q4/6 results from the photometry.
- C_{ws} – water-soluble carbon, indicating the quality of primary SOM (bio available carbon for soil microorganism). Laboratory determination consists of an hour sample extraction using 0.01mol/L CaCl2 solution (1:5 w/V) and the determination of oxidizable carbon in the filtrate evaporation residue by heating the filtrate with chromium sulphuric acid and subsequent titration with Mohr's salt.
- C_{hws} – hot water-soluble carbon, being similar for the assessment purpose to water-soluble carbon. After the soil sample was boild for 1 hour in 0.01mol/L CaCl2 solution (1:5 w/V), the oxidizable carbon in the filtrate evaporation residue through the heating of filtrate with chromium sulphuric acid and subsequent titration with Mohr's salt is determined.

The contents of potentially potentially toxic elements were analysed in sediment samples in RISWC:

- As, Cd, Co, Cr, Cu, Hg, Ni, Pb a Zn in the extract of Aqua regia (ČSN EN, 13346), Hg was analysed by AMA 254 method (Advanced mercury analyser, total content).
- As, Cd, Cu, Pb and Zn in the extract of 1mol/L NH4NO3 (mobile contents). The samples were prepared according to ISO, 11464.

The analysis of the elements in the samples were conducted by the AAS method (AAS Varian), flame and hydride technique.

Persistent organic pollutants were analysed in commercial accredited laboratories Aquatest a.s.:

- BTEX (benzene, toluene, e-benzene and xylene), gas chromatography with mass spectrometry (GS/MS), EPA Method, 8260 B.
- PAHs – polycyclic aromatic hydrocarbons, the contents of 16 substances following EPA, liquid chromatography with fluorescence detector (HPLC), methodology TNV, 75 8055.
- PCB_7 – polychlorinated biphenyls, seven indicator congeners (28, 52, 101, 118, 138, 153, 180), gas chromatography with ECD detector (GC/ECD), EPA Method, 8082.
- DDT sum – sum of DDT, DDE and DDD, gas chromatography with ECD detector (GC/ECD), EPA Method, 8082.
- C_{10} – C_{40} hydrocarbons, gas chromatography with flame-ionisation detector (GC/FID), CSN EN, 14039.

The evaluation of sediment characteristics and the contents of potentially potentially toxic elements and persistent organic pollutants in the sediments separating on the base of their origin and type were done by the use of elementary statistics where median, maximum, minimum, average, standard deviation are presented (Excel). The correlations (Pearson correlation coefficients) between selected sediment properties (pH, CEC, content and quality of soil organic matter) significant at the 0.01 and 0.05 level were processed (SPSS Statistics 17.0).

3. Results and discussion

3.1 Sewage sludge results

The values of POPs (Polycyclic aromatic hydrocarbons – PAHs, monocyclic aromatic hydrocarbons – MAHs, Chlorinated hydrocarbons – ClHs and Petroleum hydrocarbons – PHs) contents are demonstrated in table 7. The sludge samples differentiation follows the type and range of studied area. The overview of POPs contents in sludge in individual years presented table 8.

On the example of tested set of sludge samples it was concluded that fluoranthene reaches the highest average concentrations among PAHs. This finding corresponds with the fact that fluoranthene concentrations in the environment belong to the highest from PAHs group (Holoubek et al., 2003). The phenanthrene concentration with highest maximum values follows fluoranthene. The variability of the values of concentrations of these two substances is the highest among PAHs group. Opposite naphtalene reaches the lowest values of all investigated substances.

The highest average and maximum values from the monocyclic aromatic hydrocarbons (MAHs) were detected in the case of toluene. Contents of toluene in the set of sludge samples were characterised by the highest variability, too. Toluene concentrations influenced predominantly the contents of the sum of MAHs because of very low concentrations of all the other substances.

The contents of chlorinated substances reach relatively low level. The values of PCBs concentrations are characterised by maximum variability. The concentrations of DDE are increased in comparison with DDD and DDT. The persistence of decomposition products of DDT in the environment is still detected (Holoubek et al., 2003; Poláková et al., 2003; Vácha et al., 2003).

Generally the highest contents were found in the case of petroleum hydrocarbons (PHs). The evaluation of the contents is complicated by difficult resolution of substances originated from petroleum contamination and of the substances from the decomposition of organic matter in the sludge.

The comparison between the values of sum of PAHs and values of sum of their toxic equivalent factors (TEF) in 25 samples presents figure 1. Good agreement between these values is evident. It could be concluded increased rate of more nuclei substances respectively substances with higher carcinogenic risk (table 9). This findings confirm the need of PAHs monitoring in sludge used for application on agricultural soils.

The data of the contents of POPs in the set of sewage sludge were processed for the assessment of their "background values". The 90% percentile was used after elimination of outlying values. These background values (table 10) are compared with background values of POPs in agricultural soils (Němeček et al., 1996) in table 11. If we compare obtained "background values" of the content of POPs in the set of sewage sludge with the limit values of POPs in sludge in Czech and European legislative norms we get following results. The value of the sum of 6 congeners of PCBs is suitable from the viewpoint of Czech (0.6 mg/kg) and European (0.8 mg/kg for 7 congeners) legislation. More problematic seems to be content of the sum of PAHs (9.37 mg/kg) where the overcome of proposed limit of EU directive (6 mg/kg) was observed. No limit value regarding PAHs is included in Czech directive No. 382/2001.

On the base of comparison of background values of POPs in sewage sludge and soil (table 5) emerged following findings. Toluene (MAHs group) shows the maximum difference between the content in the soil and in the sludge from all POPs substances. The concentration in the sludge is cca 243-fold higher than the concentration in the soil. The difference of these contents is significantly lower in the group of PAHs with the maximal difference in the case of benzo(ghi)perylene where sludge content represented 13.7-fold higher value as compared to soil. The contents of PCBs in the sludge are cca 10-fold higher in sludge compared to soil while the contents of DDT (including DDD, DDE) are comparable with the contents in the soil.

The values of I-TEQ PCDD/F fluctuated in the range from 9.2 to 280.2 ng/kg. The value of 280.2 ng/kg was eliminated as outlying by statistic procedure. Resulting average I-TEQ PCDD/F is than 22.5 ng/kg in the set of sludge samples. For 90% percentile I-TEQ PCDD/F reaches the value 37.7 ng/kg. The values of I-TEQ PCDD/F fulfil safely the proposed limit of EU order (100 ng/kg I-TEQ PCDD/F).

The assessment of sludge load on the base of congener analysis of PCDD/F indicates regional differences (with the dominance of octo-chlorinated dibenzodioxins in sludge), which are depending on the wastewater load from the different sources very probably (the rate of communal and industrial wastewater of different type). The data are according with the finding that octo-chlorinated (OCDD) and hepta-chlorinated (HpCDDs) congeners are dominant in the sewage sludge (Holoubek et al., 2002). In spite of this fact the definition of typical general congener pattern of the load of set of sludge samples seems to be complicated considering to regional differences. Congener patterns of individual sludge samples could be used for the localisation of sources of wastewater contamination by PCDD/F (Holoubek et al., 2002).

The proposal of recommended limit values of elected POPs in sludge for the application on agricultural soils (table 12) was derived from the following:

- The background values of selected POPs in set of sludge samples from the wastewater factories of the areas of regional, district and industrial towns and smaller settlements were determined.
- Vegetation experiments did not confirm that sludge application in the dose of 5t/ha of dry matter on the soil influenced POPs contents in the soil and tested plants. Together with these findings we respect the results of the other authors following from long-term experiments about the accumulation of some POPs substances in the soil.
- The proposed limit values in "Working Document on Sludge" were observed.
- The substances from POPs group included in Czech Directive of Soil Protection No. 13/1994 Sb. were selected for the observation.
- Theoretical and simplified balance sheet of the input of POPs into soil by sludge application resulted that the background values of most selected POPs in the soil will be multiplied two times after period of 300 years by sludge application. This balance was not used for PCDD/F.
- Increased limit value was proposed for PAHs in comparison with primary proposal in "Working Document on Sludge". EU primary proposal seems to be not relevant in view of load by Czech sludge by PAHs and from the viewpoint of the strictness of PAHs limit against the other limits of the substances (PCDD/F, PCB_7) in EU primary proposal. The presence of PAHs in the environment in Czech conditions does not correspond with primary EU proposal of PAHs in the sludge and majority of sludge production will be excluded respecting the limit 6 mg/kg. We could not find the explanation for the respecting of this limit by the comparison of limit values of PAHs and PCDDs/Fs in the sludge and their background values in the soil for example. The content of PAHs in sludge is 6 times higher as in the soil but the content of PCDDs/Fs is 100 times higher as in the soil regarding the primary EU proposal.
- The extent of selected POPs substances was adapted for Czech legislative for soil protection (Directive No.13/1994 Sb.). The use of results of the research for the Czech legislation is depending on the confrontation of soil protection and sludge application needs respecting economical site of the problem. The difficulty of this process was documented by the refusal of "Working Document on Sludge" for EU legislation.

The results were derived from the set of sludge samples collected in the territory of the Czech Republic. The international validity could be assumed for European countries thanks to connected markets resulting to similar load of municipal waste waters by potentially toxic substances.

3.2 Dredged sediments results

The limit values of POPs in soil for sediment use in Czech legislation (No. 257/2009 Sb.) shows table 13 where only two POPs groups are limited.

The limit values of POPs in sediments in Czech legislation (No. 257/2009 Sb.) shows table 14 where six POPs groups are limited. The existence of national limits of pollutants in sediments for agricultural use in European countries is recommended

The basic physio-chemical properties of dredged sediments are presented in table 15. The content of dry matter, organic matter, sediment reaction, exchangeable H+ content and adsorption characteristics are defined for the set of sediment samples. The wide range of values of observed parameters is clearly visible in table 15. The differences between individual sediment groups can be observed when the separation of sediments with respect to their origin (the sediments of field, forest and village pounds) is carried out. The differences between sediment acidity were detected primarily. Forest sediments are characterised by higher acidity than the others. The lower values of the saturation of adsorption complex by basic ions (S value) and the values of the rate of adsorption complex saturation (V value) consecutively display an increase in sediment acidity.

The sediments were separated based on the sediment storage method (bottom, heap) due to the tendency to increase acidity during storage, and the comparison of the acidity of separated sediments and adsorption characteristics were observed. The prevailing separate sources (field, village and forest) were accepted also but village sediments were not calculated using this procedure due to missing data (only 1 sample of heap sediment was from a village pond). The results are presented in table 16. The storage of sediments on the heaps before application on agricultural soils is generally used methods in many countries.

The results confirm the trend of sediment acidification during sediment storage in the category of both sediment groups (field, forest). The forest sediments show sharper differences between the reaction of bottom and heap sediments. It was surprising to see, however, that the bottom forest sediments reached the highest pH value. The results demonstrate that decreasing pH value influences the values of adsorption characteristics markedly (S and V values).

The values of content and quality of sediment organic matter are presented in table 17.

The wide range of organic matter content in the set of sediment samples is evident; the sediment application with minimal C_{ox} content seems to not provide economical benefit from the viewpoint of organic matter inputs into agricultural soils. Conversely, the application of sediments with maximal C_{ox} content in a set of sediment samples will lead to increased organic matter input into soils. The lower values of organic matter contents are displayed in village pond sediments. Some countries (Slovakia for example) use minimal limit values of organic matter for sediment use in agriculture.

The quality of primary organic matter (the carbon ability for microbial utilization) when compared by water-soluble and hot-water soluble carbon contents (C_{ws} and C_{hws} values that characterise easily available carbon) reached the highest values in forest pond sediments following by field pond sediments. The lowest values in these parameters were observed in village pond sediments again. The same order can be observed by the evaluation of the content of humus substances where the rise of carbon content of total humus substances in forest pond sediments is distinctly increased. The quality of humus substances compared with the ratio of the carbon of humic and fulvic acid is higher in the field pond sediments compared with forest pond sediments. The lowest values of humus substances quality were observed in village pond sediments. From the comparison of carbon contents of primary organic matter and humus substances it follows that the highest humification degree in organic matter is observed in forest pond sediments. This parameter is comparable in field and village pond sediments. It could be generally resulted that forest sediments are very suitable for application on agricultural soils from the viewpoint of their organic matter quality.

The medians and maximums of POPs contents in field, village and forest sediments are presented in table 18 where the comparison with the Direction No. 257/2009 Sb. is available also.

The median values of PAHs indicate an increased load of village pond sediments and a similar trend can be found in the case of DDT. The contents of the others POPs are comparable between individual sediment types. The maximum limits of PAHs were exceeded in all three sediment types. Very probably, PAHs will be the most problematic of the observed POPs group in the sediments. This trend could be expected generally and the proposed limits for sludge in European proposal (Working Document of Sludge) confirm this fact. From the comparison of sediment load by PAHs with the proposal of PAHs limit values in Czech agricultural soils (Němeček et al., 1996) it was concluded that increased persistence of more nuclei compounds in the sediments was found. The tendency of the substances to accumulate in the sediments was observed in the order benzo(ghi)perylene>benzo(b)fluoranthene, benzo(k)fluoranthene, pyrene>benzo(a)pyrene, benzo(a)anthracene, fluoranthene and chrysene. The order was assessed on the basis of the rate between the individual PAHs substances content in the sediment and proposed soil limit value, and the comparison of the sum of PAHs in the sediments and soil limit value.

Despite the findings of DDT it remains that the increased contents in agricultural soils (Vácha et al., 2001; Čupr et al., 2009) did not exceed limits in sediment samples. The existence of the limit for BTEX in the sediments in Direction No. 257/2009 Sb. must be supported with more data collected, especially from river sediments. The limit for C_{10} – C_{40} hydrocarbons will eliminate their increased contents in sediments for agricultural use from local leaks of petroleum hydrocarbons.

The correlation between the contents of observed POPs groups (except of C_{10} – C_{40} hydrocarbons where a dominant number of values were under detection limit) and content and quality of organic matter was assessed. The data in table 19 confirm only sporadic correlation surprisingly.

The trend of PCB and BTEX accumulation in the dependency on content and quality of humus substances is presented. The PAHs groups did not show any trend of accumulation regarding their properties and affinity to organic carbon. Some authors (Cave et al., 2010) measured bioaccessible PAHs fraction in the soil (varied from 10 – 60%) and the multiple regression showed that the PAHs bioaccessible fraction could be explained using the PAHs compound, the soil type and the total PAHs to soil organic carbon content.

It could be assumed that the sources of the contamination by POPs determined in most POPs groups, except for BTEX, influenced the sediments load stronger than the selected sediment properties in an observed set of sediment samples.

The inputs of potentially toxic substances by sludge and sediment application can play important role in soil hygiene. The easy balance of POPs inputs into soil by sludge and sediments application in accordance with Czech legislative is presented in table 20. It must be accepted that the application of sewage sludge and dredged sediments runs under different conditions. The sludge can be applied once in 3 years in maximal dose of 5 tons of dry matter per hectare. The sediments can be applied once in 10 years in maximal dose of 750 tons of dry matter per hectare. The table presented the dose of sludge and sediments in 10 years. This balance could differ between individual countries following national legislative standards.

The maximum possible increase of POPs content in the soil after sludge and sediment application was derived from their possible maximum inputs (table 21). The values are only tentative because no process of POPs decomposition and migration in the soil was reflected.

4. Conclusion

It is evident that legislative regulation of sewage sludge and dredged sediment application on agricultural soils limits the inputs of risky substances into soils and the other parts of the environment. The uncontrolled application of these materials as well as the other biosolids could lead to serious damage of the soils and their functions. The problem with the limiting of POPs in sewage sludge is still continuing not only in the Czech Republic where only PCB$_7$ and AOX are limited but in European context especially. The refusal of Working Document on Sludge extended the validity of EU directive 86/278 with the absence for limit values of any POPs substances. At the same time it is known that sludge application significantly increased inputs of PAHs and chlorinated substances (PCBs, PCDDs/Fs) into agricultural soils.

The comparison of POPs inputs by sediment and sludge application demonstrated that the application of dredged sediments loads the agricultural soils more by POPs inputs thanks to use of high possible sediment doses. The European legislative is not available on the field of sediment use in agriculture in present time and the existence of national legislative regulations for sediment application can be highly recommended. The experiences of the practical use of limits application in individual countries can be utilized in the process of European legislative assessment.

5. Annex

5.1 Tables

Organic substances	The value (mg/kg dm)
AOX	500
LAS	2600
DEHP	100
NPE	50
PAHs	6
PCB$_7$	0,8
Dioxins	The value (ng TE/kg dm)
PCDDs/Fs	100

AOX - Sum of halogenated organic compounds
LAS - Linear alkylbenzene sulphonates
DEHP - Di(2-ethylhexyl)phthalate
NPE - Nonylphenol and nonylphenolethoxylates
PAHs - Sum of polycyclic aromatic hydrocarbons
PCB$_7$ - Sum of seven indication PCB congeners (28, 52, 101, 118, 138, 153, 180)
PCDDs/Fs - Polychlorinated dibenzodioxins/dibenzofurans

Table 1. The proposed limit values of EU directive 86/278.

Anaerobic and aerobic stabilisation (microbial activity stimulation), sludge dehydration and pressing

No.	Potential use in agriculture	Characterisation
1	yes	agglomeration, different wastewaters, high technological level of wastewater factory - WF
2	yes	Small area, municipal wastewater, lower technological level of WF
3	yes	Small area, municipal wastewater, lower technological level of WF
4	yes (in use)	Regional town up to 35 000 inhabitants., municipal wastewater predominantly, good technological level of WF
8	yes	Regional town up to 55 000 inhabitants, municipal and industrial wastewater (glass, ceramic), high technological level of WF
9	-	Regional town up to 100 000 inhabitants, municipal and industrial wastewater (food production, chemistry – pre-treatment of wastewater), high technological level of WF
10	yes	Regional town up to 100 000 inhabitants, municipal and industrial wastewater (food and paper production), high technological level of WF
11	yes	settlement up to 7 000 inhabitants, municipal wastewater, good technological level of WF
12	yes	Regional town up to 40 000 inhabitants, municipal and industrial wastewater (food production), high technological level of WF
13	yes	Town up to 15 000 inhabitants, municipal wastewater, lower technological level of WF
14	yes	Regional town up to 170 000 inhabitants, municipal and industrial wastewater (food production), high technological level of WF
15	yes	Regional town up to 20 000 inhabitants, municipal wastewater predominantly, high technological level of WF
16	yes (in use)	Regional town up to 50 000 inhabitants, municipal and industrial wastewater (car production), high technological level of WF
17	yes	Regional town up to 50 000 inhabitants, municipal and industrial wastewater (car production), high technological level of WF
18	yes	Industrial town up to 20 000 inhabitants, municipal and industrial wastewater 50/50 (chemistry), high technological level of WF
19	yes	Regional town up to 80 000 inhabitants, municipal wastewater only, high technological level of WF
20	yes	Town up to 20 000 inhabitants, municipal wastewater, high technological level of WF

21	yes	Town up to 20 000 inhabitants, municipal and industrial wastewater, good technological level of WF
22	no	Regional town up to 100 000 inhabitants, industrial WF, high technological level
23	yes	Regional town up to 100 000 inhabitants, municipal and industrial (lower rate) wastewater, high technological level of WF
24	yes	Settlement up to 5 000 inhabitants, municipal wastewater, lower technological level of WF
25	no	Industrial town up to 20 000 inhabitants, increased rate of industrial wastewater (chemistry), high technological level of WF
Mechanical filtration, cold sludge maturation		
5	yes	Spa town up to 15 000 inhabitants, municipal wastewater
6	yes	Settlement up to 5 000 inhabitants, municipal wastewater
7	yes	Central WF for few small settlements, municipal wastewater

Table 2. The characteristics of selected wastewater factories.

Analyse	Samples
pH, Cox,Ca,Mg, P, K	45 samples
As, Be, Cd, Co, Cr, Cu, Hg, Mn, Ni, Pb, V, Zn (extract of aqua regia)	45 samples
Monocyclic aromatic hydrocarbons benzene, toluene, xylene, ethylbenzene Polycyclic aromatic hydrocarbons naphtalene, anthracene, pyrene, phluoranthene, phenanthrene, chrysen, benzo(b)phluoranthene, benzo(k)phluoranthene, benzo(a)anthracene, benzo(a)pyrene, indeno(c,d)pyrene, benzo(ghi)perylene chlorinated hydrocarbons PCB, HCB, α-HCH, β-HCH, γ-HCH Pesticides DDT, DDD, DDE styrene, petroleum hydrocarbons	45 samples
PCDF 2,3,7,8 TeCDF, 1,2,3,7,8 PeCDF, 2,3,4,7,8 PeCDF, 1,2,3,4,7,8 HxCDF, 1,2,3,6,7,8 HxCDF, 1,2,3,7,8,9 HxCDF, 2,3,4,6,7,8 HxCDF, 1,2,3,4,6,7,8 HpCDF, 1,2,3,4,7,8,9 HpCDF, OCDF PCB 189, PCB 170, PCB 180 PCDD 2,3,7,8 TeCDD, 1,2,3,7,8 PeCDD, 1,2,3,4,7,8 HxCDD, 1,2,3,6,7,8 HxCDD, 1,2,3,7,8,9 HxCDD, 1,2,3,4,6,7,8 HpCDD,OCDD PCB PCB 77, PCB 126, PCB 169, PCB 105, PCB 114, PCB 118+123, PCB 156, PCB 157, PCB 167	16 samples

Table 3. The analyses in sludge samples.

PAHs											
A	N	P	Ch	Ph	F	B(a)P	B(b)F	B(k)F	B(a)A	B(ghi)P	I(cd)P	PAHs
S 1	1440	3400	2520	1420	8340	7990	3630	4190	1820	1890	1930	1740	40310
S 2	851	50	2950	2590	7220	9520	6640	7490	3360	3150	3690	2830	50341

	MAHs					ChHs						
	B	T	X	Eb	MAHs	PCB6	αHCH	βHCH	γHCH	HCB	DDT	DDD	DDE
S 1	120	830	7	2300	1043	1090	1.00	1.00	1.00	1.00	1.00	1.00	1.00
S 2	14	90	3	3800	111	57	1.00	1.00	1.00	1.25	1.26	2.08	21.5

S 1 – sludge 1, S 2 – sludge 2
A – anthracene, N – naphthalene, P – pyrene, Ch – chrysene, Ph – phenanthrene, F – fluoranthene,
B(a)P – benzo(a)pyrene, B(b)F – benzo(k)fluoranthene, B(a)A – benzo(a)anthracene, B(ghi)P –
benzo(ghi)pyrene, I(cd)P – indeno(c,d)pyrene, PAHs – polycyclic aromatic hydrocarbons, B – benzene,
T – toluene, X – xylene, EB – ethylbenzene, MAHs – monocyclic aromatic hydrocarbons, PCB6 – sum of
6 polychlorinated biphenyls congeners, HCH – hexachlorcyclohexane, HCB – hexachlorbenzene, DDT –
dichlordiphenyltrichloethane, DDD – dichlordiphenyldichlorethane, DDE – dichlordiphenylethane,
ChHs – chlorinated hydrocarbons

Table 4. POPs contents in sewage sludge used in pot trial (μg/kg).

Soil type	District of origin	pH (KCl)	Cox (%)	Trial
Arenic Cambisol	Melnik	7.05	1.02	pot
Modal Cambisol	Benesov	6.15	1.29	pot
Modal Chernozem	Nymburk	6.93	2.18	pot
Modal Cambisol	Jihlava	5.85	0.8	field

Table 5. The characteristics of soils used in the experiments.

	Field ponds	Forest ponds	Village ponds	Total
Bottom	6	4	3	13
Heap	7	7	2	16
Total	13	11	5	29

Table 6. The numbers and types of sediment samples.

		PAHs											
		A	N	P	Ch	Ph	Fl	B(a)P	B(b)F	B(k)F	B(a)A	B(ghi)P	I(cd)P	PAHs
Industrial towns	AM	299	279	1176	659	1748	941	316	290	168	481	202	158	6718
	GM	245	110	976	526	1484	783	238	222	129	375	154	124	5580
	std.	144	383	506	301	760	451	165	149	80	223	139	76	2851
	max.	477	1185	1870	984	2570	1500	539	463	245	754	506	254	9714
	min.	49	15	169	75	337	168	28	29	15	45	29	17	976
	med.	343	147	1100	791	2100	851	291	289	191	486	154	189	6771
Regional towns	AM	501	77	1706	1026	2280	1821	606	650	316	802	413	367	10564
	GM	282	26	1338	844	1599	1296	454	486	238	648	324	264	7970
	std.	551	67	1241	676	2074	1588	494	528	254	547	302	306	8454
	max.	1710	203	3850	2190	6700	4880	1410	1500	724	1670	940	918	26528
	min.	96	1	507	371	502	464	216	233	114	310	132	113	3071
	med.	165	64	1130	676	1260	1010	334	342	176	532	260	233	6165
Settlements	AM	215	20	1768	826	1399	1187	415	393	201	685	346	207	7662
	GM	181	5	1394	761	1159	1025	360	353	179	624	268	191	6624
	std.	134	25	1616	338	938	658	218	174	89	317	314	91	4705
	max.	548	69	6490	1570	3810	2810	900	732	368	1460	1240	434	20431
	min.	88	1	638	352	453	345	120	133	59	329	110	113	2741
	med.	188	1	1395	736	1100	1085	338	352	183	631	212	181	6328

PAHs – polycyclic aromatic hydrocarbon

Table 7a. The POPs contents in individual groups of sludge samples - PAHs (µg/kg).

		MAHs (µg/kg)					ChlH (µg/kg)								Sty-rene	PHs	Te
		B	T	X	Eb	MAU	PCB	α-HCH	β-HCH	γ-HCH	HCB	DDT	DDD	DDE			
Industrial towns	AM	39.6	494.9	154.5	856.4	1545.3	336	2.5	62.5	1.0	30.6	25.6	27.6	26.5	12.6	14571	2.6
	GM	10.8	138.0	93.0	79.6	481.9	240	1.8	5.7	1.0	16.0	21.9	5.4	24.9	2.1	11043	1.9
	std.	62.4	603.2	125.7	1977.6	2173.4	254	2.6	98.4	0.0	44.2	13.3	57.8	9.4	12.7	9322	1.6
	max.	191.0	1860.0	370.0	5700.0	6548.1	738	8.7	254.0	1.0	138.0	43.4	169.0	43.9	32.5	29000	6.1
	min.	0.1	5.1	19.9	16.1	61.1	98	1.0	1.0	1.0	4.3	10.6	1.0	14.2	0.1	2200	0.2
	med.	17.9	392.0	164.0	70.8	644.7	183	1.0	1.0	1.0	16.6	25.4	4.6	24.9	8.7	11000	2.3
Regional towns	AM	17.9	1543.8	38.2	25.3	1625.1	144	2.0	1.0	1.2	4.1	10.3	2.7	15.2	0.1	8943	2.3
	GM	7.7	452.6	2.6	6.1	630.5	119	1.7	1.0	1.1	2.5	9.5	2.2	10.6	0.1	8616	2.2
	std.	14.5	2804.5	57.3	22.5	2794.1	99	1.3	0.0	0.6	3.9	4.8	2.0	10.0	0.0	2496	0.7
	max.	44.1	8380.0	154.0	70.7	8437.4	358	4.5	1.0	2.7	11.6	21.6	7.1	35.3	0.1	13000	3.2
	min.	0.1	48.4	0.1	0.1	168.3	64	1.0	1.0	1.0	1.0	6.3	1.0	1.0	0.1	6200	1.1
	med.	13.6	314.0	5.7	20.9	369.9	99	1.5	1.0	1.0	1.7	9.2	2.1	14.6	0.1	7700	2.4
Settle-ments	AM	14.3	2784.4	13.4	20.5	2832.6	1566	15.1	1.6	1.0	11.4	21.2	11.4	35.9	0.1	11810	3.4
	GM	2.3	878.1	0.7	2.9	1141.6	170	2.4	1.2	1.0	7.1	15.2	6.5	25.1	0.1	10517	2.2
	std.	15.2	3463.8	30.7	22.6	3452.0	4345	40.0	1.9	0.0	7.5	16.9	12.3	34.8	0.0	5700	2.5
	max.	38.8	9330.0	104.0	59.8	9369.0	14600	135.0	7.2	1.0	20.1	58.6	43.0	134.0	0.1	21000	8.4
	min.	0.1	55.8	0.1	0.1	151.1	31	1.0	1.0	1.0	1.0	3.4	1.0	4.7	0.1	6000	0.2
	med.	8.3	800.0	0.1	14.2	830.0	123	1.6	1.0	1.0	13.9	16.5	7.3	25.3	0.1	9050	3.2

MAHs – mococyclic aromatic hydrocarbons; ClHs – chlorinated hydrocarbons
PHs – petroleum hydrocarbons Te – tenzides

Table 7b. The POPs contents in individual groups of sludge samples - MAHs, ClHs (µg/kg), Te and PHs (mg/kg)

		PAHs												
		A	N	P	Ch	Ph	Fl	B(a)P	B(b)F	B(k)F	B(a)A	B(ghi)P	I(cd)P	PAHs
2002 (20 sampl.)	AM	164	47	1442	616	1131	2499	709	906	409	595	508	398	9930
	GM	122	26	1219	527	940	2131	634	810	368	517	453	355	8488
	std.	120	43	814	363	704	1511	333	435	187	368	255	190	5964
	max	484	140	3330	1490	2880	6860	1440	1930	806	1900	1210	762	26434
	min.	16	2	363	189	240	596	206	334	141	193	161	156	2631
	med	160	30	1250	552	947	2115	635	798	370	560	413	333	8516
2001 (25 sampl.)	AM	252	45	1096	692	1479	904	316	329	173	504	216	165	5912
	GM	173	11	852	535	1067	709	241	250	133	392	163	123	4646
	std.	188	55	474	333	977	427	156	166	85	223	118	73	2571
	max	791	203	1870	1570	3910	1710	596	732	368	927	506	305	9714
	min.	3	1	14	11	17	18	7	6	4	10	3	1	95
	med	199	23	1115	676	1230	889	319	322	176	497	201	172	6170
Sum (45 sampl.)	AM	201	50	1159	685	1330	1405	540	617	294	508	352	283	6566
	GM	143	17	949	547	1010	1088	400	438	217	425	264	205	5497
	std.	144	55	536	383	887	815	364	454	200	205	229	192	2632
	max	548	207	2550	1750	3910	3630	1440	1930	806	927	940	762	11218
	min.	3	1	14	11	17	18	7	6	4	10	3	1	95
	med	164	30	1170	632	1070	1220	467	455	240	518	311	213	6525

PAHs – polycyclic aromatic hydrocarbons

Table 8a. Elementary statistic of the POPs in sludge samples – PAHs ($\mu g/kg$)

		MAHs					ChlHs					PHs
		B	T	X	Eb	MAU	PCB	HCB	DDT	DDD	DDE	(mg/kg)
2002 (20 sampl.)	AM	37.9	3815.3	39.8	12.5	3950	110	6.96	3.72	6.11	17.15	5845
	GM	27.6	1431.3	13.5	7.3	1680	98	5.54	3.13	4.94	13.75	5303
	std.	33.8	4315.9	47.8	10.6	4366	49	4.00	1.97	3.38	9.24	2310
	max.	120.0	16400.0	170.0	40.0	16820	201	15.40	7.85	11.50	36.30	8800
	min.	12.0	70.0	0.1	0.1	111	33	1.00	1.00	1.00	1.00	2300
	med.	20.5	2900.0	26.0	8.0	2974	103	6.70	3.51	6.18	15.50	6450
2001 (25 sampl.)	AM	14.8	498.8	36.5	28.2	608	122	9.13	16.91	4.24	21.54	11309
	GM	3.8	163.1	2.6	6.0	288	98	5.44	12.43	2.98	15.87	7848
	std.	13.8	553.0	59.0	27.3	614	75	7.13	12.56	3.74	12.38	6987
	max.	44.1	2040.0	192.0	82.1	2324	358	20.10	45.00	16.60	45.80	29000
	min.	0.1	0.1	0.1	0.1	0	7	1.00	1.00	1.00	1.00	20
	med.	11.1	350.0	5.7	20.2	360	104	8.23	11.40	3.24	20.75	9100
together (45 sampl.)	AM	22.4	2209.1	38.0	16.5	2293	110	8.17	8.00	5.13	20.33	6827
	GM	8.3	522.4	5.3	5.4	739	95	5.49	5.62	3.79	15.32	6219
	std.	22.8	2998.3	54.4	16.1	2998	52	6.05	6.54	3.69	12.17	2631
	max.	120.0	10200.0	192.0	59.8	10355	234	20.10	25.40	16.60	51.80	13000
	min.	0.1	0.1	0.1	0.1	0	7	1.00	1.00	1.00	1.00	2200
	med.	17.0	617.0	13.0	9.5	675	103	7.53	6.24	4.13	19.80	7000

MAHs – mococyclic aromatic hydrocarbons; ClHs – chlorinated hydrocarbons
PHs – Petroleum hydrocarbons

Table 8b. Elementary statistic of the POPs in sludge samples – MAHs, ClHs ($\mu g/kg$) and PHs (mg/kg)

Compound	The toxic equivalent value	Compound	The toxic equivalent value
Benzo(a)pyrene	1	Benzo(k)fluoranthene	0.01
Benzo(a)anthracene	0.1	Dibenzo(a,h)anthracene	1
Benzo(b)fluoranthene	0.1	Indeno(1,2,3-cd)pyrene	0.1

Table 9. The overview regarding the toxic equivalent value for individual PAHs compounds

	PAHs (µg/kg)												
	Fl	P	Ph	B(b)F	B(a)A	A	B(a)P	I(cd)P	B(k)F	B(ghi)P	Ch	N	ΣPAU
90 percentil	2412	1626	2407	1316	759	433	949	535	572	686	1148	132	9371

	MAHs (µg.kg-1)					ChlHs (µg/kg)					PHs mg/kg
	B	T	X	Eb	ΣMAU	PCB	HCB	DDT	DDE	DDD	
90 percentil	50	7300	150	37	7342	183	17.8	19.6	36.1	9.8	9440

Table 10. Background values of POPs in sludge collection, 90 percentil = backgroun value

	PAHs (µg/kg)											
	Fl	P	Ph	B(b)F	B(a)A	A	B(a)P	I(cd)P	B(k)F	B(ghi)P	Ch	N
Background - soil	300	200	150	100	100	50	100	100	50	50	100	50
Background - sludge	2412	1626	2407	1316	759	433	949	535	572	686	1148	132
difference in %	804	813	1605	1316	759	866	949	535	1144	1372	1148	264

	MAHs (µg/kg)				ChlHs (µg/kg)					PHs(mg/kg)
	B	T	X	Eb	PCB	HCB	DDT	DDE	DDD	
Background - soil	30	30	30	40	20	20	15	10	10	100
Background - sludge	50	7300	150	37	183	18	20	36	10	9440
difference in %	167	24333	500	92	917	89	130	361	98	9440

Table 11. The comparison of background values of POPs in sludge and soils.

Parameter	Content (µg/kg)							
	Sum MAHs	Sum PAHs	PCB$_7$	HCB	DDT	DDE	DDD	I-TEQ* PCDDs/Fs
Recommended limit	10 000	10 000	600	60	60	60	30	80
EU proposal	6000	-	800	-	-	-	-	100
Soil reference value (Czech)	1000	130	20	20	30	25	20	1

PAHs- polyaromatic hydrocarbons,
MAHs-monoaromatic hydrocarbons,
PCB7-sum of 7 congeners of polychlorinated biphenyls,
HCB-hexachlorbenzene,
DDT-dichlordiphenyltrichlorethane,
DDD-dichlordiphenyldichlorethane,
DDE-dichlordifenyldichlorethen,
I-TEQ PCDD/F-toxic equivalent of polychlorinated dibenzo-p-dioxins and dibenzofurans
I-TEQ PCDDs/Fs (ng/kg)

Table. 12. Recommended limit values of elected POPs in sludge, primary EU proposal and reference values in soils of the Czech Republic.

	Content (mg/kg)	
Limited substance	Middle and heavy texture soils	Light texture soils
PAHs	1.0	1.0
PCB$_7$	0.02	0.02

PAHs – polycyclic aromatic hydrocarbons
PCB7 – seven indication congeners of polychlorinated biphenyls

Table 13. Directive No. 257/2009, sediment use on agricultural soils, POPs limit values in soil.

Limited substance	Content (mg/kg)
PAHs	6
PCB$_7$	0.2
BTEX	0.4
DDT	1
C$_{10}$-C$_{40}$	300

PAHs – polycyclic aromatic hydrocarbons
PCB7 – seven indication congeners of polychlorinated biphenyls (28, 52, 101, 118, 138, 153, 180)
BTEX – sum of benzene, toluene, ethylbenzene and xylene
DDT – sum of DDT, DDD and DDE;
C10-C40 – sum of hydrocarbons - indication of petroleum hydrocarbons

Table 14. Directive No. 257/2009, sediment use on agricultural soils, POPs limit values in sediments (mg/kg).

	Dry matter %	Organic matter %	pH H_2O	pH KCl	Exchangeable H+ (mmol/100g)	CEC (mmol/100g)	BS %
Field (medians, 13 samples)	85.73	8.40	5.22	5.04	8.0	20.01	52.5
Village (medians, 5 samples)	98.06	7.07	5.42	5.23	6.5	18.91	63.5
Forest (medians, 11 samples)	80.08	8.44	4.09	3.82	16	20.88	48
Together (medians, 29 samples)	81.24	8.68	5.13	4.91	11.0	20.9	57.0
Together (A. means)	79.27	9.89	5.17	4.91	13.12	21.33	56.12
Together (St. deviation)	17.1	4.71	1.01	1.06	9.66	7.05	16.41
Maximum	99.06	22.5	7.1	6.95	42.0	41.19	100
Minimum	46.05	2.73	2.86	2.84	<0.5	8.81	28

CEC – cation exchange capacity
pH H2O – sediment pH measured in the extract of H2O
BS – the rate of complex saturation adsorption
pH KCl - sediment pH measured in the extract of 1M KCl

Table 15. Sediment characteristics in the set of 29 samples

	pH H_2O	pH KCl	Exchangeable H+ (mmol/100g)	CEC (mmol/100g)	BS %
Field-bottom	5.26	5.12	5.5	17.99	55.5
Field-heap	5.21	4.94	9.5	22.19	52.5
Forest-bottom	6.28	6.13	10	17.95	71
Forest-heap	3.81	3.63	26.5	25.24	31

CEC – cation exchange capacity
pH H2O – sediment pH measured in the extract of H2O
BS – the rate of complex saturation adsorption
pH KCl - sediment pH measured in the extract of 1M KCl

Table 16. The medians of sediment characteristics separated into sediment groups based on sediment type and storage method.

	C_{ox} %	C_{ws} mg.kg-1	C_{hws} mg.kg-1	HA %	FA %	HS %	Q4/6	HA:FA	HS:C_{ox}
Field	2.54	167.5	404.5	0.41	0.25	0.61	5.50	1.44	0.25
Village	1.72	108.5	324	0.21	0.23	0.47	6.1	0.92	0.27
Forest	2.53	199	558	0.49	0.61	0.97	5.4	1.18	0.39
Together	2.67	175	458.5	0.41	0.29	0.77	5.35	1.1	0.29
Maximum	8.29	515.0	1738.0	2.33	1.67	3.27	24.8	4.04	0.49
Minimum	0.52	72.0	76.0	0.07	0.04	0.11	3.4	0.23	0.16

Cox- organic carbon HA- carbon of humic acids
Cws- water-soluble carbon FA- carbon of fulvic acids
Chws- hot-water-soluble carbon HS- carbon of humus substances Q4/6- colour quotient

Table 17. The medians, maximum and minimum of organic matter content and quality in the sediments

		PAHs 2n	PAHs 3-4n	PAHs 5-6n	PAHs sum	PCB$_7$	DDT sum	BTEX	$C_{10}-C_{40}$*
Field	median	48	494	147	694	15.1	9.19	31.2	100
	max	210	4762	1290	6143	40.8	14.5	71.5	580
Village	median	77	2396	780	3386	14.2	15	30.35	105
	max	210	6842	2133	9052	36.9	32.3	43.2	110
Forest	median	41	326	54	517	15.4	8.83	66.3	100
	max	228	10347	2961	13536	1010	16.7	96.2	200
Limit 257/2009		-	-	-	6000	200	100	400	300
Limits exceeded. Field/Village/Forest					1/2/1	0/0/1	0/0/0	0/0/0	1/0/0

* C10-C40 – sum of hydrocarbons, content in mg/kg
PAHs 3-4n - the sum of PAHs with 3 and 4 rings
PAHs 2n - the sum of PAHs with 2 rings
PAHs 5-6n - the sum of PAHs with 5 and 6 rings
PCB7 – sum of seven indication congeners
BTEX – sum of benzene, toluene, ethylbenzene and xylene
DDT sum – sum of DDT, DDD and DDE

Table 18. The medians and maximums of POPs contents in the sediments (µg/kg) and values exceeding limits (Direction No. 257/2009 Sb.)

	PAHs 2n	PAHs 3-4n	PAHs 5-6n	PAHs sum	PCB$_7$	DDT	BTEX	C$_{10}$-C$_{40}$
C$_{ox}$	0.209	-0.194	-0.196	-0.187	0.347	0.085	**0.749***	0.287
C$_{hws}$	0.196	0.071	0.069	0.075	0.246	0.183	**0.468***	-0.015
C$_{ws}$	**0.418**	0.225	0.235	0.237	0.06	0.148	0.268	**0.419**
HS	0.170	-0.151	-0.165	-0.148	**0.462**	0.067	**0.727***	0.180
FA	0.351	-0.041	-0.51	-0.034	0.286	0.064	**0.723***	0.163
HA	0.051	-0.202	-0.217	-0.202	**0.524***	0.063	**0.632***	0.196
HA/FA	-0.299	-0.199	-0.203	-0.202	0.186	0.045	0.298	-0.034
C$_{ox}$/HS	0.121	0.092	0.083	0.091	**0.414**	0.094	0.268	-0.088

Cox - organic carbon HA - carbon of humic acids PCB7 – sum of seven indication congeners
Cws - water-soluble carbon FA - carbon of fulvic acids DDT – sum of DDT, DDD and DDD
Chws - hot-water-soluble carbon HS - carbon of humus substances PAHs 2n - the sum of PAHs with 2 rings PAHs 3-4n - the sum of PAHs with 3 and 4 rings PAHs 5-6n - the sum of PAHs with 5 and 6 rings C10-C40 – sum of hydrocarbons, content in mg/kg

Table 19. Pearson correlation coefficients between the contents of individual POPs groups and content and quality of organic matter, correlation significant at the 0,01 level (bold*) and 0,05 level (bold).

	POPs inputs (g/ha)	
Limited substance	Sewage sludge, application of 15t d.m. once in 10 years	sediments, application of 750t d.m. once in 10 years
PCB$_7$	9	150
PAHs	non limited	4500
BTEX	non limited	300
DDT	non limited	75
C$_{10}$-C$_{40}$	non limited	225000

(g/ha) PAHs – polycyclic aromatic hydrocarbons PCB7 – seven indication congeners of polychlorinated biphenyls BTEX – sum of benzene, toluene, e-benzene and xylene DDT – sum of DDT, DDD and DDE C10-C40 – sum of hydrocarbons - indication of petroleum hydrocarbons

Table 20. The comparison of POPs inputs by sewage sludge and dredged sediments application into agricultural soils.

	Sewage sludge		Dredged sediments	
Limited substance	Concentration increase (mg/kg)	% of soil background values CR	Concentration increase (mg/kg)	% of soil background values CR
PCB$_7$	0.002	10	0.03	150
PAHs	-	-	1	100
BTEX	-	-	0.07	54
DDT	-	-	0.02	27
C$_{10}$-C$_{40}$	-	-	50	50

PAHs – polycyclic aromatic hydrocarbons PCB7 – seven indication congeners of polychlorinated biphenyls BTEX – sum of benzene, toluene, e-benzene and xylene DDT – sum of DDT, DDD and DDE C10-C40 – sum of hydrocarbons - indication of petroleum hydrocarbons

Table 21. Maximum possible increase of POPs in soil after sewage sludge and dredged sediments application.

5.2 Figures

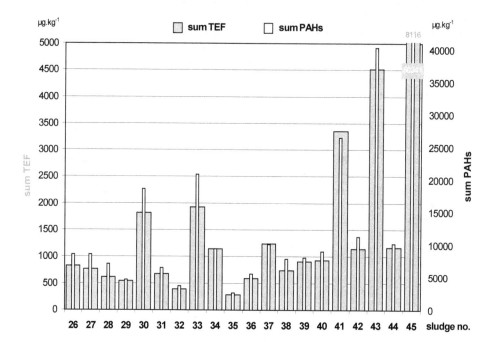

Fig. 1. The comparison of sum TEF PAHs and sum PAHs in sludge (µg/kg).

6. Acknowledgment

The chapter was prepared by the support of the Project of Ministry of Agriculture MZE0002704902 and Project No. QH 82083.

7. References

Benešová, L. & Gergel, J. (2003). Circumstances and relations of sediments agricultural use. *Odpadové fórum*, Vol.9, (September 2003), pp. 14-16, ISSN 1212-7779

Cave M.R.; Wragg, J.; Harrison, I.; Vane, C.H. ; Van de Wiele, T.; De Groeve, E.; Nathanail, C.P.; Ashmore, M.; Thomas, R.; Robinson, J. & Daly, P. (2010). Comparison of Batch Mode and Dynamic Physiologically Based Bioaccessibility Tests for PAHs in Soil Samples. *Environmental Science & Technology*, Vol.44, No.7, (April 201), pp. 2654-2660, ISSN 0013-936X

Clarke, B.; Porter, N.; Symons, R.; Marriott, P.; Ades, P.; Stevenson, G. & Blackbeard, J. (2008). Polybrominated diphenyl ethers and polybrominated biphenyls in Australian sewage sludge. *Chemosphere*, Vol.73, No.6, (October 2008), pp. 980-989, ISSN 0045-6535

Čermák, P.; Budňáková, M. & Kunzová, E. (2009). The utilization of sediments on agricultural farm land in the Czech Republic. *Book of abstracts of 18th CIEC International Symposium*, November 8 - 12th, Roma, Italy.

Default, D.A.; Gerino, M.; Laplanche, C.; Julien, F.; Winterton, P.; Merlina, G.; Delmas, F.; Lim, P.; Sanchez-Perez, J.M. & Pinelli E. (2009). Herbicide accumulation and evaluation in reservoir sediments. *Science of the Total Environment*, Vol.407, No.8, (April 2009), pp. 2659-2665, ISSN 0048-9697

Domene, X.; Alcaniz, J.M. & Andres, P. (2007). Ecotoxicological assessment of organic wastes using the soil collembolan Folsomia candida. *Applied Soil Ecology*, Vol.35, No.3, (March 2007), pp. 461-472, ISSN 0929-1393

EN 14735 (2005). Characterization of waste - Preparation of waste samples for ecotoxicity tests. Available from
http://shop.bsigroup.com/ProductDetail/?pid=000000000030152823

European Parlament (2003). Thematic Strategy for Soil Protection. European Parliament resolution on the Commission communication 'Towards a Thematic Strategy for Soil Protection' (COM(2002) 179 - C5-0328/2002 - 2002/2172(COS)). 19/11/2003.

European Union (2000). Working Document on Sludge: An EU-initiative to improve the present situation for sludge management, Brussels, ENV.E.3/LM, p. 19.

Fuentes, A.; Llorens, M.; Saez, J.; Aguilar, M.I.; Ortuno, J.F. & Meseguer, V.F. (2008). Comparative study od six different sludges by sequential speciation of heavy metals. *Bioresources Technology*, Vol.99, No.3, (February 2008), pp. 517-525, ISSN 0960-8524

Gergel, J. (1995). The extraction and the use of sediments from small water reservoirs. *Methodological approach*, 18/1995, Research Institute for Soil and Water Conservation, Prague, 45 p. (in Czech)

Gomez-Alvarez, A.; Valenzuela-Garcia, J.L.; Aquayo-Salinas, S.; Meza-Figueroa, D.; Ramirez-Hernandez, J. & Ochoa-Ortega, G. (2007). Chemical partitioning of sediment contamination by heavy metals in the Pedro River, Sonora, Mexico. *Chemical Speciation and Bioavailability*, Vol.19, No.1, (January 2007), pp. 25-35, ISSN 0954-2299

Heininger, P.; Pelzer, J.; Claus, E. & Pfitzner, S. (2004). Results of long-term sediment quality studies on the river Elbe. *Acta Hydrochimia et Hydrobiologica*, Vol.31, No.4-5, (February 2004), pp. 356-367, ISSN 0323-4320

Heise, S.; Claus E.; Heininger, P.; Krämmer, Th.; Krüger, F.; Schwarz, R. & Förstner, U. (2005). Studie zur Schadstoffbelastung der Sedimente im Elbeeinzugsgebiet - Ursachen und Trends. Im Auftrag von Hamburg Port Authority. Abschlussbericht (Dezember 2005), 169 p.

Hilschnerová, K.; Dušek, L.; Kubík, V.; Klánová, J. & Holoubek, I. (2007). Distribution of organic pollutants in sediments and alluvial soils after major floods. *Journal of Soils & Sediments*, Vol.7, No.3, (June 2007), pp.167-177, ISSN 1439-0108

Holoubek, I.; Čupr, P.; Škarek, M.; Černá, M. & Sáňka M. (2002). The conclusion of analyses of contamination of vicinity of Spolana Neratovice factory by PCDD/F and biphenyl after floods in 2002. TOCOEN, s.r.o. Brno. *TOCOEN REPORT No. 236*, 62 p. (In Czech), Available from http://www.tocoen.cz/zpravy.htm

Holoubek, I.; Adamec, V.; Bartoš, M.; Černá, M.; Čupr, P.; Bláha, K.; Bláha, L.; Demnerová, K.; Drápal, J.; Hajšlová, J.; Holoubková, I.; Jech, L.; Klánová, J.; Kohoutek, J.; Kužílek, V.; Machálek, P.; Matějů, V.; Matoušek, J.; Matoušek, M.; Mejstřík, V.; Novák, J.; Ocelka, T.; Pekárek, V.; Petira, O.; Punčochář, M.; Rieder, M.; Ruprich, J.; Sáňka, M.; Vácha, R. & Zbíral, J. (2003). National stocktaking of Persistent organic pollutants in the Czech Republic. Project GF/CEH/01/003 Enabling activities to facilitate early action on the implementation of the Stockholm Convention on Persistent organic pollutants (POPs) in the Czech Republic. TOCOEN REPORT No. 249. Available from
http://www.genasis.cz/stockholm-stockholmska_umluva-inventura_pops_2007/
Hraško, J.; Červenka, L.; Facek, Z.; Komár, J.; Němeček, J.; Pospíšil, F. & Sirový, V. (1962). Soil analysis. SVPL Bratislava, 342 p. (in Czech)
ISO 15799 (2003). Soil quality – Guidance on the ecotoxicological characterization of soils and soil materials. International Organization for Standardization. Geneve, Switzerland. Available from
http://www.iso.org/iso/iso_catalogue/catalogue_tc/catalogue_detail.htm?csnum ber=29085
Janeček, M.; Bohuslávek, J.; Dumbrovský, M.; Gergel, J.; Hrádek, F.; Kovář, P.; Kubátová, E.; Pasák, V.; Pivcová, J.; Tippl, M.; Toman, F.; Tomanová, O. & Váška, J. (2005). The protection of agricultural soil against erosion. ISV Press Praha, 2nd edition, 195 p. ISBN 80-86642-38-0
Liu, X.; Sun L.; Yin, X. & Wang, Y. (2008). Heavy Metal Distributions and Source Tracing in the Lacustrine Sediments of Dongdao Island, South China Sea. Acta Geologica Sinica-English Edition, Vol.82, No.5, Special Issue 4, (2008), pp. 1002-1014, ISSN 1000-9515
Ministry of agriculture and Ministry of Environment of Czech Republic (2009). Direction No. 257/2009 Sb. for sediment use on agricultural soils.
Markard, C. (1988). Organic contaminants in sexage sludge – do they constitute a danger for the food chain. Korrespondenz Abwasser, Vol.35, No.5, (May 1988), pp. 449-452, ISSN 0341-1540
Melcer, H.; Monteith, H. & Nutt, S. G. (1988). Variability of toxic trace contaminants in municipal sewage treatments plants. Water Science and Technology, Vol.20, No.4-5, (1988), pp. 275-284, ISSN 0273-1223
Ministry of Environment of Czech Republic (1994). The notice of the Ministry of Environment for the management of the soil protection, No. 13/1994 Sb. (in Czech)
Ministry of Environment of Czech Republic (2001). The notice of Ministry of Environment for the sludge application on agricultural soil, No. 382/2001 Sb. (in Czech)
Natal-da-Luz, T.; Tidona, S.; Jesus, B.; Morais, P.V. & Sousa, J. P. (2009). The use of sewage sludge as soil amendment. The need for an ecotoxicological evaluation. Journal of Soils and Sediments, Vol.9, No.3, (June 2009), pp. 246-260, ISSN 1439-0108
Němeček, J.; Podlešáková, E. & Pastuszková, M. (1996). Proposal of soil contamination limits for persistent organic xenobiotic substances in the Czech Republic. Rostlinna Vyroba, Vol.42, No.2, (February 1996), pp. 49-53, ISSN 0370-663X
Netzband, A.; Reincke, H. & Bergemann, M. (2002). The river Elbe – a case study for the ecological and economical chain of sediments. Journal of Soils and Sediments, Vol.2, No.3, (June 2002), pp. 112-116, ISSN 1439-0108

Oleszczuk, P. (2007). Organic pollutants in sewage sludge-amended soil part I. General remarks. *Ecological Chemistry and Engineering-Chemia I Inzynieria Ekologiczna*, Vol.14, No.S1, (2007), pp. 65-76, ISSN 1231-7098

Pandard, P.; Devillers, J.; Charissou, A.M.; Poulen, V.; Jourdain, M.J.; Férard, J.F.; Grand, C. & Bispo, A. (2006). Selecting a battery of bioassays for ecotoxicological characterization of wastes. *Science of the Total Environment*, Vol.363, No.1-3, (June 2006), pp. 114-125, ISSN 0048-9697

Passuello, A.; Mari, M.; Nadal, M.; Schuhmacher, M. & Domingo, J. L. (2010). POP accumulation in the food chain: Integrated risk model for sewage sludge application on agricultural soils. *Environment International.*, Vol.36, No.6, (August 2010), pp. 577-583, ISSN 0160-4120

Podlešáková, E.; Němeček, J. & Hálová, G. (1994). Contamination of fluvisols on Labe floodplains by hazardous elements. *Rostlinna Vyroba*, Vol.40, No.1, (January 1994), pp. 69-80, ISSN 0370-663X

Podlešáková, E.; Němeček, J. & Hálová, G. (1996). Proposal of soil contamination limits for potentially hazardous trace elements in the Czech Republic. *Rostlinna Vyroba*, Vol.42, No.3, (March 1996), pp. 119-125, ISSN 0370-663X

Podlešáková, E.; Němeček, J. & Vácha, R. (2000). Contamination of agricultural soils with polychlorinated dibenzo-p-dioxines and dibenzofurans. *Rostlinna Vyroba*, Vol.46, No.8, (August 2000), pp. 349-354, ISSN 0370-663X

Poláková, Š.; Tieffová, P. & Provazník, K. (2003): DDT and the residues in agricultural soils of the Czech Republic. *Bulletin of Central Institute for Supervising and Testing in Agriculturale*. Brno, Vol.11, No.3 (June 2003), pp. 73-75, ISSN 1212-5458

Riba, I.; DelValls, TA.; Reynoldson, T.B. & Milani, D. (2006). Sediment quality in Rio Guadiamar (SW, Spain) after a tailing dam collapse: Contamination, toxicity and bioavailability. *Environment International*, Vol.32, No.7, (September 2006), pp. 891-900, ISSN 0160-4120

Rosik-Dulewska, C.; Ciesielczuk, T. & Karwaczynska, U. (2009). Polycyclic Aromatic Hydrocarbons (PAHs) Degradation During Compost Maturation Process. *Rocznik Ochrona Srodowiska*, Vol.11, Part 1, (2009), pp. 133-142, ISSN 1506-218X

Santin, C.; de la Rosa, J.M.; Knicker, H.; Otero, X.L.; Alvarez, M.A. & Gonzales-Vila, F.J. (2009). Effects of reclamation and regeneration processes on organic matter from estuarine soils and sediments. *Organic Geochemistry*, Vol.40, No.9, (September 2009), pp. 931-941, ISSN 0146-6380

Starke, U.; Herbert, M. & Einsele, G. (1991). Polyzyklische aromatische Kohlenwasserstoffe (PAK) in Boden und Grundwasser, Teil I Grundlage zur Beurteilung von Schadenfällen. 1680 BOS 9 Lfg., 10: pp. 1-38.

Tripathy, S. & Praharaj, T. (2006). Delineation of water and sediment contamination in river near a coal ash pond in Orissa, India, In: Sajwan, K.S.; Twardowska, I.; Punshon, T.; Ashok, K. & Alva, A.K. (Eds.), *Coal Combustion Byproducts and Environmental Issues*, Springer, New York, pp. 41-49, ISBN 978-0-387-25865-2

Urbaniak, M.; Zielinski, M.; Weselowski, W. & Zalewski, M. (2009). Polychlorinated dibenzo-p-dioxins (PCDDs) and polychlorinated dibenzofurans (PCDFs) compounds in sediments of two shallow reservoirs in central Poland. *Archives of Environmental Protection*, Vol.35, No.2 (2009), pp. 125-132, ISSN 0324-8461

Vácha, R.; Poláček, O. & Horváthová, V. (2003). State of contamination of agricultural soils after floods in August 2002. *Plant, Soil and Environment*, Vol.49, No.7, (July 2003), pp. 307-313, ISSN 1214-1178

Vácha, R.; Horváthová, V. & Vysloužilová, M. (2005a). The application of sludge on agriculturally used soils and the problem of persistent organic pollutants. *Plant, Soil and Environment*, Vol.51, No.1, (January 2005), pp. 12-18. ISSN 1214-1178

Vácha, R.; Vysloužilová, M. & Horváthová, V. (2005b): Polychlorinated dibenzo-p-dioxines and dibenzofurans in agricultural soils of Czech Republic. *Plant, Soil and Environment*, Vol.51, No.10 (October 2005), pp. 464-468, ISSN 1214-1178

Part 2

Environmental Fate, Effects and Analysis
of Organic Pollutants

4

Exposure Assessment to Persistent Organic Pollutants in Wildlife: The Case Study of Coatzacoalcos, Veracruz, Mexico

Guillermo Espinosa-Reyes, Donaji J. González-Mille,
César A. Ilizaliturri-Hernández, Fernando Díaz-Barríga Martínez
and Jesús Mejía-Saavedra
Universidad Autónoma de San Luis Potosí,
Facultad de Medicina-Departamento de Toxicologia Ambiental,
Mexico

1. Introduction

Until the early 70s, it was thought that pollution was a phenomenon circumscribed to zones where pollutants were generated. Because of that, in each country concern was limited to regions where pollutant concentration was higher or its danger was greater. However, it has gradually become aware that pollution is a problem that affects everybody and, because of that, everybody is responsible to control it, regardless of the sites distance where pollutants are produced. Therefore, the problem of pollution has become a global phenomenon. Mankind has always depended on natural resources located in the region where they dwell. Nevertheless, the fast population growth coupled with a fast agricultural and industrial development as well as life style changes have increased emissions of pollutants in different ecosystems.

Persistent organic pollutants (POPs) is a group of compounds chemically very stable, able to travel considerable distances and it is resistant to natural degradation processes, most of them were produced to be used as pesticides and certain chemicals to be used as industrial processes, and others are generated as by-products unintentionally from human activities, such as combustion processes or power generation (PNUMA, 2005). Most of these compounds are highly toxic; they bioaccumulate in human and animal tissue, mainly in the fatty tissues, and can damage different organs and systemic targets such as the liver, kidney, hormonal system, nervous system, etc., of both humans and wildlife. According to the Stockholm Convention held in 2001, there are twelve compounds known as POPs: pesticides (DDT, aldrin, chlordane, dieldrin, endrin, mirex, toxaphene and heptachlor), industrial chemicals (hexachlorobenzene and polychlorinated biphenyls -PCB-) and unintentional compounds (dioxins, furans, PCDD-and PCDF-) [Albert, 2004]. In May of 2009 nine new Chemicals were added to the POPs list: alpha hexachlorocyclohexane, beta hexachlorocyclohexane; hexabromodiphenyl heptabromodiphenyl ether and ether tetrabromodiphenyl pentabromodiphenyl ether and ether chlordecone, hexabromobiphenyl, lindane, pentachlorobenzene, perfluorooctane sulfonic acid, its salts and perfluorooctane sulfonyl fluoride.

POPs main route of entry into the organism is food. However, we cannot ignore environmental exposure (inhalation) and dermal exposure (accidents). Because of their properties, POPs are classed as persistent, bioaccumulative and toxic. Therefore, POPs are to be considered as one of the most harmful groups of Toxic environmental pollutants to humans and wildlife. In countries where these compounds have been used are frequently found residuals in food. They are a problem because of their persistence in the environment and characteristics of bioaccumulation and biomagnification along the food chain and because these compounds generate toxic effects in both human population and in biota.

POPs, mainly organochlorine compounds make up a big part of hazardous waste. Mexico annually generates approximately 8 million tons of hazardous waste. Of this amount, only 12 percent is handled properly. The question is: Where does the remaining 88% go? To make matters worse, Mexico's infrastructure for hazardous waste is by far insufficient.

Coatzacoalcos, Veracruz is one of the most commercial and industrialized ports in Mexico. Presently, the Coatzacoalcos River and the areas surrounding it are regarded by many as some of the most heavily polluted sites in Mexico. Several of Mexico's chief petrochemical complexes, such as Cangrejera, Morelos and Pajaritos are based in the region. Furthermore, there have been various toxic substances present in the area which have been stored inside environmental and biological compartments, including persistent organic pollutants (Espinosa-Reyes et al., 2010; Gonzalez-Mille et al., 2010; Stringer et al., 2001), polycyclic aromatic hydrocarbons, volatile organic compounds (Riojas-Rodriguez et al., 2008), polybrominated compounds (Blake 2005), dioxins and metals (Petrlink & DiGangi 2005; Rosales 2005; Vázquez-Botello 2004).

A number of POPs have been registered in Coatzacoalcos. One of these, Hexachlorocyclohexane (HCH), is a manufactured chemical from which there are, theoretically, eight chemical forms or isomers. The three most common isomers are α-HCH, β-HCH and γ-HCH (commonly called lindane). Lindane is used as a pesticide on fruit and vegetable crops as well as on forest plantations; it is also found in medications to treat diseases such as scabies and pediculosis. There are no records indicating that lindane has ever been manufactured in Mexico; however, approximately 20 tons of these compounds are imported and subsequently used in Mexico each year. At present, lindane is authorized for use in Mexico for ectoparasite control in livestock for ticks, fleas, and common fly larvae. It is also registered for use as a seed treatment for oats, barley, beans, corn, sorghum and wheat. Pharmaceutical uses of lindane in Mexico include the formulation of creams and shampoos for scabies and lice treatment (CEC, 2006; ATSDR, 2005). In 1994, the Canadian Environmental Protection Act (CEPA) proclaimed hexachlorobenzene (HCB) as highly toxic. HCB is a manufactured chemical (which was) used as a wood preservative, as a fungicide for treating seeds and as intermediary in organic syntheses. Additionally, hexachlorobenzene can be formed as an unwanted byproduct in preparation processes like in the synthesis of organochlorines from high-temperature sources (Sala et al., 1999; Newhook & Meek 1994). Dichlorodiphenyldichloroethane (DDT) is a synthetic organochloride which is relatively stable with slow degradation rates through sunlight or oxidation, and possesses good absorption capacity and resistance to biodegradation in sediments and soils. It is also insoluble in water (CEC, 2001). In 1945, DDT was used for the first time in Mexico for the control of Malaria, and was widely used in agriculture between the 50s and 70s (CEC, 1997). The use of DDT in the Malaria Control Program was abandoned in the year 2000, when it was replaced by pyrethroids. Polychlorinated biphenyls (PCBs) are allowed in "totally enclosed uses" such as coolants and lubricants, in transformers and capacitors. Dioxins are produced during the combustion of organic materials containing chlorine as well as during the manufacture of various chlorine-containing chemicals, such as ethylene dichloride. Existing involuntary sources of intake include electric arc furnaces, shredders, sinter plants, cement plants, cremation facilities, and coal-based power plants (Lutharddt et al., 2002).

Biomonitoring wildlife can be used to detect chemical pollution and to evaluate the ecosystem's health, using test species as systematic models in the evaluation of risks associated to paths of real exposure. Wildlife species residing in polluted sites are exposed to complex mixtures of pollutants through multiple pathways which could hardly be evaluated in lab studies. The main purpose of this research was to pinpoint exposure levels to POPs in wildlife from different sets of ecosystems throughout the industrial area of Coatzacoalcos, Veracruz, Mexico to obtain a baseline of the ecological condition of this region.

2. Materials and methods

2.1 Test area, sampling sites and species selection

The Coatzacoalcos region is located in the South eastern State of Veracruz, Mexico, in the municipality under the same name, at 18° 8' 56 " N and 94° 24' 41" W. The average altitude is 14 m.a.s.l. The predominant climate is tropical rain [Am (i') gw"], the average annual temperature is 24.5°C and average annual rainfall is 2780.1 mm (García, 2004). The main inland body of water is the Coatzacoalcos River, which has an area of 322 km. It originates above 2000 m in elevation in the State of Oaxaca and over its course is fed by countless other rivers (Jaltepec, Coachapa, Uxpanapa, Calzada) and streams (Teapa, Tepeyac, San Francisco) which inflows contribute to the discharge of pollutants (Páez-Osuna et al., 1986.; Rosales-Hoz & Carranza-Edwards, 1998). The region is comprised of urban, industrial, livestock, riparian and wetland areas. However, its main activity is chemical, namely petrochemical (Ruelas-Inzunza et al., 2007).

In October 2006, six sampling stations for biological sampling were set up at the lower basin of the Coatzacoalcos River (Fig. 1). The selection of sampling sites was based on wind direction, location of industrial zones and urban areas, the presence of organisms and the influence of riparian systems as well as on previous investigations within the area (Páez-Osuna et al., 1986; Rosales-Hoz & Carranza-Edwards 1998; Stringer et al., 2001; Bahena-Manjarrez et al., 2002.).

In this research we selected earthworms, crabs, fish, toads, turtles, iguanas, and crocodiles to measure levels of POPs in muscle or blood. These groups are critical species because they have an important role in the ecosystems dynamics and/or an importance (economic, cultural and scientific) for man. Species were selected according to the following criteria: The kind of pollutant located in the study area. Based on literature, a revision on the pollutant environmental behavior was conducted, considering their physiochemical characteristics as well as environmental parameters (humidity, temperature, pH, type of soil, etc.) that can influence in the environment pollutant levels. Once pollutant groups to evaluate were determined, potential pathways and exposure routes were established. An important criterion that was also considered when selecting animal groups was their biology, as it should be well documented; finally groups that are relatively easy to capture and handle were selected.

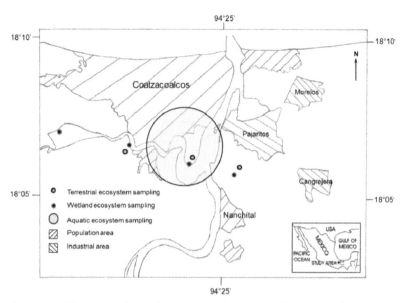

Fig. 1. Study area and location of sampling sites in the region of Coatzacoalcos, Ver.

2.2 Ecological importance of selected species
2.2.1 Earthworms
Earthworms' ecological importance is that they are decomposer organisms (important in biogeochemical cycles) and because of that, they have an important role when adding nutrients to soil (they favor the availability of nitrogen, phosphorus and sulfur) which can be used by vegetable species (Reines et al., 1998; Legall, 2006); they are also an important link in the trophic chain, mainly for some bird species and it has been proven that earthworms can accumulate important metal concentrations (Sánchez-Hernández, 2006). Earthworms can be found in many types of soil and are vulnerable to impacts occurred in soil, their small size represents an advantage to be handled; their distribution is ubiquitous in edaphic horizons with detritus, they are easy to capture, have a close contact with soil, and have a short life-cycle favoring the study of several generations (Ogunseitan, 2002). Persisting organic pollutants have the capacity to bioaccumulate and biomagnify along the trophic chain, as well as animals that belong to decomposers or detritivores levels are very important for the ecosystem functioning, therefore, if animals that are part of soil are affected by pollutants that may show in the ecosystems' health.

2.2.2 Crabs
Crabs are ubiquitous in all temperate and tropical regions in the world. In the wetland ecosystems they are an ecologically important kind, because they play a primary role in the decomposition of organic material and the addition of nutriment to soil. Due to the fact that they build up their galleries by the rivers' basin, lakes or seas, they spend part of their life on land surface and when tides rise they shelter in their burrows (≈ 30 cm. deep) under water. They mainly feed themselves with detritus, so they are excellent filtering organisms, capable to accumulate great pollutant quantities.

2.2.3 Iguanas

There are different kinds of iguanas, however the most common ones in the Coatzacoalcos region are the green iguana (*Iguana iguana*) and the black iguana (*Ctenosaura* spp.); Distribution of both goes from the south of Mexico to South America. According to Lara-Lopez and Gonzalez-Romero, (2002), iguanas are herbivores, the diet of the green iguana is composed mainly as follows: leaves (57.36%), flowers (24.15%) and fruits (3.43%), most of their life is spent on top of the trees, because of the previously said, this specie may be used as a POPs bio-monitor in air. In many coastal communities in Mexico, it is usual to eat iguana as an important source of protein, and it is believed that their blood and eggs contain a lot of energy and help to prevent certain diseases such as anemia. Furthermore people use the skin of this reptile and dissected to sell as ornaments (Alvarez del Toro, 1982). In southern Mexico (Leon & Montiel, 2008), in Central America (FAO, 1997) and in parts of South America, the iguana is one of the most consumed wildlife species. Iguanas have been an important source of protein for humans for over 7000 years FAO (1997). Many of the rural inhabitants of Central America still rely on iguana as a protein source; however, consumption of these species are not uniform during the year because in most cases eating frequency depends on the availability in certain times or seasons that are usually 3 to 4 months per year (Pers. Obs.).

2.2.4 Fish and invertebrates

Fish and aquatic invertebrates are commonly used to monitor pollutants because they bioaccumulate toxic substances and are wile spread, coupled with the diversity and importance of these environments. It has been observed that they are highly sensible to changes in the aquatic environment as well as to low concentrations of environmental pollutants, (Russo et al., 2004; Klobučar et al., 2010). On the other hand, fish have a crucial position in the toxicological field, due to the fact that they have been widely used in studies related to human and ecological health. Fish study includes a wide variety of approaches to detect aquatic pollution impacts from direct measures of mortality, to the analysis of demographical dynamics and the community structure, and to the detection of measures of sub cellular changes (Di Giulio & Hinton, 2008).

At the same time aquatic invertebrates and specifically crustaceans are organisms that have a wide distribution (example: marine, terrestrial and freshwater environment), they are organisms that are in close contact with pollutants in sediment, so they have been used in countless eco-toxicological studies. They have proven to be useful to evaluate effects on different pollutants. They have also served as aquatic pollution indicators. In addition, they may be a source or exposure for local consumers (Nacci et al., 1996; Rinderhagen et al., 2000; Rigonato et al., 2005; Külköylüolu 2004; Regoli et al., 2006). Otherwise, spatial distribution of pollutants on sediments and biota in aquatic ecosystems have been related to a great variety of biological answers in populations and fish/invertebrates communities, with the purpose to determine a possible relation between pollutants in the environment and health in the organisms (Adams et al., 1999).

2.2.5 Giant toads

The giant toad (*Rhinella marina*, after *Bufo marinus*) is a native and geographically widespread species in Mexico and Central America (Zug & Zug 1979). It is an omnivorous and opportunistic species (Zug & Zug 1979), which indicates that toads would integrate different exposure paths due to the ingestion of a wide variety of food items and amphibious living habits. The giant toad is one of the largest amphibians in Mexico (adult body length ranges from 10 to 17 cm), with a life expectancy from 10 to 15 years in the wild. The high lipid-somatic index (2 to 10% compared to less than 0.1% in most anuran species after the spawning period) and the elevated hepatosomatic index (Feder & Burggren 1992) along with its breeding biology make this species prone to bioaccumulation of organic and inorganic pollutants and their toxicological effects (Sparling et al., 2010; Linder et al., 2003). Recently, the giant toad has been used as an aquatic ecosystem biomonitor in the evaluation of air pollution (Dohm et al., 2008), infectious diseases (Zupanovic et al., 1998), organochlorine pesticides (Linzey et al., 2003) and endocrine disruptors (McCoy et al., 2008).

2.2.6 Turtles
Slider turtle (*Trachemys scripta*) is geographically widespread across Mexico and Central America (Burger and Gibbons, 1998). They are eligible species because they have several characteristics associated with their metabolism, life history and ecology (Overmann & Krajicek, 1995). These turtles are generally omnivorous and have temperature-dependent of sex determination and studies have investigated contaminant effects on this process (Selcer 2006), this makes them ideal for studies of chronic exposure of local pollutants. This species has been employed for exposure assessment to metals, radiation, organochloride pesticides and polybrominated biphenyls (Bergeron, et al., 1994; Bickham, et al., 1998; Burger and Gibbons, 1998; Lovelette & Wrigth, 1996; Meyers-Schöne & Walton 1994; Willingham et al., 1999; Willingham et al., 2000). In Mexico, slider turtles are considered an endangered species and are protected by Mexican laws.

2.2.7 Crocodiles
Swamp crocodiles (*Crocodylus moreletii*) are aquatic reptiles living in Mexico's tropical regions. They have a life strategy based on late maturation; they are extremely long-lived animals, show parental care and determine sex depending on temperature (Selcer, 2006). Crocodiles reach the highest levels of food chains, so they are useful for the evaluation of persistent and biomagnifying pollutants. Around the world, their populations are endangered (including Mexico), and because of that the concern on effects (mainly reproductive ones) of pollutants in their populations has increased (Guillette et al., 1999); it must be mentioned that the highest DDE levels registered in wild reptile were found in these organisms (De Solla, 2010). Different Crocodylia species have been used to evaluate heavy metals as well as persistent organic compounds.

2.3 Biological sampling techniques
Wild Earthworms (*Eisenia* sp) were collected by excavation. Crabs were harvested using pitfall traps (*Uca* sp) and traditional fishing gear (*Callinectes* sp). fish (*Aplodinotus* sp, *Ariopsis felis*, *Centropomus parallelus*, *Eucinostomus* sp, *Eugerres axillaris*, *Gobiomorus* sp, *Menticirrhus* sp, *Mugil cephalus* and *Oreochromis* sp) were caught using traditional fishing gear (i.e. cast net) with the help of fishermen. Giant toads (*Rhinella marina*) were collected from each site using nets in nocturnal transects within an area of 10,000 m². Crocodiles (*Cocodrylus moreletti*) and Iguanas (*Iguana iguana*) were caught using a noose trap. Turtles (*Trachemys scripta*) were captured with a baited piper trap placed near fallen trees and along the edge of the river during the afternoon and checked early the following morning.
Immediately after capture, organisms were measured, weighed and sorted by type of species; type of ecosystem (terrestrial, aquatic and wetland) and by feeding behaviours (carnivores, omnivores, detritivores and herbivores). Blood samples drawn were obtained using

heparinized syringes on endangered animals (turtles and crocodiles). All organisms were subsequently released. Samples were stored at 4°C for transport and subsequent laboratory analysis. Dissection was performed on each of the specimens from the rest of different species to extract the muscle tissue. The tissue was placed in amber glass containers and frozen at -20°C until analysis. All organisms were collected with a Scientific Collector's Permit (Wild Fauna and Flora Scientific Collector) issued by México's SEMARNAT (Ministry of Environment) or Secretaría de Medioambiente y Recursos Naturales-No. FAUT-0133.

2.4 Analysis of blood and tissue residues

Concentrations of the following compounds were tested for on biological samples: α-, β-, γ-hexachlorocyclohexane (HCH), hexachlorobenzene (HCB), aldrin, dieldrin, mirex, α-, γ-chlordane, oxychlordane, trans-, cis-nonachlor, heptachlor epoxide, p, p'-DDT, p, p'-DDE, polychlorinated biphenyls (PCBs, IUPAC No 28, 52, 99, 101, 105, 118, 128, 138, 153, 156, 187, 180, 183, and 170) and polybrominated diphenyl ethers (PBDE, only on some species of fish). The method of extraction, separation and cleaning of muscle tissue was carried out according to the method established by Jensen et al., (2003) with slight modifications (Gonzalez-Mille et al., 2010) and Dallaire et al., (2006) for blood samples. The endrin-C13 and PCB 14-C13 were used as internal standards and were added to all samples. The chromatographic method (Gas chromatography–mass spectrometry GC-MS) was carried out according to that reported by Trejo-Acevedo et al., (2009). The detection limit for POPs was approximately 0.3 mg/L.

3. Results and discussion

3.1 Terrestrial ecosystem

In Table 1, it can be observed that earthworms have the highest concentrations of polychlorinated compounds biphenyls (PCBs) y persistent organic pollutants (POPs), followed by iguanas and finally by crabs. PCBs congeners that were analyzed were (PCBs 105, 128, 138, 153, 156, 170, 180 y 183).

ECOSYSTEM	SPECIES	α-HCH	β-HCH	γ-HCH	DDT	DDE	Mirex	ΣPCBs	ΣPOPs
	Eisp	12.8		106.3	2.5	13.2		13.4	146.2
	(n=6)*	(5.8 - 31.4)	N.D.	(39.4 - 196.0)	(2.5 - 2.5)	(0.3 - 57.2)	N.D.	(2.3 - 39.2)	(50.3 - 323.8)
	Igig	4.76	0.59	0.55		0.04	0.36	0.11	6.42
TERRESTRIAL	(n=3)	(4.33 – 5.41)	(0.48 – 0.71)	(0.45 – 0.66)	N.D.	(N.D. – 0.06)	(0.27 – 0.49)	(N.D. - 0.22)	(5.97 – 6.82)
	Ucsp	0.24	0.37	0.06	0.02	0.04	0.73	0.04	1.52
	(n=2)*	(0.17 – 0.31)	(N.D. - 0.74)	(0.05 – 0.08)	(0.020- 0.026)	(0.04 – 0.05)	(0.72 – 0.74)	(N.D. - 0.09)	(1.25 – 1.78)

Values represent the mean and range. Eisp: *Eisenia* sp., Igig: *Iguana iguana*, Ucsp: *Uca* sp., * Pool samples, N.D.: non detected

Table 1. Concentrations of persistent organic pollutants (ng/g tissue) from terrestrial wildlife collected in Coatzacoalcos, Veracruz.

Registered results of iguanas (*Iguana iguana*) are relevant because until now there are no studies showing POPs exposure background. In addition, because they are a basic part of rural communities' diet, meat intake may be a potential route of exposure to organic pollutants persistent for humans. Furthermore, because it is a tree species, it may be used to indirectly monitor air quality of some volatile organic compounds (VOCs) and semi-volatile ones as DDT and their metabolites, as well as some congeners of PCBs.

With crabs (*Uca* sp.) there are a few exposure studies to POPs. De Sousa et al., (2008) that registered POPs concentrations in crab eggs (*Chasmagnathus granulata*) in different Brasil stereos. Concentrations of DDE, DDT, γ-HCH, PCBs y total POPs are higher (35.95; 0.38; 3.52; 286.27 and 339.68 ng/g) than those reported in the present work. The previously said may be due to the fact that the matrix analyzed by de Sousa et al., contains a greater quantity of lipids, and crabs evaluated in this study (*Uca* sp.) were not in reproductive stage. Bayen et al., (2005) a study was conducted in a Singapure's mangrove swamp where a thropic web was established and the thunder crab (*Myomenippe hardwicki*) was one of the species presenting high concentrations of POPs. Falandysz et al., (2001) also used crabs (*Carcinus means*) to evaluate the exposure to organochlorine pesticides. It is complicated to make a comparison between crab species because they have different etiology and habitants, however, because most crab species are detritivores, we consider them as a good option to do studies related to POPs exposure.

There are several studies on earthworms with ecotoxicological background mainly focused on effects at population levels (lethality), at both laboratory controlled conditions (Heimbach, 1984; Ma & Bodt, 1993; Kula, 1995, Morrison et al., 2000) as in field ones (Thompson, 1970; Tomlin, 1981; Edwards & Brown, 1982; Haque & Ebing, 1983; Potter et al., 1994; Espinosa-Reyes et al., 2010), however, there are a few studies where exposure to POPs are evaluated, Jones and Hart (1998) revised works related to exposure of different earthworms species to various pesticides–Benomyl, Carbaryl, Carbendazim, Carbofuran, Chlordane, Methiocarb, Parathion, Pentachlorophenol, Phorate, Propoxur, Thiophanateme-. These same authors mentioned that earthworms of the *Eisenia* type are the most resistent to the pesticides previously mentioned. Morrison et al., (2000) evaluated the bioavailibility of DDT, DDE, DDD y Dieldrin in soil samples with different age of use of the mentioned pesticides. They exposed *Eisenia* earthworms. The concentrations registered by Morrison et al., (2000) are high when compared with the registered in this study (DDT 28.3±8.4 & DDE 3.77±0.48 mg/Kg tissue). Based on these backgrounds and results registered in this study, it is possible to postulate the Eisenia earthworms as POPs biomonitors in terrestrial ecosystems because earthworms play a major role in facilitating pivotal interactions within ecosystems through the mixing and translocation of soil constituents, or serving as a conduit for contaminants to predators at higher trophic levels (Harris et al., 2000; Langdon et al., 2003). Earthworms have been used extensively in ecotoxicology like biomonitors (Fitzpatrick et al., 1992; Goven et al., 1993; Reinecke & Reinecke, 1998; Espinosa-Reyes et al., 2010) to assess the effects of diffuse contaminants present in soils.

Finally, when doing a POPs biomonitoring in terrestrial ecosystems, it is important to take into account the following factors: a) species susceptibility difference; b) soil type; c) species behavior; d) exposure time; e) pollutant toxicity; f) when using pesticides consider the applying method.

3.2 Aquatic ecosystem

Because fish are organisms with a wide movement range, the sample was taken within an approximate area of 8 Km². Thirty-one fish from five species were caught: *Centropomus parallelus*, *Mugil cephalus*, *Eugerres axillaris*, *Oreochromis* sp, *Ariopsis felis* and thirty organisms of one crustacean species *Callinectes* sp., for quantification of POPs.

From de 29 quantified compounds in the simple, only were detected concentrations of HCB, α-, β-, γ-HCH, DDT, DDE, mirex and 6 congeners of PCBs (52, 101, 105, 118, 138, 153) (Table 2). Most concentrations registered were of β-HCH, α-HCH and mirex. HCB was only detected in 26% of the samples and just on species *Mugil cephalus y Callinectes* sp. El γ-HCH and DDT were registered in a 32% and 10% of the samples respectively. Pollutants α-HCH, β-HCH, DDE, mirex and PCBs were detected in a 100% of the samples. Species where a higher number of compounds were found are *Callinectes* sp, *Ariopsis felis y Eugerres axillaris* (8, 7 y 6 respectively), however, most concentrations were found in *Callinectes* sp and *Eugerres axillaris*. According to these results, POPs concentrations in muscular tissue by species decrease in this order β-HCH > α-HCH > mirex > DDE > Total PCBs > γ-HCH > DDT > HCB.

ECOSYSTEM	SPECIES	HCB	α-HCH	β-HCH	γ-HCH	DDT	DDE	Mirex	ΣPCBs	ΣPOPs
AQUATIC	Cepa	N.D.	1.2	0.1	N.D.	N.D.	0.1	0.3	0.2	1,8
	(n=9)		(0.05-5.3)	(0.05-0.5)			(0.05-0.5)	(0.05-0.6)	(0.05-0.4)	(0.5-5,9)
	Muce	0.1	0.2	0.3	N.D.	N.D.	0.2	0.3	0.1	1.0
	(n=7)	(0.05-0.2)	(0.05-0.3)	(0.05-0.7)			(0.05-0.5)	(0.2-0.4)	(0.05-0.2)	(0.6-1,5)
	Euax	N.D.	0.4	1.7	0.1	N.D.	0.1	0.6	0.1	2.9
	(n=7)		(0.03-1.9)	(0.05-3.8)	(0.05-0.3)		(0.05-0.1)	(0.1-0.9)	(0.1-0.2)	(0.9-4.7)
	Orsp	N.D.	0.3	N.D.	N.D.	N.D.	0.2	0.4	N.D.	0.8
	(n=5)		(0.2-0.4)				(0.1-0.2)	(0.3-0.5)		(0.7-1.0)
	Arfe	N.D.	0.3	N.D.	0.1	0.1	0.1	0.5	0.3	1.3
	(n=3)		(0.2-0.5)		(0.05-0.1)	(0.05-0.1)	(0.05-0.2)	(0.4-0.7)	(0.1-0.6)	(1.1-1.5)
	All fishes	0.1	0.6	0.9	0.1	0.06	0.2	0.4	0.2	1.7
	(n=31)	(0.05-0.2)	(0.03-5.3)	(0.05-3.8)	(0.05-0.3)	(0.05-0.1)	(0.05-0.5)	(0.05-0.9)	(0.05-0.6)	(0.5-5.9)
	Casp	0.2	1.2	1.6	0.2	0.1	0.4	1.3	0.3	14.8
	(n=4)*	(0.05-0.5)	(0.3-2.4)	(0.05-5.2)	(0.05-0.2)	(0.05-0.1)	(0.1-0.9)	(0.9-1.7)	(0.1-0.7)	(13.3-17.6)

Values represent the mean and range. Cepa: *Centropomus parallelus*, Muce: *Mugil cephalus*, Euax: *Eugerres axillaris*, Orsp: *Oreochromis* sp, Arfe: *Ariopsis felis*, Casp: *Callinectes* sp. N.D.: non detected

Table 2. Concentrations of persistent organic pollutants in muscle tissue (ng/g wet weight) from aquatic wildlife collected in Coatzacoalcos, Veracruz.

Out of the *Ariopsis felis* fish species the highest concentrations of PCBs were registered and it is the only one that presented DDT, *Mugil cephalus* is the only species in which HCB was detected and it presented the highest concentrations of DDE, *Eugerres axillaris* showed the highest concentrations of lindane and α, β-HCH's, finally *Oreochromis* sp registered the lowest load of POPs in comparison to the other species.

Concentrations of β- HCH and α-HCH were the most abundant among the evaluated species, which matches with what was registered by Lee et al., (1997) and Yim et al., (2005) in other fish species. This can be explained because β-HCH and α-HCH have a greater bio-concentration factor (log BCF 2.8 and 2.5 respectively) in aquatic animals and they are more persistent than γ-HCH (log BCF 1.2) (Willett et al., 1998). Even though technical grade HCH has a greater constitution of α-HCH (60-70%), concentrations of β-HCH and α-HCH were very similar, this can be due to the fact that α-HCH has a high volatility (Henry' law constant 6.68 x 10-6 atm-3/mol) and environment degradation, which can cause low persistency (Yim et al., 2005). Another possible explanation is that HCH metabolized and excreted faster than β-HCH (Takazawa et al., 2005).

DDT was only detected in *Ariopsis Felis* while DDE was detected in all species. Most concentrations were registered in detritivores fish. As it was mentioned, DDT was widely used in Mexico for malaria control and because of its toxic effects, its use was eliminated. Chemical and biological processes transformed DDT into DDD and DDE, particularly DDE has been the most registered in biota (Takazawa et al., 2005). Registered concentrations in fish from the zone indicate that they are mainly exposed to residual DDT and to its degradation products. However, as it was seen, DDT/DDE relation in sediment suggests current use, reason why it is expected to have greater concentrations in the tissues, which suggests that fish from the zone have the capacity to eliminate DDT fastly.

Furthermore, quantification of polybrominated diphenyl ethers (PBDE) was done in 30 fish that belong to different species (*Aplodinotus* sp, *Centropomus parallelus*, *Mugil cephalus*, *Eucinostomus* sp, *Gobiomorus* sp, *Menticirrhus* sp and Torombolo –common name-), in these organisms, there were concentrations of 6 congeners registered (PBDE 47, PBDE 100, PBDE 99, PBDE 154, PBDE 153, PBDE 209) (Table 3). Considering all species, the order of concentration of congeners found was 47>41>154>209>153>100, being *Eucinostomus* sp species where the highest concentrations were found and in Torombolo species the lowest ones.

Within this context, aquatic ecosystems are highly vulnerable because of their tendency to accumulate concentrations relatively greater of pollutants coming from terrestrial ecosystems surrounding them, as well as those from direct entries (downloads), in that, regardless of its source of entry into the environment, aquatic systems are frequently deposits for a great variety of chemicals. Pollution in these environments can have negative effects over aquatic life (example: alteration of reproduction and decreased of species) as well as directly or indirectly affect human health and threat food safety. (Jha, 2008).

3.3 Wetland ecosystem

In Table 4 are presented the levels of persistent organic compounds detected in the different animal species from the Coatzacoalcos wetland. When monitoring them, 6 organochlorine pesticides out of the 14 ones analyzed were detected and 13 polychlorinated compounds biphenyls (PCBs) out of 21. All captured organisms from the different species presented detectable levels of at least 3 persistent organic compounds: DDE, Lindane y PCBs (data not shown, found in giant toads' adipose and hepatic tissue). Generally, the pattern of pollutants presence in giant toads' muscular tissue was $\sum$DDT> $\sum$HCH> Mirex > HCB; and for the turtles and crocodiles' case, the exposure levels of PCBs and DDE in serum were similar. It must be said that lindane and DDT concentrations in crocodiles were higher.

SPECIES	PBDE 47	PBDE 100	PBDE 99	PBDE 154	PBDE 153	PBDE 209	Σ PBDEs
Mesp (n=7)	3.2 (0.5-6.5)	1.9 (0.5-4.3)	4.5 (1.0-8.0)	0.3 (0.04-0.5)	0.7 (0.3-1.2)	0.7 (0.3-1.2)	11.3 (3.8-19.7)
Eusp (n=2)	12.03 (7,7-16,4)	3.7 (0,8-6,5)	2.6 (1,7-3,5)	18.0 (9,5-26,5)	10.2 (1,5-9,2)	13.2 (8,6-17,7)	59.7 (43,1-76,2)
Muce (n=7)	5.5 (0,2-24,8)	1.6 (0,1-2,9)	5.4 (0,5-10,6)	1.3 (0,03-4,2)	1.1 (0,1-2,0)	0.9 (0,1-2,2)	15.8 (1,0-45,1)
Gosp (n=1)	2.9	2.3	8.8	0.8	1.2	1.4	17.3
Apsp (n=3)	1.9 (1,1-2,7)	0.7 (0,5-1,0)	2.4 (1,21-3,7)	0.2 (0,1-0,2)	0.3 (0,1-0,4)	0.3 (0,1-0,6)	5.7 (3,2-8,7)
Torombolo (n=2)	2.0 (1,3-2,8)	0.4 (0,2-0,5)	1.3 (0,8-1,9)	0.1 (0,1-0,2)	0.2 (0,1-0,3)	0.2 (0,1-0,3)	4.3 (2,6-6,0)
Cepa (n=8)	2.2 (1,0-4,0)	1.0 (0,5-1,4)	3.7 (0,8-5,5)	0.2 (0,1-0,5)	0.5 (0,2-0,6)	0.5 (0,1-0,8)	8.1 (3,1-12,6)
All fish (n=30)	4.2 (0.2-24.8)	1.7 (0.1-6.5)	4.1 (0.5-10.6)	3.0 (0.03-26.5)	2.0 (0.1-9.2)	2.5 (0.1-17.7)	17.5 (1.0-76.2)

Values represent the mean and range. Mesp: *Menticirrhus sp.*, Eusp: *Eucinostomus sp*, Muce: *Mugil cephalus*, Gosp: *Gobiomorus s*, Apsp: *Aplodinotus sp*, Torombolo: -common name-, Cepa: *Centropomus parallelus*.

Table 3. Concentrations of polybrominated diphenyl ethers (ng/g lipid) in fish muscle tissue from species collected in Coatzacoalcos, Veracruz.

ECOSYSTEM	SPECIES	HCB	α-HCH	β-HCH	γ-HCH	DDT	DDE	Mirex	ΣPCBs	ΣPOPs
WETLAND	Rhma	0.08	0.75	0.31	0.26	ND	1.51	0.39	ND	3.1
	(n=12)	(N.D.-0.15)	(0.41-2.05)	(N.D.-1.24)	(N.D.-0.53)		(0.05-7.8)	(N.D.0.67)		(0.84-9.56)
	Crmo‡	N.D.	N.D.	N.D.	64.5	59.2	11.5	N.D.	2.7	76.0
	(n=2)				(N.D. -64.5)	(N.D. - 59.2)	(N.D. - 11.5)		(2.5 - 2.7)	(72.7 - 79.2)
	Trsc‡	N.D.	N.D.	N.D.	3.5	ND	8.8	N.D.	3.0	7.0
	(n=4)				(N.D. - 4.58)		(N.D. - 8.78)		(N.D. - 5.83)	N.D. - 14.61

Values represent the mean and range. Rhma: *Rhinella marina*, Crmo: *Crocodylus moreletii*, Trsc: *Trachemys scripta*. * Pool samples, ‡Blood samples, N.D.: non detected

Table 4. Concentrations of persistent organic pollutants in muscle tissue (ng/g wet weight) and blood serum (ng/ml) from wildlife collected in Coatzacoalcos, Veracruz.

The DDT/DDE relation obtained in toads' tissue was lower than 1.0 which suggests the organism's ability to metabolize the parental compound to DDE. Studies on amphibians, especially the toads' family (Bufonidae), has proven these organisms capacity to accumulate high concentrations of DDT and its metabolites. It should be said that these studies have also shown that proportions of DDE and DDT were higher than those of the parental compound. In our case, DDD concentrations in tissues were not quantified, so exposure of toads from the Coatzacoalcos region to DDT and its metabolites may be underestimated. Registered DDE concentrations in this research are comparable with those found in adult anuran coming from other polluted sites (Table 4 and 5). There are records of $\sum$DDT concentrations and their metabolites (mainly DDE) up to 3480 µg/g of fat in samples of *R. clamitans* coming from the wetlands from the south of Ontario (Harris et al 1998); in our study, we found concentrations that go up to 3094.5 µg/g of fat in samples of giant toad's liver (data not shown). POPs exposure works in fresh water turtles' populations are scarce; studies done in Terrapin (*Malaclemys terrapin*), Snapping turtle (*Macrochelys temminckii*) and Common musk (*Sernotherus odoratus*) have determined levels of DDE of 0.73-21.7 ng/g wet mass (Basile, 2010, Moss et al., 2009, De Solla et al,. 1998). Several studies on marine turtles have proven levels of DDE of 0.06- 0.73 ng/g wet mass (Stewart et al 2011). Our study shows comparable levels between species of terrestrial and fresh water turtles. Crocodiles present the highest levels of DDT and DDE in plasma; be noted that higher organochlorine pesticides concentrations in the world (mainly DDE) have been found in the Crocodilia species due to their position in the trophic chain. Guillette et al., (1999) Reported DDE levels of 0.9- 17.9 ng/ml and DDT levels of 0.45 to 0.70 in plasma of alligator's residents at Apopka Lake in florida; our study shows comparable levels; even greater to the ones found in Apopka Lake.

The distribution general pattern of Hexaclorociclohexanos (HCH's) observed in toads' tissues was α-HCH>β-HCH>γ-HCH and for the crocodiles and turtles' case, levels of γ-HCH were detected. In accordance with literature, general distribution patterns of HCH's in mammals, birds and fish are β-HCH>α-HCH>γ-HCH; this distribution pattern is mainly determined by the compound persistency, the exposure path, the species metabolism, and the trophic position (Willet et al., 1998). Our data contrast with the distribution pattern observed in other studies of environmental and biological matrix. Some possible explanations in order to interpret the presence of a higher proportion of γ-HCH is the chronic exposure to isomer of different routes, which can be noted with the found values in crocodile and turtle's plasma where isomer γ-HCH is the only detectable one (Table 4). El γ-HCH is more volatile in comparison with the other isomers, which implies an important transport by air, it is also the isomer with most solubility in water (Walker et al 1999). Coupled to the previously said, it is possible that other important exposure routes exist towards HCH's which have not been explored yet and which may significantly contribute to corporal load and its distribution to these animals. Even though differences in isomers proportions may indicate different sources, routes and times of exposure, for many species it is not clearly understood the influence of processes such as intake, distribution, metabolism and storage in the differences of isomer distribution in tissues (Willet et al., 1998). HCH's concentrations detected in this study are lower than those observed in other studies done on wild amphibians living in agricultural sites (Table 4 and 5). In the case of turtles and crocodiles, there was no useful information found to compare data with.

SPECIE	COMPOUND	CONCENTRATION	TISSUE	REFERENCE
Pseudacris crucifer	DDE	1001	Whole	Russell et al 1995
	DDT	160.6		
	α- HCH	0.37		
	β- HCH	1.37		
	γ- HCH	<DL		
Rana clamitans	DDE	0.58-45.0	Whole	Russell et al 1997
	HCB	0.08-0.49		
	∑DDT	1.24		Guilliland 2001
	∑HCH	0.12		
Rana perezi	DDE	<DL-190	Muscle	Rico et al 1987
	∑DDT	50-550		
	γ- HCH	<DL-10		
	∑DDT	35.4	Whole	Pastor et al 2004
	HCB	2.7		
	α, γ- HCH	0.5		
Rana pretiosa	DDE	91-173	Whole	Kirk 1988
	DDT	563-1750		
Rana mucosa	DDE	17-100	Whole	Fellers et al 2004
	α- HCH	<DL-4.9		
	γ- HCH	<DL-0.7		
Necurus maculosus	DDE	0.3-90.0	Whole	Bonin et al 1995
	DDT	<DL-8.3		
	∑HCH	<DL-10.1		
Necturus lewisi	DDE	60	Whole	Hall et al 1985
Chaunus arenarum	DDE	ND-4.5	Whole	Jofre et al 2008
	α- HCH	ND-5.6		
	β- HCH	ND-2.3		
	γ- HCH	ND-2.7		
Hypsiboas cordobae	DDE	1.1-1.7	Whole	
	α- HCH	3.9-5.4		
	β- HCH	3.0-7.3		
	γ- HCH	4.9-7.2		
Leptodactylus mystacinus	DDE	ND-6.0	Whole	
	α- HCH	3.5-6.9		
	β- HCH	0-8.9		
	γ- HCH	ND-5.7		
Melanophryniscus stelzneri	DDE	10.6	Whole	
	α- HCH	6.99		
	β- HCH	ND		
	γ- HCH	ND		
Odontophrynus occidentalis	DDE	0.8-1.8	Whole	
	α- HCH	1.3-4.1		
	β- HCH	1.0-2.7		
	γ- HCH	1.5-3		
Pleurodema tucumanum	α- HCH	4.7	Whole	
	β- HCH	4.1		
	γ- HCH	4.4		

DL-Detection limit, ND-Not detected

Table 5. Persistent Organic Pollutants concentrations (mg/Kg wet weight) measured in various amphibian species from different studies.

Mirex and HCB were pollutants found in less proportion in giant toads' tissues. There are a few studies related to Mirex and HCB exposure on adult amphibians. Russell y collaborators (2002) reported average concentrations of 0.26-1 ng/g tissue of HCB in cricket frogs (*Acris crepitans*) coming from 5 agricultural locations in Ohio USA; in table (Table 5) other studies are presented with levels of HCB which are higher when compared to the ones obtained in this study (0.14-0.67 ng/g of tissue). Detectable levels of Mirex and HCB in turtles and crocodiles were not found.

PCBs congeners detected in blood and tissue sample were 52,101, 105, 118, 138, 153, 156, 170 and 180, which corresponds with the reported in other studies as the most common for human and biological samples (Table 4). The PCBs congeners presence pattern observed in the giant toad's tissue is consistent with other studies on amphibians from other regions around the world (Loveridge et al., 2007, Russell et al., 1997); as well as in the found concentrations (Table 6). Studies conducted in diverse species of fresh water and terrestrial turtles have reported levels of PCBs totals of 5-414.8 ng/g wet mass (Basile, 2010, Moss et al., 2009, De Solla et al,. 1998) and for alligator populations there have been found levels of 1.54 ± 0.12 ng/ml (Guillette el al 1999); these levels are comparable with the ones obtained in this study. Detected congeners (except 52) correspond in greater proportion (>30%) to aroclor 1254, one of the most sold commercial mixtures in the world, which suggests that the origin of these compounds probably is related to the use of these oils in the region's industrial areas. Detected congeners are characterized for being some of the most persistent ones in the environment and for being absorbed in greater proportion in the organisms.

SPECIE	TISSUE	CONCENTRATION	REFERENCE
Rana clamitans	Carcass	2.8	Loveridge et al 2007
Various anurans	Carcass	151-4470	DeGarady y Halbrook 2003
Rana clamitans	-	7.51	Rusell et al 1997
Necturus maculosus	Carcass	113-1082	Bonin et al 1995
Rana pipiens and *Rana clamitans*	Carcass	50-112	Phaneuf et al 1995
Rana perezi	Muscle	50-1080	Rico et al 1987

Table 6. Total PCB concentrations (ng/g wet weight) measured in various amphibian species from different studies.

Amphibian and reptiles populations are declining at an alarming way in the world (Alford, 2020; Todd et al., 2010); some of the determining concomitant causes in this phenomenon is the exposure to toxic agents; let us note that reptile and amphibian toxicological information is growing, however, it is yet limited in comparison with other vertebrate groups (Sparling et al., 2010). Amphibians and reptiles may be exposed to a wide spectrum of toxic substances; pollutant accumulation in these organisms may be influenced by many factors (physiological, trophic, behavioral, etc.) at the same time, exposure may occur by different routes and in different environments during their life-time. It is known that these organisms can accumulate significant pollutant loads in their tissues, mainly of heavy metals and organic compounds. The caused effects by exposure to POPs of greater concern in reptile and amphibian populations are the endocrine disruption, DNA damage, and development abnormalities; some of the studies of greater impact over these effects have been found in these organisms. Ecological importance to maintain viable reptile and amphibian populations is determinant because these organisms are the link between terrestrial and aquatic ecosystems; at the same time they are mainly placed between intermediate links of trophic chains; they have also diversified and occupied a wide spectrum of ecological niches in different types of ecosystems; the presence of highly persistent, bioaccumulative and biomagnifying pollutants is a potentially dangerous situation worth to be evaluated in the Coatzacoalcos Veracruz ecosystems.

3.4 Trophic levels and bioaccumulation of POPs

In Figure 2, the integration of exposure to POPs in tissue (ng/g lip) of the evaluated species in each one of the ecosystems present in Coatzacoalcos; a trophic hypothetical web was established in the region. Results were classified according to the trophic level that species belong to. It can be observed that herbivores are the ones presenting the lowest POPs concentrations followed by carnivorous, omnivorous, and finally the ones with the highest concentrations are the omnivorous. This is explained because detritivores organisms are found in grater contact with contaminated matrix (as soil and sediment), while omnivorous organisms include a greater number of exposure routes (environment and food) and are, therefore, the ones presenting greater POP's concentrations in their tissues.

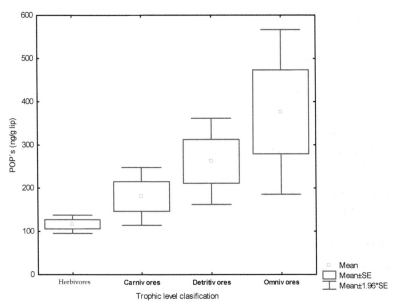

Fig. 2. Concentration Levels of POPs* in tissue taking into account the feeding habits of some species collected at Coatzacoalcos, Ver. *Corresponds to the sum of chloride compounds. Herbivores (iguana); Carnivores (Fish–Cep-); Detritivores (Fish –Arfe, Muce-Crab –Ucsp- and–Casp-); Omnivores (Fish –Tilapia-, Amphibian–Rhma-)

4. Conclusion

With the noted exception of HCB which was not found in the terrestrial ecosystem, traces of all other POPs were identified in species from all three ecosystems. Performing an analysis of feeding habits, it was learned that concentrations of POPs rise as follows: herbivores < carnivores < detritivores < omnivores. With these results it may be established that the region's biota is in fact exposed to diverse Persistent Organic Pollutants. In similar studies it has been found that exposure to POPs may cause several effects (Espinosa-Reyes et al., 2010; González-Mille et al., 2010). In addition, one must consider that Coatzacoalcos organisms are exposed to other types of pollutants in addition to POPs whose collective action could well increase the magnitude of the effects. On the other hand, as far as brominated compounds go, this is the first ever study on fish in the region. The described scenario leads us to ask ourselves what exactly is the risk to humans and wildlife in the region. The Outlook is not promising if you consider that these organisms are of paramount importance in the food chain both to humans and other species of wildlife.

Available data on concentrations of chemical substances in the environment and human beings as well as over effects of exposure to chemicals complex combinations is still scarce. Chemicals generally pose a risk for the environment; they are a rapidly growing pollution load including chemical compounds increasingly complex from which potential effects on public health and environment are probably known. It is estimated that between 70,000 and 100,000 chemicals are available in the market and this number is growing fast. Around 5,000 of these substances are being produced in high volumes, over one million tons a year. The biggest chemical producers are countries members of the OECD, but countries such as India, China, Brazil, South Africa and Indonesia are rapidly increasing their productions. One of the main reasons for the development and adoption of the REACH Regulation is that a great number of substances have been produced and marketed in Europe during many years, at times in very large quantities yet there is not enough information over the hazards for human health and the environment. The European Commission has estimated that in order to fill in information gaps related to toxic effects of that large number of substances it may be required the use of 9 million of lab animals with an approximate cost of 1.3 billion € to perform the necessary tests. A later estimate suggested that required tests would imply 54 million vertebrate animals and costs would grow up to 9.5 billion €. Within this context, wildlife bio-monitoring may be used to detect chemical pollution and evaluate the health of ecosystems, with species as systematic test models during the evaluation of risks associated to actual exposure routes. Wildlife species residing in polluted zones are exposed to pollutant complex mixtures through multiple paths that could hardly be evaluated in laboratory tests.

5. Acknowledgments

This work was supported by a grant from the National Institute of Ecology, SEMARNAT (DGICUR-INE) [No. de convenio INE/A1-047/2007]. We also thank the University of Veracruz, campus Coatzacoalcos. Special thanks to Prof. Jesús Guerrero Cabrera for English language editing of the manuscript.

6. References

Adams, S.M., Bevelhimer, M.S., Greeley, M.S., Levine, D.A., Teh, S.J. 1999. Ecological risk assessment in a large river reservoir: 6. Bioindicators of fish population health. *Environ. Toxicol. Chem.* 18:628–640.

Alford, R.A, 2010. Declines and the global status of amphibians, D.W. Ecotoxicology of Organic Contaminants in amphibians. In Sparling, D.W. Linder, G., Bishop, C.A., Krest; S.K, editors. Ecotoxicology of amphibians and reptiles 2nd ed. Pensacola (FL): SETAC Press p 13-46.

Álvarez del Toro, M. 1982. *Los reptiles de Chiapas.* Talleres Gráficos del Estado. Tuxtla Gutiérrez, Chiapas.

ATSDR. 2005. *Toxicological profile for alpha-, beta-, gamma- and delta-hexachlorocyclohexane.* Department of Health and Human Services. Agency for Toxic Substances and Diseases Registry. 377.

Basile, E.R., Avery, H.W., Bien, W.F., Keller, J.M. 2011. Diamondback terrapins as indicator species of persistent organic pollutants: Using Barnegat Bay, New Jersey as a case study. *Chemosphere* 82:137-44.

Bayen, S.; Wurl, O.; Karuppiah, S.; Sivasothi, N.; Kee Lee, H.; Philip Obbard, J. 2005. Persistent organic pollutants in mangrove food webs in Singapore. *Chemosphere* 61: 303 –313.

Bergeron, J.M., Crews, D., Mc Lachlan, J.A. 1994. PCBs as environmental estrogens: turtle sex determination as a biomarker of environmental contamination. *Environ. Health Perspect.* 102: 780–786.

Bickham, J.W., Hanks, B.G., Smolen, M.J., Lamb, T., Gibbons, J.W. 1988 Flow cytometric analysis of the effects of low-level radiation exposure on natural populations of slider turtles (Pseudemys scripta). *Arch. Environ. Contam. Toxicol.* 17:837–841.

Blake, A. 2005. *The next generation of POP's: PBDE's and lindane.* International POP's Elimination Network (IPEN). Washington D.C. USA. 15.

Bonin, J., DesGranges, J.L., Bishop, C. A. Rodrigue, J., A. Gendron J. Elliott, E. 1995. Comparative study of contaminants in the mudpuppy (amphibia) and the common snapping turtle (reptilia), St. Lawrence River, Canada. *Arch. Environ. Contam. Toxicol.* 28:184-194.

Burger, J., Gibbons, J.W. 1998 Trace Elements in Egg Contents and Egg Shells of Slider Turtles (*Trachemys scripta*) from the Savannah River Site. *Arch. Environ. Contam. Toxicol.* 34:382–386.

CEC. 1997. *Historia del DDT en Norteamérica.* Commission for Environmental Cooperation. Available in:
http://www.cec.org/files/PDF/POLLUTANTS/historiaDDTs_ES.PDF (accessed Jul 2011)

CEC. 2001. *Diagnostico situacional del uso de DDT y el control de la malaria.* Commission for Environmental Cooperation. Available in:
http://www.cec.org/files/PDF/POLLUTANTS/InfRegDDTb_ES_EN.pdf (accessed Jul 2011)

CEC. 2006. *The North American regional action plan (NARAP) on lindane and other hexachlorocyclohexane (HCH) isomers.* Commission for Environmental Cooperation. 51p.

De Solla, S. 2010.Organic Contaminants in reptiles. In Sparling, D.W. Linder, G., Bishop, C.A., Krest; S.K, editors. *Ecotoxicology of amphibians and reptiles.* 2nd ed. Pensacola (FL): SETAC Press p 289-324.

De Solla, S.R., Bishop, C.A., Van der Kraak, G., Brooks, R.J., 1998. Impact of organochlorine contamination on levels of sex hormones and external morphology of common snapping turtles (Chelydra serpentina serpentina) in Ontario, Canada. *Environ. Health. Perspect.* 106:253-260.

De Souza, A., Machado, J.P., Ornellas, R., Curcio, R., Souto, M., Silveira, C. 2008. Organochlorine pesticides (OCs) and polychlorinated biphenyls (PCBs) in sediments and crabs (*Chasmagnathus granulata*, Dana, 1851) from mangroves of Guanabara Bay, Río de Janeiro state, Brazil. *Chemosphere* 73(1): 186 – 192.

DeGarady, C. J. and Halbrook, R. S. 2003. Impacts from PCB Accumulation on Amphibians Inhabiting Streams Flowing from the Paducah Gaseous Diffusion Plant. *Arch. Environ. Contam. Toxicol.* 45(4): 525-532.

Di Giulio, R.T. and Hinton, D.E. 2008. The toxicology of fishes. Ed. CRC Press, Florida, 1071pp.

Dohm, M.R., Mautz, W.J., Doratt, R.E., Stevens, J.R. 2008 Ozone exposure affects feeding and locomotor behavior of adult *Bufo marinus*. *Environ. Toxicol. Chem.* 27:1209-1216.

Edwards P.J., Brown S.M. 1982. Use of grassland plots to study the effect pesticides on earthworms. *Pedobiol.* 24: 145-150.

Espinosa-Reyes, G.; Ilizaliturri, C.; González-Mille, D.; Costilla, R.; Díaz-Barriga, F.; Cuevas, M.C.; Martínez, M.A.; Mejía-Saavedra, J. DNA Damage in earthworms (*Eisenia* spp.) as indicator of environmental Stress in the industrial zone Coatzacoalcos, Veracruz, Mexico. *J Environ. Science Health A* 45: 49-55.

Falandysz, J.; Strandberg, L.; Zpuzyn, T.; Gucia, M. 2001. Chlorinated cyclodiene pesticide residues in Blue mussel, crab, and fish in the Gulf of Gdansk, Baltic Sea. *Environ. Sci. Technol.* 35: 4163 – 4169.

FAO. 1997. Lista mundial de vigilancia para la diversidad de los animales domésticos. 2a Ed. Organización de las Naciones Unidas para la Agricultura y la Alimentación. Roma, Italia. http://www.fao.org/docrep/V8300S/v8300s00.HTM (última visita diciembre de 2008)

Feder, M.E., Burggren, W.W. 1992. Environmental physiology of the amphibians. University of Chicago Press, Chicago, p 646.

Fellers, G.M., McConnell, L.L., Pratt, D., Datta, S. 2004. Pesticides in mountain yellow-legged frogs (*Rana muscosa*) from the Sierra Nevada Mountains of California, USA. *Environ. Toxicol. Chem.* 23: 2170-2177.

Fitzpatrick, L.C.; Sassani, R.; Venables, B.J.; Goven, A.J. 1992. Comparative toxicity of polychlorinated biphenyls to earthworms *Eisenia fetida* and *Lumbricus terrestris*. *Environ. Pollut.* 77: 65-69.

Gibbons, J.W. (ed). 1990. The slider turtle. In: *Life history and ecology of the slider turtle.* Smithsonian Institution Press, Washington, DC, pp 3-18

Gillilland, C.D., Summer, C.L., Gillilland, M.C., Kannan, K., Villeneuve, D.L., Coady, K., Muzzall, P., Mehne, C., Giesy, J.P. 2001 Organochlorine insecticides, polychlorinated biphenyls and metals in water, sediment and green frogs from southwestern Michigan. *Chemosphere* 44:327-339.

González-Mille, D.J.; Ilizaliturri-Hernández, C.A.; Espinosa-Reyes, G.; Costilla-Salazar, R.; Díaz-Barriga, F.; Ize-Lema, I. and Mejía-Saavedra, J. 2010. Exposure to persistent organic pollutants (POPs) and DNA damage as an indicator of environmental stress in fish of different feeding habits of Coatzacoalcos, Veracruz, Mexico. *Ecotoxicology* 19:1238-1248.

Goven, A.J.; Eyambe, G.S.; Fitzpatrick, L.C.; Venables, B.J.; Cooper, E.L. 1993. Cellular biomarkers for measuring toxicity of xenobiotics: effects of polychlorinated biphenyls on earthworm *Lumbricus terrestris* coelomocytes. *Environ. Toxicol. Chem.* 12: 863-870.

Guillette, L.J., Brock, J.W., Rooney, A.A., Woodward, A.R. 1999. Serum concentrations of various environmental contaminants and their relationship to sex steroid concentrations and phallus size in juvenile American alligators *Arch. Environ. Contam. Toxicol.* 36:447-455.

Hall, R.J, Driscoll, C.T., Likens, G.E., Pratt, J.M.1985. Physical, chemical and biological consequences of episodic aluminium addition to a stream. *Limnol. Oceanogr.* 30, 212-220.

Haque, A. Ebing, W. 1983. Toxicity determinationof pesticides to earthworms in the soil substrate. *J. Plant Dis. Prot.* 90: 395-408.

Harris, M.L., Bishop, C.A., Struger, J., Van Den Heuvel, M.R., Van Den Kraak, M.R., Dixon, G.J., Ripley, B., Bogart, J.P. 1998. The functional integrity of northern leopard frog (*Rana pipiens*) populations in orchard wetlands. I. Genetics, physiology, and biochemistry of breeding adults and young-of-the-year. *Environ. Toxicol. Chem.* 17:1338-1350.

Harris, M.L.; Wilson, L.K.; Elliott, J.E.; Bishop, C.A.; Tomlin, A.D.; Henning, K.V. 2000. Transfer of DDT and Metabolites from Fruit Orchard Soils to American Robins (*Turdus migratorius*) Twenty Years after Agricultural Use of DDT in Canada. *Arch. Environ. Contam. Toxicol.* 39, 205-220.

Heimbach, F. 1984. Correlations between three methods for determining the toxicityof chemicals to earthworms. *Pest. Sci.* 15: 605-611

Jha, A.N. 2008. Ecotoxicological applications and significance of the comet assay. *Mutagenesis* 23:207-221.

Jofré, M.B., Antón, I.R. and Caviedes-Vidal, E. 2008. Organochlorine Contamination in Anuran Amphibians of an Artificial Lake in the Semiarid Midwest of Argentina. *Arch. Environ. Contam. Toxicol.* 55 (3): 471-480.

Jones, A. and Hart, D.M. 1998. Comparison of laboratory toxicity tests for pesticides with field effects on earthworm population: a review. Pp: 247-267. In: Sheppard, S.C; Bembridge J.D.; Holmstrup, M.; Posthuma, L. 1998. *Advances in earthworm ecotoxicology.* Proceedings from the second international Workshop on earthworm ecotoxicology. 2 – 5 april 1997. Amsterdam, The Netherlands. Pensacola FL: Society of environmental Toxicology and chemistry (SETAC) 472 p.

Kirk, J.J. 1988. Western spotted frog (*Rana pretiosa*) mortality following forest spraying of DDT. *Herp. Review* 19:51-53.

Klobučar, G. I. V., Štambuk, A., Pavlica, M., Sertić, P.M. Kutuzović, H., Hylland. 2010. Genotoxicity monitoring of freshwater environments using caged carp (*Cyprinus carpio*). *Ecotoxicology* 19:77–84.

Kula, H. 1995. Comparision of laboratory and field testing for the assessment of pesticide side effects on earthworms. *Acta Zool. Fennica*, 196: 338-341.

Külköylüolu, O. 2004. On the usage of ostracods (Crustacea) as bioindicator species in different aquatic habitats in the Bolu region, Turkey. *Ecological Indicators* 4: 139-147

Langdon, C.J., Piearce, T.G., Meharg, A.A., Semple, K.T. 2003. Interactions between earthworms and arsenic in the soil environment: a review. *Environ. Pollut.* 124, 361-373.

Lara-López, M.S., González-Romero, A. 2002. Alimentación de la iguana verde *Iguana iguana* (Squamata: Iguanidae) en la Mancha, Veracruz, México. *Acta Zool Mex* 85: 139-152.

Lee R.F., Steinert S. 2003. Use of the single cell gel electrophoresis/comet assay for detecting DNA damage in aquatic (marine and freshwater) animals. *Mutation Res.* 544:43-64.

Legall, J.R.; L.E., Dicovskiy, Valenzuela, Z.I. 2006. *Manual Básico de lombricultura para condiciones tropicales.* Escuela de Agricultura y Ganadería de Estelí. Estela, Nicaragua. 16 p.

León, P., Montiel, S. 2008. Wild Meat Use and Traditional Hunting Practices in a Rural Mayan Community of the Yucatan Peninsula, Mexico. *Hum. Ecol.* 36: 249–257.

Linzey, D., Burroughs, J., Hudson, L., Marini, M., Robertson, J., Bacon, J., Nagarkatti, M., Nagarkatti, P. 2003. Role of environmental pollutants on immune functions, parasitic infections and limb malformations in marine toads and whistling frogs from Bermuda. *Int. J. Environ. Health Res.* 13:125-148.

Lutharddt, P.; Mayer, J.; Fuchs, J. 2002. Total TEQ emissions (PCDD/F and PCB) from industrial sources. *Chemosphere.* 46, 1303-1308.

Ma, W., Bodt, J. 1993. Differences in toxicity of the insecticide chlorpyrifos to six species of earthworms (oligochaeta, lumbricidae) in standardized soils tests. *Bull. Environ. Cont. Toxicol.* 50: 864-870.

McCoy, K.A., Bortnick, L.J., Campbell, C.M., Hamlin, H.J., Guillette, L.J., St Mary, C.M. 2008. Agriculture alters gonadal form and function in the toad *Bufo marinus*. *Environ. Health Perspect.* 116:1526-1532.

Morrison, D.E.; Robertson, B.K.; Alexander, M. 2000. Bioavailability to earthworms of aged DDT, DDE, DDD, and Dieldrin in soil. *Environ. Sci. Technol.* 34: 709 – 713.

Moss, S., Keller, J.M., Richards, S., Wilson, T.P. 2009. Concentrations of persistant organic pollutants in plasma from two species of turtle from the Tennessee River Gorge. Chemosphere 76:194-204.

Nacci, D.E., Cayulab S., Jackim E. 1996. Detection of DNA damage in individual cells from marine organisms using the single cell gel assay. *Aquat. Toxicol.* 35:197-210.

Newhook, R., Meek, M.E. 1994. Hexachlorobenzene: evaluation of risks to health from environmental exposure in Canada. *Environ. Carcin. Ecotox. Rev.* 12(2), 345-360.

Ogunseitan, O.A. 2002. *Microbial proteins as biomarkers of ecosystem health.* 217-232 pp. In: Integrated Assessment of Ecosystem Health. Editado por: Scow, K. M.; Fogg, G.E.; Hinton, D.E.; Jonson, M.L. Lewis Publishers. Boca Raton, Florida, U.S.A. 340 p.

Overman, S.R., Krajicek, J.J. 1995. Snapping turtles (Chelydra serpentina) as biomonitors of lead contamination of the Big River in Missouri old lead belt. *Environ. Toxicol. Chem.* 14:689–695.

Pastor, D., Sanpera, C., González-Solís, J., Ruiz, X., Albaigés, J. 2004. Factors affecting the organochlorine pollutant load in biota of a rice field ecosystem (Ebro Delta, NE Spain). *Chemosphere* 55:567–576.

Petrlink, J., J. DiGangi. 2005. *The egg report.* International POPs Elimination Network (IPEN). Washington D.C. USA. 52.

Phaneuf, D., DesGranges, J.L., Plante, N., Rodrigue, J. 1995 Contamination of local wildlife following a fire at a polychlorinated biphenyls warehouse in St. Basile le Grand, Quebec, Canada. Arch Environ Contam Toxicol 28:145–153.

Portter, D.A., Spicer, P.G., Redmond, C.T., Powel,l A.L. 1994. Toxicity of pesticides to earthwormsin Kentucky bluegrass turf. *Bull. Environ. Cont. Toxicol.* 52: 176-181.

Regoli, F., Gorbi. S., Fattorini D., Tedesco S., Notti A., et al.,. 2006 Use of the land snail *Helix aspersa* as sentinel organism for monitoring ecotoxicologic effects of urban pollution: an integrated approach. *Environ. Health Perspect.* 114:63-69.

Reinecke, A.J. and Reinecke S.A. 1998. *The use of earthworms in ecotoxicological evaluation and risk assessment: new approaches* In: Earthworm ecology. Edwards, C.A. (Ed.) St. Lucie Press. Boca Raton. U.S. 273-293

Reines, M., Rodríguez, C.,. Sierra, A., Vázquez, M. 1998. *Lombrices de tierra con valor comercial: Biología y técnicas de cultivo*. Universidad de Quintana Roo. Chetumal, Quintana Roo, México. 60 p.

Rico, M.C., Hernandez, L.M., Gonzalez, M.J., Fernandez, M.A., Montero, M.C. 1987. Organochlorine and metal pollution in aquatic organisms sampled in the Doñana National Park during the period 1983–1986. *Bull. Environ. Contam. Toxicol.* 39:1076–1083.

Rigonato, J., Mantovani, M.S., Quinzani, J.B. 2005. Comet assay comparasion if different *Corbicula fluminea* (Mollusca) tissues for detection of genotoxicity. *Genet. Mol. Biol.* 28: 464-468.

Rinderhagen, M., Ritterhoff, J. & Zauke, G.P. 2000. Crustaceans as bioindicators. In *Biomonitoring of polluted water: reviews on actual topics* (A. Gerhardt, ed.). Trans Tech Publications; Environmental Research Forum, Uetikon, p. 161-194.

Riojas-Rodriguez, H., Baltazar-Reyes, M.C., Meneses, F. 2008. Volatile organic compound presence in environmental samples near a petrochemical complex in Mexico. *Abstracts Epidemiology* 19 (1), S219.

Rosales, L. Carranza, E. Estudio geoquímico de metales en el estuario del río Coatzacoalcos. *In: Golfo de México contaminación e impacto ambiental: diagnóstico y tendencias.* Vázquez-Botello, A.; Rendón-Von Osten, J.; Gold-Bouchot, G., Agraz-Hernández, C., Eds.; Universidad Autónoma de Campeche, Universidad Autónoma de México, Instituto de Ecología. 2005, 389-406.

Russell, R., Lipps, G., Hecnar, S., Haffner, D. 2002. Persistent Organic Pollutants in Blanchard's Cricket Frogs (Acris crepitans blanchardi) from Ohio. *Ohio J Sci.* 102 (5):119-122.

Russell, R.W., Gillan, K.A., Haffner, G.D. 1997 Polychlorinated biphenyls and chlorinated pesticides in southern Ontario, Canada, green frogs. *Environ. Toxicol. Chem.* 1:2258–2263.

Russell, R.W., Hecnar, S.J., Haffner, G.D. 1995. Organochlorine pesticide residues in southern Ontario spring peppers. *Environ. Toxicol. Chem.* 14:815–817.

Russo, C., Rocco, L., Morescalchi, M.A., Stingo, V. 2004. Assessment of environmental stress by the micronucleus test and the comet assay on the genome of teleost populations from two natural environments. *Ecotoxicol. Environm. Saf.* 57:168-174.

Sala, M.; Sunyer, O.; Otero, R.; Santiago-Silva, M.; Ozalla, D.; Herrero, C.; To-Figueras, J.; Kogevinas, M.; Anto, J.; Camps, C.; Grimalt, J. 1999. Health effects of chronic high exposure to hexachlorobenzene in a general population sample. *Arch. Environ. Health.* 54(2), 102-109.

Sánchez-Hernández, J.C. 2006. Earthworm biomarkers in ecological risk assessment. *Rev. Environ. Contam. Toxicol.* 188: 85-126

Selcer, K.W. 2006.Reptile ecotoxicology: studying the effects of contaminants on populations . In Gardner S.C, Oberdösrter, E., editors. *Toxicology of reptiles.* Boca raton (FL): Taylor and Francis Press p 267-297.

Sparling, D.W. Linder, G., Bishop, C.A., Krest; S.K,. 2010. Recent advancements in amphibian and reptile ecotoxicology. In Sparling, D.W. Linder, G., Bishop, C.A., Krest; S.K, editors. *Ecotoxicology of amphibians and reptiles* 2nd ed. Pensacola (FL): SETAC Press p 1-12.

Stringer, R.; Labunska, I.; Bridgen, K. 2001. *Organochlorine and heavy metals contaminants in the environmental around the Complejo Petroquimicos Paharitos, Coatzacoalcos, México.* Technique note Greenpeace. University of Exeter. U.K. 60.

Takazawa, Y., Kitamura, K., Yoshikane, M., Shibata, Y., Morita, M., Tanaka, A. 2005. Distribution patterns of hexachlorocyclohexanes and other organochlorine compounds in muscles of fish from a japanese remote lake during 2002-2003. *Bull. Environ. Contam. Toxicol.* 74:652–659. Thompson AR. 1970. Effects of nine insecticides on numbers and biomass of earthorms in pasture. *Bull. Environ. Cont. Toxicol.* 5: 577-585.

Vázquez-Botello, A.; Villanueva-Fragoso, S.; Rosales-Hoz, L. 2004. *Distribución y contaminación por metales en el Golfo de México.* In: Diagnostico ambiental del Golfo de México. Caso, M. Pisanty, I.; Ezcurra, E., Eds.; SEMARNAT-INE. 682-712.

Venne, L., Anderson, T., Zhang, B., Smith, L., McMurry, S. 2008 Organochlorine pesticide concentrations in sediment and amphibian tissue in playa wetlands in the southern high plains, USA. *Bull. Environ. Toxicol.* 80: 497-501.

Willett K.L., Ulrich E.M., Hites R.A. 1998. Differential toxicity and environmental fates of hexachlorocyclohexane isomers. *Environ. Sci. Technol.* 32:2197-2207.

Willett, K.L., Ulrich, E.M., Hites, R.A. 1998 Differential toxicity and environmental fates of hexachlorocyclohexane isomers. *Environ. Sci. Technol.* 32:2197-2207.

Willingham, E.J., Crews, D. 1999. Organismal effects of environmentally relevant pesticide concentrations on the red-eared slider turtle. *Gen. Comp. Endocrinol.* 113:429–435.

Willingham, E.J., Crews, D. 2000. The red-eared slider turtle: an animal model for the study of low doses and mixtures. American Zool. 40:421–428.

Yim, U.H., Hong, S.H., Shim, W.J., Oh, J.R. 2005 Levels of persistent organochlorine contaminants in fish from Korea and their potential health risk. *Arch. Environ. Contam. Toxicol.* 48:358–366.

Zug, G.R., Zug, P.B. 1979. The marine toad, *Bufo marinus:* a natural history resumé of native populations. Smithsonian Institution Press, Washington, D.C. 284 pp. Available in: http://hdl.handle.net/10088/5188

Zupanovic, Z., Musso, C., Lopez, G., Louriero, C.L., Hyatt, A.D., Hengstberger, S., Robinson, A.J. 1998. Isolation and characterization of iridoviruses from the giant toad *Bufo marinus* in Venezuela. *Dis. Aquat. Org.* 33:1-9.

5

Bioavailability of Polycyclic Aromatic Hydrocarbons Studied Through Single-Species Ecotoxicity Tests and Laboratory Microcosm Assays

Bernard Clément

LEHNA-IPE, Université de Lyon-ENTPE
France

1. Introduction

The bioavailability of a pollutant is defined as its capacity to transfer from the surrounding environment (water and sediments) to organisms and is one of the key factors governing its bioaccumulation and toxicity. Knowledge of the bioavailable fraction of a compound is therefore vital in order to evaluate environmental risks and in particular to study its effects on living organisms. We began works on this topic at the laboratory Transfert et Effets des Polluants sur l'Environnement (TEPE) of Université de Savoie by studying the toxicity of a mixture of three polycyclic aromatic hydrocarbons (PAHs) (phenanthrene, fluoranthene and benzo(k)fluoranthene) in the framework of a PNETOX (programme national d'écotoxicologie) 1998-2001 program (Verrhiest, 2001; Verrhiest *et al.* 2000, 2001, 2002a, 2002b). The aim of this program, which brought together three research laboratories (TEPE of the Université de Savoie, Centre d'Etudes du Machinisme Agricole, du Génie Rural, des Eaux et Forêts (CEMAGREF) Lyon, INERIS, Institut Français pour la Recherche et les Etudes en Mer (IFREMER)), was to contribute to the development of a method of evaluating the risks of freshwater sediments contaminated by PAHs *via* in-depth changes to methods and better knowledge of the fate and effects of PAHs in aquatic ecosystems. The choice of PAHs was motivated by the fact that they are:

- ubiquitous organic contaminants (**table 1**), stemming from incomplete combustion (pyrolitic origin) or from the slow maturation of organic matter (petroleum hydrocarbons), and mostly emitted by anthropic activities (industry, transport),
- hydrophobic compounds which, after being introduced in a water column, rapidly become linked to colloidal and suspended matter, thus they enter the sediment where their effects depend on their bioavailability and fate, in relation with the physicochemical and microbiological properties of the sediment and interstitial water,
- toxic (**tables 2 and 3**) or/and mutagenic and carcinogenic compounds.

2-L microcosm assays (28 days) were performed on different types of sediment (artificial and natural) spiked with a mixture of PAHs (fluoranthene, phenanthrene, benzo(k)fluoranthene, **table 4**). The effects were evaluated on higher organisms and on the bacterial compartment of the sediment. Details on the protocols can be found in Verrhiest *et* *al.* (2000, 2001, 2002a, 2002b). The microcosms consisted in glass beakers filled with artificial sediment (300 g sand + kaolinite + alpha-cellulose + TetraMin®) or natural sediment, synthetic water (2000 mL), and inoculated with daphnids (*Daphnia magna*), micro-algae (*Pseudokirchneriella subcapitata*), duckweeds (*Lemna minor*), chironomids (*Chironomus riparius*) and amphipods (*Hyalella azteca*). The organisms were exposed 28 days to PAH-spiked sediment, measurements were performed on PAH contents, organism development (survival, growth, reproduction), bacterial exo-enzymatic activities involved in the nitrogen (leucine-aminopeptidase) and carbon (β-glucosidase) cycles, and bacterial density (direct counting of bacteria by epi-fluorescence or Colony Forming Units after spreading on agar). The fate and the effects of PAHs were analysed as a function of the sediments' characteristics, especially organic carbon content and sediment type (artificial sediment conditioned or not, natural sediment).Studying the bioavailability of PAHs has been extended through a program on the impact of pyrene (**table 4**) on the organisms of an aquatic ecosystem (Clément *et al.*, 2005b). This study, performed in the framework of the Centre National de la Recherche Scientifique (CNRS) Life and Societies Environment Program (Ecosystems and Environment), was performed in 2001-2002 (Jouanneau *et al.*, 2003). It brought together four laboratories: the Laboratory of Biochemistry and Biophysics of Integrated Systems, the Department of Molecular and Structural Biology, Unité Mixte de Recherche (UMR) 5092, Commissariat à l'Energie Atomique (CEA) Grenoble (Yves Jouanneau), the TEPE Laboratory, Université de Savoie (Pr Gérard Blake), the Laboratoire Chimie Moléculaire et Environnement (LCME) Laboratory, Université de Savoie (Emmanuel Naffrechoux), and the Laboratoire des Sciences de l'Environnement (L.S.E.) (Bernard Clément). During the first phase (2001), the toxicity of pyrene was evaluated at the L.S.E. by single-species assays (daphnia, algae, duckweed, chironomidae, amphipods), and a bioaccumulation measurement protocol for *D. magna* and *H. azteca* was formulated. During the second year, 2-litre microcosm tests were performed in the presence of two lacustrine sediments. As in the case of Verrhiest's (2001) study of PAH mixtures, we completed monitoring of the higher organisms of the microcosm by measuring enzymatic activity and bacterial density in the bacterial compartment of the sediments

	Place	Area studied	PHE (mg/kg)	FLU (mg/kg)	BKFLU (mg/kg)	Pyrene (mg/kg)	Sum of 16 PAHs (mg/kg)	References
Marine sediments	West of the sea Méditerranée	N+C	0.06 to 1.86	0.001 to 3.18	–	dna	1.18 to 20.44	Baumard *et al.*, 1998
	Bassin d'Arcachon	C	–	–	–	dna	5 to 10	Geffard *et al.*, 1999 Raymond *et al.*, 1999
	West of the Beaufort Sea (Alaska)	N	–	–	–	dna	0.16 to 1.1	Valette-Silver *et al.*, 1999
	Baltic Sea	N	0.03 to 0.08	0.05 to 0.27	0.05 to 0.16	dna	0.72 to 1.9	Witt, 1995

	Ain	N *	0.05 to 0.12	0.25 (mean)	0.07 to 0.3	dna	–	Verrhiest, 2001
	Neyrieux	N *	<0.05	<0.05	0.03	dna	–	Verrhiest, 2001
	Confluence of Seine-Marne	N	0.5 (mean)	0.82 (mean)	–	dna	1.5 to 7.4	Garban and Ollivon (1995)
	Rhone, Saone and Ain	N+C	<0.05 to 0.55	< 0.04 to 0.88	< 0.01 to 0.67	dna	0.52 to 14.64	Bonnet, 2000
	Moselle and Meurthe	N+C	<0.53 to 3.44	0.75 to 6.3	0.38 to 1.5	dna	6.68 to 32	Bonnet, 2000
	Seine	N C	– –	– –	– –	dna	2 to 4 60	Ollivon et al., 1995
	Lake Annecy	C	–	–	–	dna	1.43	Naffrechoux et al., 1999
	Niagara river (EU)	N C	0.01 0.4	0.04 0.9	–	dna	0.4 (mean) 3.3 to 5.4	Eisler, 2000
Fresh water sediments	Black river (EU)	C	52 (mean)	33 (mean)	1.5 (mean)	dna	–	Eisler, 2000
	CEBS code B2	C*	0.51	0.87	0.26	0.95	5.86	Bray et al., 2001
	CEBS code B13	C*	1.3	1.6	0.48	1.5	9.58	Bray et al., 2001
	CEBS code B22	C*	0.84	1.45	0.46	1.33	9.59	Bray et al., 2001
	North 13990	C*	0.53	0.68	0.89	1.84	10.09	Bray et al., 2003
	North 17000	C*	3.96	2.53	0.15	7.00	19.11	Bray et al., 2003
	North 12570	C*	1.23	1.64	2.04	3.31	17.04	Bray et al., 2003
	North 12730	C*	2.29	7.36	4.43	6.34	37.64	Bray et al., 2003
	North 12800	C*	2.52	4.67	3.71	5.59	62.54	Bray et al., 2003

Table 1. PAH concentrations of several sediments (according to Verrhiest 2001 and completed by the author) (PHE: phenanthrene; FLU: fluoranthene; BKFLU: benzo(k)fluoranthene; N: normal; C: contaminated; CEBS: Canal de l'Est Branche Sud; dna: data non available; * sediments used in our studies).

PAH	Organism	Type	EC50/LC50 (µg/l overlaying water or mg/kg dry sediment)	Duration of exposure/ Endpoint
	Hydra *Hydra sp.*	Water	96 µg/l	_/mortality
	Daphnia *Daphnia magna* *Daphnia pulex*	Water	700 µg/l 734 µg/l	_/mortality _/mortality
Phenanthrene	Amphipod *Gammarus pseudolimnaeus*	Water & sediment (epibenthic)	126 µg/l	_/mortality
	Insect chironomid *Chironomus tentans*	Sediment (benthic)	490 µg/l	_/mortality
	Annelid *Lumbriculus variegatus*	Sediment (benthic)	> 419 µg/l	_/mortality
	Hydra *Hydra americana*	Water	70.06 µg/l	_/mortality
	Daphnia *Ceriodaphnia dubia* *Daphnia magna*	Water	45 µg/l 102.8 µg/l 43 to 92 µg/l 4.2 to 15 mg/kg	_/mortality _/mortality 10 days/mortality 10 days /mortality
Fluoranthene	Amphipod *Hyalella azteca*	Water & sediment (epibenthic)	97 to 114 µg/l 32 to 54 µg/l 2.3 to 7.4 mg/kg	10 days/mortality 10 days/mortality 10 days/mortality
	Insect *Chironomus tentans* *Chironomus riparius*	Sediment (benthic)	30 to 61 µg/l 3 to 8.7 mg/kg 29 to 41 µg/l 170 mg/kg	10 days/mortality 10 days/mortality 11 days/mortality 28 days /emergence
	Annelid *Lumbriculus variegatus*	Sediment (benthic)	> 178.5 µg/l	mortality

Table 2. Acute toxicity data for phenanthrene and fluoranthene towards water invertebrates (Verrhiest, 2001) (EC/LC50: concentration which produces 50% effect on a given end point or kills half of the organisms initially present).

Bioavailability of Polycyclic Aromatic Hydrocarbons Studied Through Single-Species Ecotoxicity Tests and
Laboratory Microcosm Assays

115

2. Results and main findings

2.1 Difficulties specific to experiments on PAHs

PAHs are difficult to handle due to their properties. Indeed, their strong affinity for particular phases and low hydrosolubility complicate assays when spiking sediments and studying the fate of these substances in experimental systems. The sediment has to be spiked so that the PAHs introduced become homogenously distributed within the matrix. However, their poor solubility does not allow spiking with an aqueous solution. This makes it necessary to use an apolar solvent which makes it possible to concentrate PAHs, although this may prove toxic for organisms. We used the "wall-coating" method (Ditsworth et al., 1990) for all the assays. This method consists in distributing the PAHs on the wall of a glass flask after adding an acetone solution, evaporating the solvent by rotating the flask horizontally, and then adding the wet sediment that comes into contact with the PAHs inside the flask as it is being rotated. High spiking yields were always obtained during the assays performed on fluoranthene, phenanthrene, benzo(k)fluoranthene and their mixtures (>80%) (Verrhiest et al., 2000, 2001, 2002a). On the contrary, the assays on pyrene provided generally variable and much lower yields (**table 5**). This may have been due to the different characteristics of the sediments used in these assays, as the efficiency of the spiking depends on the way in which sediments come into contact with the flask wall. Chevron (2004) proposed another more classical method that consists in dissolving pyrene in DMSO (dimethylsulfoxide) at a given concentration, then mixing it with an organic matrix. The low toxicity of this solvent to microorganisms means that it can be used, as she did, in biodegradation assays. It remains to be determined, however, whether DMSO has no effect on higher organisms in contact with the sediment.

µg/L	Daphnia		Algae		Ceriodaphnia	
	LC50-48 h survival	IC95%	EC10 growth	IC95%	EC10 reprod 7 d	IC95%
Benzo(k)fluoranthene	> 1.1		> 1.5		> 1.5	
Benzo(a)anthracene	> 9.1		4.1	3.8-4.6	> 13	
Benzo(b)fluoranthene	> 1.1		> 1.5		> 1.5	
Benzo(ghi)perylene	> 0.2		> 0.26		0.124	0-0.17
Benzo(a)pyrene	> 2.7		1.54	1.52-1.57	0.77	0.03-2.2
Dibenzo(a,h)anthracene	> 0.35		0.73	0.52-1.31	> 0.046	
Acenaphtene	958	916-994	308	266-371	64	50-100
Acenaphtylene	1800	1731-1956	595	505-703	95	47-231
Anthracene	> 25		23.3	18.6-27.8	> 4.9	
Fluorene	408	368-449	485	411-540	33	28-45
Chrysene	> 1.3		> 3.9		> 0.13	
Fluoranthene	> 112		33.1	27.8-37.5	1.2	0.2-4.9
Indeno(1,2,3,cd)pyrene	> 357		5.7	3.6-9.8	0.38	0-9.49
Naphtalene	1664	1441-1902	> 8024		999	761-1331
Phenanthrene	> 400		123.5	80.1-170.4	15	4.7-19
Pyrene	24.6	21.6-28.4	12.4	6.7-17.9	2.1	1.3-3.1

Table 3. Toxicity data of PAHs in aqueous phase for three pelagic organisms (Vindimian et al., 2000) (EC10: concentration which produces 10% effect on a given end point; LC50 concentration which kills half of the organisms initially present; IC95: confidence interval at 95%).

The affinity of PAHs for solid surfaces can also bias results when studying their effects in the aqueous phase. PAHs dissolved in water tend to adsorb onto the walls of beakers, even glass ones, as already observed (Clément et al., 2005b), thus confirming the results of other experiments (McCarthy, 1983; Gauthier et al., 1986; Pelletier et al., 1997; Miller, 1999). Therefore a solution of 10 µg/L pyrene contains only 4 µg/L pyrene after 30 days in the absence of light to avoid any photodegradation. Finally, the same problem exists when conserving aqueous samples before analysis. Since we were unable to carry out the first assays on pyrene ourselves, we had to subcontract this task to another laboratory which, due to the lead times given, provided results that greatly underestimated the real values that we were able to check against solutions spiked at known concentration. We strongly recommend strict control over the analytical part (extraction and analyses) when working on PAHs, otherwise there is a risk of coming up against the same problems.

PAH	Structure	Hydrosolubility (µg/L) to 25°C	log Kow	Photosensitivity
Phenanthrene $C_{14}H_{10}$		1000 (May, 1980)	4.29 (May, 1980)	Non absorbent (Newsted and Giesy, 1987) UV_C (Huovinen et al., 2001)
Fluoranthene $C_{16}H_{10}$		206 (May, 1980)	5 (May, 1980)	UV_A, UV_B (Newsted and Giesy, 1987)
Benzo(k)fluoranthene $C_{20}H_{12}$		1.5 (Swartz et al., 1995)	6.72 (Pelletier et al., 1997)	UV_A, UV_B and visible (Pelletier et al., 1997)
Pyrene $C_{16}H_{10}$		160	5.18	UV_A and above all UV_B (Huovinen et al., 2001)

Table 4. Physicochemical properties of phenanthrene, fluoranthene, benzo(k)fluoranthene, and pyrene (according to Verrhiest et al. (2001) completed by data on pyrene) (Kow: water/octanol partition coefficient; UV: ultra-violet).

Bioavailability of Polycyclic Aromatic Hydrocarbons Studied Through Single-Species Ecotoxicity Tests and
Laboratory Microcosm Assays

117

μg/g nominal f.w.	Test	Carbonate sediment		Peat sediment	
		μg/g measured f.w.	Spiking rate, %	μg/g measured f.w.	Spiking rate, %
2	ss	4.9 [a]	243	1.5 [a]	73.5
20	ss	15.1 [a]	75.7	11.5 [a]	57.7
50	sp	31.6 [b] ± 7.9	63.2 ± 15.8	23.6 [b] ± 7.6	47.3 ± 15.2
50	m1	24.1 [a]	48.2	1.4 [a]	2.7
50	m2	33.6 [a]	67.2	3.9 [a]	7.8
50	m3	29.1 [b]	58.2	7.2 [b]	14.4
200	ss	175.3 [a]	87.7	194.6 [a]	97.3

Table 5. Nominal and measured concentrations of sediments from lake Aiguebelette spiked with pyrene, with spiking rates (p.f.: fresh weight; ss: amphipod test; sp: spiking assay; m1: 1st microcosm assay; m2: 2nd microcosm assay; m3: microcosm assay without organisms; carbonate sediment; peat sediment; [a] concentrations measured on D0, after 7 days underwater ; [b] concentrations measured on 3 samples immediately after spiking the sediment) (source: Clément et al., 2005b).

2.2 Toxicity in the aqueous phase towards microcosm organisms exposed during single-specific bioassays and modulating factors

The results of single-species bioassays in the aqueous phase on *Daphnia magna* and *Hyalella azteca* (table 6) confirm the acute toxicity of the PAH studies on these crustaceans, except for benzo(k)fluoranthene, which has very weak hydrosolubility (1.5 μg/L) (Verrhiest et al., 2001; Clément et al., 2000; Clément et al., 2005b).

The absence of acute toxicity of benzo(k)fluoranthene was confirmed by the results of Vindimian et al. (2000), who also showed the absence of chronic toxicity (table 3). When organisms are exposed in obscurity, toxicity at 24 and 48 h is observed at contents close to solubility. Conversely, in the presence of the fluorescent light usually used in the laboratory, particularly in all microcosm assays, toxicity is higher, though the induction of toxicity is more significant for fluoranthene than for phenanthrene, generally recognised as non phototoxic (Newsted and Giesy, 1987; Swartz et al., 1997; Boese et al., 1998), and pyrene. This type of light (*cool-white fluorescent*) contains a fraction of UV (Clément et al., 2000) which probably explains the increase in toxicity. Phototoxicity mechanisms are recalled in **figure 1**. For the most part, phototoxicity can be attributed to photosensitisation (Arfsten et al., 1996). Furthermore, the phototoxicity of phenanthrene and fluoranthene has been confirmed by the post-exposure of organisms for 2 hours to pure UV radiation (A or C) (according to the method of Wernersson and Dave, 1997). This post-exposure could not be performed with pyrene. Our results strengthen the hypothesis that, although many PAHs are not acutely toxic to aquatic organisms at concentrations corresponding to their range of solubility (National Research Council of Canada (NRCC), 1983), the presence of natural light containing UV rays represents a factor that considerably increases the risk of toxicity (Lyons et al., 2002). This is especially the case for pelagic invertebrates such as *Daphnia magna*, which are positively heliotropic and are thus more exposed in the presence of light and PAHs in the water column (Wernersson et al., 1999).

	24 h lux	48 h lux	48 h lux + UV	24 h dark	48 h dark	48 h dark + UV
	Daphnia magna					
PHE	2500 lux: 678	2500 lux: 604	2500 lux: 273 [a]	854	731	725 [a]
FLU	1500 lux: 56	1500 lux: 36	1500 lux: 29 [b]	>200	201	80 [b]
	2500 lux: 63	2500 lux: 34	2500 lux: <18 [a]	>180	>180	20 [a]
BKFLU			non toxic			
PYR	1500 lux: 139	1500 lux: 74				
	2500 lux: 105	2500 lux: 48	/	167	68	/
	6000 lux: 161	6000 lux: 45				
	Hyalella azteca					
PYR	1500 lux: 80	1500 lux: 41				
	2500 lux: 134	2500 lux: 41	/	82	63	/
	6000 lux: 77	6000 lux: 26				

Table 6. Toxicity in the aqueous phase of PAHs to *Daphnia magna* and *Hyalella azteca* (according to Verrhiest *et al.* (2001), Clément *et al.* (2000), and Clément *et al.* (2005b); lux: exposure to 1500 and 2500 lux 16h/day; dark: obscurity; lux+UV [a]: post-exposure 2 h to UV-A (365 nm, 247 µW/cm²); lux+UV [b]: post-exposure 2 h to UV-C (365 nm, 247 µW/cm²).

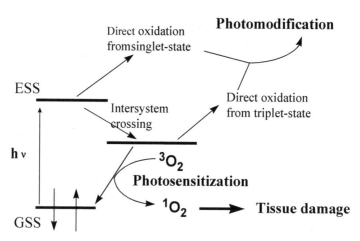

Fig. 1. Phototoxicity mechanisms (according to Krylov *et al.*, 1997).

The absence of phototoxicity of benzo(k)fluoranthène to *D. magna* (Verrhiest *et al.*, 2001) is probably due to the fact that the weak hydrosolubility of this PAH (1.5 µg/L) does not allow sufficient bioaccumulation to cause effects, as suggested by Boese *et al.* (1998) for benzo(b)fluoranthene (hydrosolubility: 6 µg/L).

Although fluoranthene has been shown to be stable in the aqueous phase for 48 h in light or obscurity (Clément *et al.*, 2000), under the same conditions, pyrene is much more sensitive to light (30 to 44% degradation over 24 h).

When algae were added in the beakers in which the daphnia were exposed, significant adsorption of fluoranthene occurred on the algae over 48 h, and there was a considerable reduction of toxicity in the presence of light. This phenomenon of reduced toxicity in the presence of algae is not attributed to adsorption, which remains limited; rather we presume that a possible protective effect may be due to algal pigments of type β-carotene (Bennett *et al.*, 1986), or more simply, that the algae absorb part of the light, though this absorption could not be quantified.

Algae are highly sensitive to pyrene (Clément *et al.*, 2005b), with an EC-72 h < 10 µg/L. According to Vindimian *et al.* (2000), they are also more sensitive to fluoranthene and phenanthrene than *Daphnia magna*. Conditions of exposure to very intense light (6000 lux) to ensure sufficient algal growth can induce significant phototoxicity, thereby explaining this higher sensitivity. Nonetheless, Warshawsky *et al.* (1995) obtained only weak inhibition of the growth of *Selenastrum capricornutum*, another species used by us, exposed to 400 µg pyrene/L for 4 days under fluorescent light similar to ours. According to Lei *et al.* (2001), *Selenastrum capricornutum* is resistant to pyrene. Furthermore, pyrene stimulates its glutathione-S-transferase (GST) activity during exposures of several days to concentrations ranging from 0.1 to 1 mg/L (Lei *et al.*, 2003). This enzymatic activity involves an enzyme, GST, that catalyses the binding reaction between an endogenous bio-molecule and the toxic substance, thereby detoxifying the contaminants. The authors associate the stimulation of GST activity with the degradation of pyrene observed, to suggest that the GST activity is responsible for the metabolization of pyrene. It is possible that the differences in culture and assay conditions between the experiments performed by Lei *et al.* (2001, 2003) and us resulted in very different expressions of GST activity. The same authors observed that pyrene inhibited the growth of another species of Chlorophyceae, *Scenedesmus quadricauda* (Lei *et al.*, 2003). The type of light, containing more or less UV, could have an impact on the cytotoxicity of PAHs to algae, due to its importance in biotransformation, highlighted with benzo(a)pyrene and *Selenastrum capricornutum* (Warshawsky *et al.*, 1995).

Contrary to algae, duckweed is not or only slightly sensitive to PAHs. We observed a value close to the solubility limit for pyrene, which proved to be non toxic in a single-species bioassay for an exposure of 6 days at 180 µg/L (3500 lux 16 h/day). An absence of sensitivity was also observed in another species of duckweed, *Lemna gibba*, by different authors (Huang *et al.*, 1995; Ren *et al.*, 1994). This resistance may be linked to the strong metabolization related to GST activity.

2.3 Toxicity to pelagic organisms in microcosms

In microcosms containing sediments contaminated by PAHs, organisms of the water column were exposed to PAHs present in the latter in dissolved, colloidal and particulate form, and some were also exposed to PAHs of the sediment with which they can be in contact for long periods (sedimented algae) or occasionally (daphnia consuming particles on the surface of the sediment). The results obtained on the daphnia exposed in beakers containing the sediment spiked with fluoranthene or phenanthrene (Verrhiest *et al.*, 2001) showed increased toxicity in comparison to bioassays in the aqueous phase, by taking into account in both cases the contents measured in the supernatant water. This increase can be attributed to different exposure routes, as the daphnia above the sediment can be brought into contact with the sediment or ingest fine particles returned to suspension. Other results take the same direction. During microcosm bioassays on a mixture of fluoranthene, phenanthrene and benzo(k)fluoranthene (Verrhiest, 2001), we observed strong toxicity after a week's exposure to *Daphnia magna*, for a natural sediment (Neyrieux) spiked at 300 mg/kg, whereas the contents measured in the supernatant water after filtration at 0.8 μm did not exceed 1 μg/L. Conversely, in microcosm bioassays on pyrene (carbonate and peat sediments from lake Aiguebelette spiked at 50 mg/kg), the response of *Daphnia magna* was likely related to contents measured in the supernatant water: a significant effect on the survival above the carbonate sediment in water containing 30 ± 9 μg pyrene/L; absence of effect above the peat in water containing 5 ± 1 μg pyrene/L (Clément *et al.*, 2005b). The same correlations were observed in single-species bioassays in small beakers (Jouanneau *et al.*, 2003), where the survival of daphnia above the contaminated sediment was reduced by about 50% to contents close to EC50.

We have never been able to evaluate the effects of PAHs on algae in microcosms for various reasons : the difficulty of taking all the algae into account (in suspension or sedimented), the influence of browsing and/or insufficient growth, and in particular high variability.

As mentioned previously, duckweed is not sensitive to PAHs and, in general, we have not observed effects in microcosms, in the presence of artificial or natural sediments spiked with contents ranging from 4 to 300 mg/kg. This can be explained easily by the fact that PAH contents in the supernatant waters could not exceed the solubility limits of these contaminants. No effect on *Lemna minor* was observed at these values in our works (single-species bioassays) or in the literature. However, there was one noteworthy exception. In a microcosm bioassay on a natural sediment (Neyrieux) spiked with a mixture of PAHs (phenanthrene, fluoranthene, and benzo(k)fluoranthene) at 300 mg/kg, significant inhibition was demonstrated from 19 days onwards (Verrhiest, 2001). However, as the sum of the PAH contents of supernatant waters did not exceed 1 μg/L during the test, it is possible that the inhibition of growth observed was due to a PAH degradation metabolite or the occurrence of an indirect effect. Verrhiest *et al.* (2002a) highlighted the effects on the bacterial compartment of a natural sediment spiked at 300 mg/kg. This disturbance perhaps concerned the sediment of Verrhiest (2001), possibly resulting in modifications of contents and flows of nutritive substances vital for *Lemna minor*. This hypothesis could not be supported by precise measurements.

2.4 Toxicity to benthic organisms in single-species and microcosm bioassays

We were able to measure the sensitivity of the species *Hyalella azteca* (crustacean-amphipod) and *Chironomus riparius* (dipterous insect) to several PAHs (fluoranthene, phenanthrene, benzo(k)fluoranthene, pyrene) alone and in mixture (for the first three), under different experimental conditions : single-species bioassays, microcosm bioassays, artificial sediments and natural sediments. Taking all the results into account, the first observation is that for exposures not exceeding one month, sediments contaminated by PAHs presented relatively weak acute toxicity as the first effects on benthic organisms (mortality and inhibited growth) were observed for contents of several mg/kg (or μg/g), higher than 30 mg/kg for the mixture "phenanthrene + fluoranthene + benzo(k)fluoranthene", and from 20 mg/kg for pyrene. Such contents are those of strongly contaminated sediments (**table 1**) and these results correspond to the threshold values provided by McDonald *et al.* (2000) and Kalf *et al.* (1997), at least if we consider the PEC (probable effect concentration) and MPC (maximum permissible concentration), thresholds from which an effect is very much probable (**table 7**). What is more, it should be also be mentioned that these thresholds result from matching contamination data with data from bioassays on benthic organisms, the effects generally being imputable to all the contaminants present, especially total PAHs, whose effects are generally additive (Munoz and

Bioavailability of Polycyclic Aromatic Hydrocarbons Studied Through Single-Species Ecotoxicity Tests and
Laboratory Microcosm Assays

121

Tarazona, 1993; Swartz *et al.*, 1995). However, it can be seen that effects are probable from 23 mg/kg in total PAHs, a value comparable to those for which we were able to show the effects in certain cases. The results obtained from microcosm bioassays were not different from those obtained from single-species tests, despite exposure being nearly twice as long. Other more sensitive sublethal effect criteria should be taken into account with these substances, some of which are known to be carcinogenic or mutagenic.

Substance	TEC (µg/g dw)	PEC (µg/g dw)	MPC (µg/g dw)
anthracene	0.0572	0.845	0.12
fluorene	0.0774	0.536	/
naphtalene	0.176	0.561	0.14
phenanthrene	0.204	1.170	0.51
benzo(a)anthracene	0.108	1.050	0.36
benzo(a)pyrene	0.150	1.450	2.70
chrysene	0.166	1.290	10.7
dibenzo(a,h)anthracene	0.033	/	/
fluoranthene	0.423	2.230	2.60
benzo(k)fluoranthene	/	/	2.40
benzo(ghi)perylene	/	/	7.50
indeno(1,2,3-cd)pyrene	/	/	5.90
pyrene	0.195	1.520	/
Total PAHs	1.610	22.800	/

Table 7. Toxicity thresholds in sediment for PAHs (TEC (threshold effect concentration) and PEC (probable effect concentration) taken from MacDonald *et al.*, 2000; MPC (maximum permissible concentration) taken from Kalf *et al.*, 1997).

The toxicity of mixtures of PAHs has rarely been studied, although natural sediments are contaminated by several substances. We approached this section with the study of the toxicity of the mixture of three PAHs (phenanthrene + fluoranthene + benzo(k)fluoranthene, Verrhiest *et al.*, 2001), and were able to highlight synergetic effects for this specific case, which contrasts with the hypothesis of additivity generally accepted.

We also showed (Verrhiest, 2001; Verrhiest *et al.*, 2001) that the effects are more significant in artificial sediments than in natural sediments, a result that we impute to a partition between the particular and aqueous phases (interstitial water) favouring the aqueous phase, the main path of exposure for the organisms studied (Di Toro *et al.*, 1991), more in artificial sediment. These differences of partition are not only due to the quantity of organic matter in the sediments (content of total organic carbon for which PAHs have great affinity), but also seem to affect the quality of organic matter. (Grathwohl, 1990; DePaolis and Kukkonen, 1997; Haitzer *et al.*, 1999).

More generally, the type of sediment (grain size, proportion of clay, sand, silt, etc.) influences the toxicity of PAHs (Landrum and Faust, 1994; Borglin *et al.*, 1996; Landrum *et al.*, 1997; Haitzer *et al.*, 1998). In the tests on the two natural sediments taken from Lake Aiguebelette ("carbonate sediment" and "peat sediment"), that differ in terms of grain size, composition, and above all the quantity of organic matter, we showed that the toxicity of the "peat" sediment was lower but that the partition coefficients between the dissolved fraction adsorbed on the organic carbon (Koc) were close, leading us to conclude that the quantity of organic matter was the main explanatory factor (Clément *et al.*, 2005b). The response of the amphipod *Hyallela azteca* to pyrene in the two types of sediment can be explained for the most part by the contents measured in the interstitial water.

Toxicity to benthic organisms is also influenced by the fate of PAHs in the sediment. A balance in the partition of PAHs between solid and liquid phases occurs in the sediment after the latter has been spiked. A study performed on this point showed that the equilibration of artificial sediments for 8 hours led to lower PAH toxicity (Verrhiest, 2001). Beyond this period, other phenomena can contribute towards modifying PAH bioavailability. In natural sediments spiked with a mixture of PAHs (Verrhiest *et al.*, 2002a) over a period of 30 days, we highlighted that PAHs degraded with the exception of benzo(k)fluoranthene, a heavier and probably more recalcitrant PAH. We attributed this degradation to endogenous bacteria of the sediments, whose β-glucosidase activity, measured in parallel, was stimulated. In the studies on pyrene, we also observed a considerable reduction of content over 30 days. This reduction was presumed to be due to ageing, although biodegradation could have been partially responsible (Clément *et al.*, 2005b). This ageing, or reduced extractability (Guthrie *et al.*, 1999; Leppänen et Kukkonen, 2000; Alexander, 2000; Conrad *et al.*, 2002), could be due to the migration of PAHs inside particle pores, making their extraction more difficult and, in parallel, reducing their bioavailability to microorganisms (biodegradation) and higher organisms (toxicity). Therefore the tests on spiked sediments raise the question of sediment conditioning to take into account both the arrival at partition balancing between the different phases, by hypothesising that such a balance exists, and the influence of biodegradation and ageing phenomena on the bioavailability of PAHs in such a way as to reproduce the conditions prevailing for contaminated sediments.

2.5 Bioaccumulation of PAHs during tests

In parallel with monitoring effects on *Daphnia magna*, we were able to measure this organism's bioaccumulation of fluoranthene (Clément *et al.*, 2000; **figure 2**) and pyrene (Jouanneau *et al.*, 2003), in single-species tests (fluoranthene and pyrene) and in microcosms (pyrene). In the studies on pyrene, measurements were also performed on other benthic organisms. The method employed consisted in an extraction procedure using acetone, followed by spectrofluorimetric analysis to identify the PAH spectrum and quantify the dose accumulated. The fact of working each time on a single PAH allowed us to use this simple method, without having to separate the compounds by chromatography.

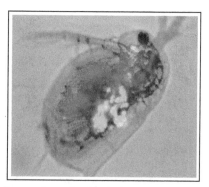

Fig. 2. Bioaccumulation of fluoranthene by *Daphnia magna*, visualised by epifluorescence microscopy (*photo B. Clément*).

In the single-species tests we showed significant bioaccumulation of fluoranthene and pyrene, directly correlated with the PAH content in the water. We found bioconcentration factors of 1000 L/kg (fresh weight) for fluoranthene and 1986 ± 445 L/kg (fresh weight) for pyrene, the latter value being close to those found in the literature (1900 to 2000 L/kg for Nikkilä and Kukkonen (2001) in *D. magna*, 1700 to 3500 L/kg for Akkanen *et al.* (2001) in *D. magna* in river water whose dissolved organic carbon content varied from 0 to 18 mg C/L and spiked at 1 µg pyrene/L, 2702 L/kg for Southworth *et al.* (1978) in *Daphnia pulex*, values expressed in each case on the basis of fresh weight).

The comparison of bioaccumulation and effect data shows good correlation between them (**figure 3**). This is in line with the hypothesis of narcotic effects occurring after a given accumulation in tissues, with narcosis resulting from physical modifications and transformations of the phospholipidic membrane by adsorption of a hydrophobic compound. Disturbances of membrane functions occur when the quantity of compound adsorbed is sufficient (Driscoll *et al.*, 1997). Acute narcosis therefore occurs as a function of the quantity of PAHs bioaccumulated by the organism. According to Landrum *et al.* (1994) and Driscoll *et al.* (1997), EC50 is obtained for different organisms (e.g. the amphipod *Diporeia*) exposed to sediments spiked with PAH for internal doses close to 6 µmol/g (fresh weight of organism). In the tests on fluoranthene and pyrène, we obtained an EC50 of about 0.66 - 0.7 µmol/g (fresh weight), which is consistent with this theory, despite this value being only a tenth of that found by these authors. On the contrary, during the tests with "carbonate sediment" (Aiguebelette) spiked with pyrene on the amphipod *Hyalella azteca* (Jouanneau *et al.*, 2003), we observed that doses leading to effects were of about the same magnitude as the internal dose necessary to achieve 50% mortality in *Diporeia* spp., i.e. 6 µmol/g fresh weight of organism. In the "peat" sediment from Aiguebelette, we also observed an effect on the growth of *Hyalella azteca* for an internal dose of about 13 µmol/g.

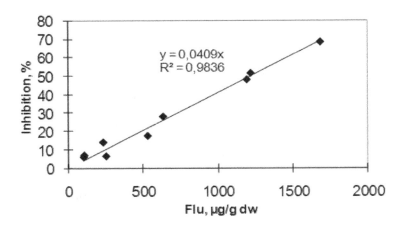

Fig. 3. Relation between the mean fluoranthene dose of *Daphnia magna* and the mean inhibition of mobility following 48 h exposure in darkness (Clément *et al.*, 2000).

During the microcosm bioassays on sediments spiked with pyrene, we were able to monitor the pyrene accumulated by daphnia introduced at the beginning of the assay and by their offspring. The mother daphnia recovered after 12 and 21 days exposure in microcosms probably containing a compound derived from pyrene, as suggested by the differences observed between the fluorescence emission spectra (**figure 4**). It is noteworthy that the measurements performed on the young born in the microcosms and exposed for a maximum of 3 days in the systems, displayed the same type of spectra, whereas there was no modification in the accumulation of pyrene in the young bred by us and introduced in the microcosms for exposure for a maximum of 3 days. These observations suggest that the modified pyrene accumulated by the mothers was transmitted to the embryos and found in the young released during hatching. As with the daphnia in the microcosms, measurements of bioaccumulation in amphipods and chironomidae larvae showed that the modifications of pyrene spectra were similar to those observed for daphnia. Therefore it was not possible to quantify the doses of pyrene accumulated for any of these organisms. The modifications of the pyrene spectra suggest biotransformation processes that could occur in all the organisms of the microcosm, namely daphnia, chironomidae and amphipods. *C. riparius* is known to develop strong pyrene biotransformation activity (Guerrero *et al.*, 2002), and the results of Gourlay *et al.* (2002) showed that although *Daphnia magna* has little effect on fluoranthene, it is capable of biotransforming pyrene and benzo(a)pyrene. As in this study, these authors obtained a different spectrum of bioaccumulated pyrene (with a shift of peaks and an increase of ratio between peaks) that did not result from a matrix effect. As in the case of benzo(a)pyrene, they also highlighted strong fluorescence of the phase recovered in water and less fluorescence of the dichloromethane phase, tending to prove that the product derived from pyrene is more polar and thus clearly the result of biological transformation. This biotransformation of pyrene by *D. magna* was demonstrated by Akkanen and Kukkonen (2003), who showed the involvement of Cytochrome P450 monooxygenases in this elimination route.

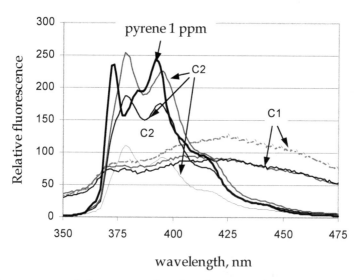

Fig. 4. Fluorescence emission spectra of methanol extracts of mother daphnia exposed for 12 days during microcosm bioassay no. 2 (carbonate sediment contaminated at 50 mg/kg) *versus* the spectrum of pure pyrene at 1 ppm in methanol.

Bioavailability of Polycyclic Aromatic Hydrocarbons Studied Through Single-Species Ecotoxicity Tests and
Laboratory Microcosm Assays

125

2.6 Responses of the bacterial compartment to PAHs

Microorganisms play a vital role in environmental dynamics as they are involved in
biogeochemical cycles acting as a medium through which matter and energy flow. As in any
ecosystem, disturbing the microbial communities of a microcosm can impact the entire
trophic chain and the balance of the environment. We also considered that it was important
to take into account the microbial compartment of sediments in the microcosm bioassays. By
using ecologically pertinent parameters (bacterial density and exoenzymatic activity), part
of our work consisted in evaluating the responses of indigenous microorganisms in
sediments to the contamination of the latter by PAHs.

Furthermore, the capacity of bacteria to biodegrade organic material as well as certain
organic compounds, such as PAHs, influences the bioavailability of hydrophobic
contaminants. This is why we chose, in addition to monitoring bacterial density, to focus on
certain enzymatic activities involved in organic matter transformation processes, namely β-
glucosidase (carbon cycle) and leucine-aminopeptidase (nitrogen cycle). In preliminary
works on an artificial sediment spiked with fluoranthene (Verrhiest et al., 2000), we also
used the activity of INT-reductase by following the protocol of Merlin et al. (1995).
Fluoranthene had no effect, even at 1000 mg/kg, on any of the parameters monitored
(bacterial density of the sediment and supernatant water, INT-reductase and β-glucosidase
activity of the sediment and the supernatant water). The study on the mixture of the three
PAHs (phenanthrene + fluoranthene + benzo(k)fluoranthene) showed effects at high
contents (300 mg/kg) on the bacterial populations of a natural sediment (Ain): reduction of
the bacterial density of the sediments, partial inhibition of leucine-aminopeptidase activity,
but stimulation of β-glucosidase activity, which it is tempting to parallel with the
considerable degradation of fluoranthene and phenanthrene observed during the same
bioassay (Verrhiest et al., 2002a). Assays were also performed with pyrene (Jouanneau et al.,
2003), first under simple conditions (spiked sediment + water; bacterial density, β-
glucosidase and leucine-aminopeptidase activities), then in microcosms (bacterial density
and β-glucosidase). Over the range 1, 10, 50, 100 and 200 mg/kg, no significant effect was
obtained in the single-species tests or in the microcosms.

In conclusion, PAHs do not appear to lead to effects on the bacterial compartment, at least
as seen through the few parameters monitored in these works, except for very high
concentrations rarely encountered in the environment. On the other hand, capacities to
degrade phenanthrene, fluoranthene and pyrene (Jouanneau et al., 2003) have been
demonstrated.

3. General discussion and conclusion on the microcosm study of PAH toxicity

The purpose of the microcosm bioassays performed on sediments contaminated by PAHs
was to evaluate the risks for lentic ecosystems related to the presence of these ubiquitous
organic contaminants in this sediment compartment. We were able to obtain reasonably
realistic results since it was possible to take into account different compartments of these
ecosystems, and the relations between the different populations represented and between
these populations and their environment.

We first confirmed the possibility of spiking initially non or only slightly contaminated
artificial and natural sediments with PAHs, drawing attention to the need to ensure good
spike rates and good distribution of PAHs in the sediment, as the physicochemical

properties of the latter can influence both rates and distribution. Once the sediment has been spiked, it is vital to observe the equilibration period, evaluated by us at being at one week (Verrhiest, 2001), which generally results in lower toxicity due to the higher adsorption of PAHs in the solid phase and which corresponds better to the situations most usually encountered in the environment (deposited, undisturbed sediment). We also looked into the possibility of extending this equilibration period to take into account ageing (migration of PAHs within particles, reducing their bioavailability) and biodegradation phenomena that we showed could be significant for certain PAHs over a few weeks (Verrhiest et al., 2002a; Clément et al., 2005b). In this framework we think that the use of natural sediments, with microflora more suitable for degrading PAHs and physicochemical and mineralogical characteristics more likely to simulate ageing, is probably preferable to using artificial sediments. The latter have other disadvantages related in particular to the impossibility of pertinently simulating the physicochemical properties impacting on the partition of PAHs between the dissolved and particular (mineral and organic) phases. On several occasions the comparative study of the fate and effects of PAHs in artificial and natural sediments showed large divergences between these two types of sediment, generally expressing an overestimation of risks of exposure and effects in artificial sediment (Verrhiest, 2001; Verrhiest et al., 2001), not only for benthic microorganisms but also for pelagic organisms in contact directly with the sediment or/and via the water column. This overestimation is related to the difficulty of representing all the potential adsorption sites, in particular linked to a generally complex, natural organic material. Although we recommend giving up the use of artificial sediment in this type of research, the question of what natural sediment model should be used remains unanswered, given the great diversity of sediments. Here again, the a priori wide distribution of physicochemical properties of sediments can lead to behaviours that vary considerably as a function of the sediment chosen. The sediment should therefore be selected according to a site or a specific property under study.

Other procedures of the microcosm test protocol have a significant influence on the fate and effects of PAHs. We showed that the lighting chosen in our tests (classical laboratory fluorescent light) favoured the phototoxicity of certain PAHs (Clément et al., 2000, 2005b). The adsorption of UV by PAHs governs photosensitisation and phototransformation phenomena, varying according to the type of radiation (UVA and UVB). Thus the choice of lighting more or less representative of the solar light spectrum influences the results observed (Wilcoxen et al., 2003). Certain authors have also shown in the laboratory that the penetration of UV down to the sediment could generate phototoxicity for benthic organisms (Ankley et al., 1994, 1995).

However, McDonald and Chapman (2002) questioned the ecological pertinence of phototoxicity, widely studied in the laboratory but, according to them, rarely expressed in situ. Indeed, many parameters have to be taken into account to estimate the probability of an organism bioaccumulating PAHs being subject to active solar radiation, as phototoxicity is essentially explained by photosensitisation. McDonald and Chapman (2002) showed that a large number of processes in situ allow organisms to avoid this exposure (reduction of bioavailability by organic and particular matter, protection mechanisms against UV in certain organisms, deep burrowing of benthic organisms, shade provided by aquatic plants, etc.). However, certain experimental parameters lead to overestimating exposure (glass flasks facilitate the passage of light in several directions, a shallow water column facilitates the passage of UVs down to the sediment, thin layers of sediment, radiation used at constant

Bioavailability of Polycyclic Aromatic Hydrocarbons Studied Through Single-Species Ecotoxicity Tests and
Laboratory Microcosm Assays

127

intensity non representative of the variations observed during the day, water saturated with oxygen favouring photosensitisation, etc.). In tests performed in the presence of daphnia and algae, we were able to demonstrate their role in reducing the phototoxicity of PAHs to daphnia.

As mentioned earlier, the use of glass recipients does not prevent the adsorption of PAHs on their walls, a phenomenon that, for the generally low contents of PAHs found in the water column, can lead to underestimating the contents expected (Clément et al., 2005b). Conversely, the use of synthetic environments generally free of dissolved organic materials leads to overestimating exposure. Regarding this, the incorporation of a sediment permits reducing this bias by enriching the water column with dissolved organic matter. Sediment also contributes particular and colloidal materials, the latter leading to an increase in the exposure of organisms in the water column to the PAHs adsorbed in it (Baumard, 1997). Similarly, the presence of micro-algae also modifies the exposure of daphnia and other consuming organisms (amphipods), though we did not have the opportunity to specify in which direction since although the algae capture some of the PAHs and thus reduce the dissolved fraction, they introduce an additional route of exposure for the organisms that consume them.

Although we studied monocontamination in most cases, we were also able to evaluate the effects of mixtures of PAHs, a situation closer to reality. Although the hypothesis of effect additivity is generally accepted, we showed that synergetic effects are possible (Verrhiest et al., 2001). It is however necessary to go further by working on a mixture of a greater number of PAHs (for example the 16 priority PAHs of the USEPA), and by incorporating other types of pollutants, such as metals.

The microcosm bioassays made it possible to diversify the exposure routes of organisms, for example, daphnia present in the water column but which can also be in contact with sediment particles. The results of two different studies failed to converge: whereas the effects on daphnia were increased by sediment contaminated by PAHs in one (Verrhiest et al., 2001), the other showed effects were essentially linked to pyrene contents in the water column. This absence of convergence can be explained by the different natures of the sediments used in these two studies and probably other parameters. Specific tests should be performed to study this point.

The toxicity criteria studied do not show that PAHs are very toxic to benthic organisms, even for exposures lasting a month. This is generally due to the high adsorption of PAHs on the particular and dissolved organic matter of the sediments which significantly reduces the bioavailability of these substances, a reduction that continues through time (ageing). It is reassuring in this initial approach to observe that chironomidae are capable of developing and emerging in nonetheless heavily contaminated sediments, a fact corroborated by results obtained on certain natural sediments also heavily contaminated by PAHs and heavy metals. We do not have data on the effects of PAHs on amphipod reproduction, due to the short time in which the tests were performed. It would be interesting to take this biological criterion into account by longer exposure or by the exposure of older individuals at the beginning of the test. Likewise, a study on several generations of chironomidae exposed to PAHs would make it possible to evaluate long term effects, by taking into account the number of hatched larvae and their capacity to pass through their life cycle. The few bioaccumulation measurements that we performed in the test on pyrene and the results in the literature encourage us to persevere along these lines, since the lower toxicity of PAHs could be explained by the capacity of pelagic (daphnia, algae) and benthic (chironomidae and amphipods) organisms to biotransform PAHs. Given the measurements performed on the microbial compartment, it appears that the presence of high PAH contents does not significantly disturb this compartment. This result is important as the functioning of the ecosystem depends in part on the biological activities of sediment which contributes to recycling organic matter and renewing the mineral elements required for primary producers.

Although we were able to perform a global study of the fate of PAHs in microcosms, we were unable to identify the role played by organisms in this fate and in the effects stemming from them, thus this could form the basis for an additional path of research. The bioturbation activity of benthic organisms can contribute towards modifying the distribution of contaminants in sediment and stimulating their biodegradation through better oxygenation, for example, of superficial layers. When studying the fate of fluoranthene in the presence of the marine worm *Capitella*, Madsen *et al.* (1997) observed that bioturbation activity contributed towards burying fluoranthene, but the total loss of fluoranthene in the sediment was higher in the presence of the worms, whose activity increased the transfer of fluoranthene to the supernatant water or/and stimulated the biodegradation of this PAH. The role played on this level by the chironomidae and the amphipods used in our tests remains to be determined. Do they contribute to greater exposure of pelagic organisms or, on the contrary, do they reduce the risks to which they are exposed ? These questions are obviously far-reaching, as answering them requires a large number of microcosm bioassays in which certain populations are present simultaneously in order to highlight the interactions described above.

4. References

Akkanen, J. & Kukkonen, J.V.K. (2003). Biotransformation and bioconcentration of pyrene in *Daphnia magna*, *Aquatic Toxicol.* 64: 53-61.

Akkanen, J.; Penttinen, S., Haitzer, M. & Kukkonen, J.V.K. (2001). Bioavailability of atrazine, pyrene and benzo(a)pyrene in European river waters, *Chemosphere* 45: 453-462.

Alexander, M. (2000). Aging, bioavailability, and overestimation of risk from environmental pollutants, *Environ. Sci. Technol.* 34: 4259-4264.

Ankley, G.T.; Collyard, S.A., Monson, P.D. & Kosian, P.A. (1994). Influence of ultraviolet light on the toxicity of sediments contaminated with polycyclic aromatic hydrocarbons, *Environ. Toxicol. Chem.* 13: 1791-1796.

Ankley, G.T.; Erickson, R.J., Phipps, G.L., Mattson, V.R., Kosian, P.A., Sheedy, B.R. & Cox, J.S. (1995). Effects of light intensity on the phototoxicity of fluoranthene to a benthic macroinvertebrate, *Environ. Sci. Technol.*, 29: 2828-2833.

Arfsten, D.P.; Schaeffer, D.J. & Mulveny, D.C. (1996). The effects of near ultraviolet light radiation on the toxic effects of polycyclic aromatic hydrocarbons in animals and plants: a review, *Ecotoxicol. Environ. Safety* 33: 1-24.

Baumard, P. (1997). Biogéochimie des composés aromatiques dans l'environnement marin, Thèse de Doctorat en Chimie Analytique et Environnement de l'Université de Bordeaux I, 145 pages.

Bioavailability of Polycyclic Aromatic Hydrocarbons Studied Through Single-Species Ecotoxicity Tests and
Laboratory Microcosm Assays

129

Baumard, P., Budzinski, H. & Garrigues, P. (1998). Polycyclic aromatic hydrocarbons in sediments and mussels of the western Mediterranean Sea. *Environ. Toxicol. Chem.* 17: 765-776.

Bennett, W.E., Maas, J.L., Sweeney, S.A. & Kagan, J. (1986). Phototoxicity in aquatic organisms: the protecting effect of beta-carotene, *Chemosphere* 15: 781-786.

Boese, B.L., Lamberson, J.O., Swartz, R.C., Ozretich, R. & Cole, F. (1998). Photoinduced toxicity of PAHs and alkylated PAHs to a marine infaunal amphipod (*Rhepoxinius abronius*), *Arch. Environ. Contam. Toxicol.* 34: 235-240.

Borglin, S., Wilke, A., Jepsen, R. & Lick, W. (1996). Parameters affecting the desorption of hydrophobic organic chemicals from suspended sediments, *Environ. Toxicol. Chem* 15: 2254-2262.

Bray, M., Babut, M., Vollat, B., Montuelle, B., Devaux, A., Bedell, J.P., Delolme, C., Durrieu, C., Clément, B., Perrodin, Y. & Triffault-Bouchet, G. (2003). Evaluation écotoxicologique de matériaux de dragage: Application to 5 sédiments du Nord-Pas de Calais, rapport Cetmef/Drast/VNF, septembre 2003, 142 pages.

Bray, M., Babut, M., Montuelle, B., Vollat, B., Devaux, A., Delolme, C., Durrieu, C., Bedell, J.P. & Clément, B. (2001). Evaluation écotoxicologique de sédiments contaminés ou de matériaux de dragage. (III). Application au Canal de l'Est Branche Sud, rapport Cetmef/Drast et VNF, avril 2001, 70 pages.

Chevron, N. (2004). Mise en évidence de la biodégradation du pyrène dans une matrice organique, rapport de DEA Sciences et Stratégies Analytiques de Lyon 1, 37 p.

Clément, B., Cauzzi, N., Godde, M., Crozet, K. & Chevron, N. (2005b). Pyrene toxicity to aquatic pelagic and benthic organisms in single-species and microcosm tests, *Polycyclic Aromatic Compounds* 25: 271-298.

Clément, B., Muller, C. & Verrhiest, G. (2000). Influence of exposure conditions on the bioavailability of fluoranthene to *Daphnia magna* (Cladocera), *Polycyclic Aromatic Compounds* 20: 259-274.

Conrad, A.U., Comber, S.D. & Simkiss, K. (2002). Pyrene bioavailability; effect of sediment-chemical contact time on routes of uptake in an oligochaete worm, *Chemosphere* 49: 447-454.

DePaolis, F. & Kukkonen, J. (1997). Binding of organic pollutants to humic and fulvic acids: influence of pH and the structure of humic material, *Chemosphere* 34: 1693-1704.

Di Toro; D.M.; Zarba, C.S., Hansen, D.J., Berry, W.J., Swartz, R.C., Cowan, C.E., Pavlou, S.P., Allen, H.E., Thomas, N.A. & Paquin, P.R. (1991). Technical basis for establishing sediment quality criteria for nonionic organic chemicals using equilibrium partitioning, *Environ. Toxicol. Chem.* 10: 1541-1583.

Ditsworth, G.R.; Schults, D.W. & Jones, J.K.P. (1990). Preparation of benthic substrates for sediment toxicity testing, *Environ. Toxicol. Chem.* 9: 1523-1529.

Driscoll, S.K., Landrum, P.F. & Tigu, E. (1997). Accumulation and toxicokinetics of fluoranthene in water-only exposures with freshwater amphipods, *Environ. Toxicol. Chem.* 16: 754-761.

Eisler, R. (2000). Polycyclic Aromatic Hydrocarbons. In: *Handbook of chemical risk assessment. Health hazards to humans, plants and animals. Organics.* Ed. R. Eisler. CRC Press. Lewis Publishers, pp. 1343-1411.

Garban, B. & Ollivon, D. (1995). Transport et devenir de polluants et micropolluants en Seine. Rôle des matières en suspension et des sédiments dans les processus de transfert, Thèse de Doctorat de l'Université de Paris VI, pp 46-49.

Gauthier, T.D.; Shane, E.C., Guerin, W.F., Seltz, W.R. & Grant, C.L. (1986). Fluorescence quenching method for determining equilibrium constants for polycyclic aromatic hydrocarbons binding to dissolved organic materials, *Environ. Sci. Technol.* 20: 1162-1166.

Geffard, O.; Budzinski, H. & His, E. (1999). The toxicity of phenanthrene (PHE), 2 methylphenanthrene (2MP) and benzo(a)pyrene (BAP) on embryogenesis and larval development oysters (*Crassostrea gigas*) and sea urchins (*Paracentrotus lividus*). Abstract. 17 th International Symposium on Polycyclic Aromatic Compounds, 25-29 October, Bordeaux, France, p 195.

Gourlay, C; Miège, C, Tusseau-Vuillemin, MH, Mouchel, JM & Garric, J. (2002). The use of spectrofluorimetry for the determination of polycyclic aromatic hydrocarbons bioaccumulation and biotransformation in *Daphnia magna. Polycyclic Aromatic Compounds* 22: 3-4, pp 501-516.

Grathwohl, P. (1990). Influence of organic matter from soils and sediments from various origins on the sorption of some chlorinated aliphatic hydrocarbons, implications on Koc correlations, *Environ. Sci. Technol.* 24: 1687-1693.

Guerrero, N.R. V.; Taylor, M.G., Davies, N.A., Lawrence, M.A.M., Edwards, P.A., Simkiss, K. & Wider, E.A. (2002). Evidence of differences in the biotransformation of organic contaminants in three species of freshwater invertebrates, *Environ. Pollut.* 117: 523-530.

Guthrie, E.A.; Bortiatynski, J.M., Van Heemst, J.D.H., Richman, J.E., Hardy, K.S., Kovach, E.M. & Hatcher, P.G. (1999). Determination of [C-13]pyrene sequestration in sediment microcosms using flash pyrolysis GC-MS and C-13 NMR, *Environ. Sci. technol.* 33: 119-125.

Haitzer, M.; Höss, S., Transpurger, W. & Steinberg, C. (1998). Effects of dissolved organic matter (DOM) on the bioconcentration of organic chemicals in aquatic organisms – a review, *Chemosphere* 37: 1335-1362.

Haitzer, M.; Höss, S., Traunspaurger, W. & Steinberg, C. (1999). Relationship between concentration of dissolved organic matter (DOM) and the effect of DOM on the bioconcentration of benzo(a)pyrene, *Aquat. Toxicol.* 45: 147-158.

Huang, X.D.; Dixon, D.G. & Greenberg, G.B.M. (1995). Increased polycyclic aromatic hydrocarbon toxicity following their photomodification in natural sunlight: impacts on the duckweed *Lemna gibba* L.G-3, *Ecotoxicol. Environ. Safety* 32: 194-200.

Huovinen, P.S.; Soimasuo, M.R. & Oikari, A.O.J. (2001). Photoinduced toxicity of retene to *Daphnia magna* under enhanced UV-B radiation, *Chemosphere* 45: 683-691.

Bioavailability of Polycyclic Aromatic Hydrocarbons Studied Through Single-Species Ecotoxicity Tests and
Laboratory Microcosm Assays

131

Jouanneau, Y.; Blake, G., Clément, B., David, B. & Naffrechoux, E. (2003). Devenir des hydrocarbures aromatiques polycycliques (HAP) dans un écosystème aquatique et impact sur les organismes vivants: exemple du pyrène. Rapport CNRS-PEVS, Ecosystèmes et Environnement, appel d'offres "Dynamique des contaminants", 126 pages.

Kalf, D.F.; Crommentuijn, T. & Van De Plassche, E.J. (1997). Environmental Quality Objectives for 10 Polycyclic Aromatic Hydrocarbons (PAHs), *Ecotoxicol. Environ. Safety* 36: 89–97.

Krylov, S.N.; Huang, X.D., Zeiler, L.F., Dixon, D.G. & Greenberg, B.M. (1997). Mechanistic quantitative structure-activity relationship model for the photoinduced toxicity of polycyclic aromatic hydrocarbons: I. Physical model based on chemical kinetics in a two-compartment system, *Environ. Toxicol. Chem.* 16: 2283-2295.

Landrum, P.F.; Dupuis, W.S. & Kukkonen, J. (1994). Toxicokinetics and toxicity of sediment-associated pyrene and phenanthrene in *Diporeia* spp.: examination of equilibrium-partitioning theory and residue-based effects for assessing hazard, *Environ. Toxicol. Chem.* 13: 1769-1780.

Landrum, P.F. & Faust, W.R. (1994). The role of sediment composition on the bioavailability of laboratory-dosed sediment-associated organic contaminants to the amphipod, *Diporeia* (spp.), *Chem. Speciation Bioavail.* 6: 85-92.

Landrum, P.F.; Gossiaux, D.C. & Kukkonen, J. (1997). Sediment characteristics influencing the bioavailability of nonpolar organic contaminants to *Diporeia* spp., *Chemical Speciation and Bioavailability* 9: 43-55.

Lei, A.P.; Wong, Y.S. & Tam, N.F.Y (2003). Pyrene-induced changes of glutathione-S-transferase activities in different microalgal species, *Chemosphere* 50: 293–301.

Lei, A.P.; Wong, Y.S. & Tam, N.F.Y. (2001). Removal of pyrene by different microalgal species. In: Proceedings of Asian Waterqual 2001, 1st Asia-Pacific Regional Conference, Fukuoka, Japan. pp. 969–974.

Leppänen, M.T. & Kukkonen, J.V.K. (2000). Effect of sediment-chemical contact time on availability of sediment-associated pyrene and benzo(a)pyrene to oligochaete worms and semi-permeable membrane devices, *Aquat. Toxicol.* 49: 227-241.

Lyons, B.P.; Pascoe, C.K. & McFadzen, I.R.B. (2002). Phototoxicity of pyrene and benzo[a]pyrene to embryo-larval stages of the pacific oyster *Crassostrea gigas*, *Marine Environ. Research* 54: 627–631.

MacDonald, D.D.; Ingersoll, C.G. & Berger, T.A. (2000). Development and evaluation of consensus-based sediment quality guidelines for freshwater ecosystems, *Arch. Environ. Contam. Toxicol.* 39: 20-31.

Madsen, S.D.; Forbes, T.L. & Forbes, V.E. (1997). Particle mixing by the polychaete *Capitella* species 1: coupling fate and effect of a particle-bound organic contaminant (fluoranthene) in a marine sediment, *Marine Ecology Progress Series* 147: 129-142.

May, W.E. (1980). The solubility behaviour of polycyclic aromatic hydrocarbons in aqueous systems, *Adv. Chem. Series.* 285 : 143-192.

McCarthy, J.F. (1983). Role of particulate organic matter in decreasing accumulation of polynuclear aromatic hydrocarbons by *Daphnia magna, Arch. Environ. Contam. Toxicol.* 12: 559-568.

McDonald, B.G. & Chapman, P.M. (2002). PAH phototoxicity - an ecologically irrelevant phenomenon ?, *Marine Pollut. Bull.* 44: 1321-1326

Merlin, G.; Lissolo, T. & Morel, V. (1995). Precautions for Routine Use of INT Reductase Activity for Measuring Biological Activities in Soil and Sediments, *Environ. Toxicol. Water Qual.* 10: 185-192.

Miller, J.S. (1999). Determination of polycyclic aromatic hydrocarbons by spectrofluorimetry, *Anal. Chim. Acta* 38 : 27-44.

Munoz, M.J. & Tarazona, J.V. (1993). Synergistic effect to two- and four-component combinations of the polycyclic aromatic hydrocarbons: Phenanthrene, Anthracene, Naphthalene and Acenaphthene on *Daphnia magna, Bull. Environ. Contam. Toxicol.* 50: 363-368.

Naffrechoux, E.; Combet, E., Fanget, B., Paturel, L. & Saber, A. (1999). Occurence and fate of PAHs from motorway runoff in the north drainage basin of Annecy lake, France. Abstract. 17 th International Symposium On Polycyclic Aromatic Compounds, 25-29 October, Bordeaux, France, p 182.

Newsted, J.L. & Giesy, J.P. (1987) Predictive models for photoinduced acute toxicity of polycyclic aromatic hydrocarbons to *Daphnia magna,* Strauss (Cladocera, crustacea), *Environ. Toxicol. Chem.* 6: 445-461.

Nikkilä, A. & Kukkonen, J.V.K. (2001). Effects of dissolved organic material on binding and toxicokinetics of pyrene in the waterflea *Daphnia magna, Arch. Environ. Contam. Toxicol.* 40: 333-338.

NRCC (1983). Polycyclic aromatic hydrocarbons in the aquatic environment: formation, sources, fate and effects on aquatic biota. NRC Associate Committee on the Scientific Criteria for Environmental Quality, National Research Council of Canada, NRCC Publication No. 18981, Ottawa, 209 pp.

Ollivon, D.; Garban, B. & Chesterikoff, A. (1995). Analysis of the distribution of some polycyclic aromatic hydrocarbons in sediments and suspended matter in the river Seine (France), *Water Air Soil Pollut.* 81: 135-152.

Pelletier, M.C.; Burgess, R.M., Ho, K.T., Kuhn, A., McKinney, R.A. & Ryba, S.A. (1997). Phototoxicity of individual polycyclic aromatic hydrocarbons and petroleum to marine invertebrate larvae and juveniles, *Environ. Toxicol. Chem.* 16: 2190-2199.

Raymond, N.; Geoffroy, L., Bourasseau, L., Budzinski, H., Nadalig, T. & Gilewicz, M. (1999). Bacterial aerobic degradation of polycyclic aromatic hydrocarbons by pure strains: characterisation of catabolic abilities using respiration studies. Abstract. 17 th International Symposium on Polycyclic Aromatic Compounds, 25-29 October, Bordeaux, France, p. 213.

Ren, L.; Huang, X.D., Mcconkey, B.J., Dixon, D.G. & Greenberg, B.M. (1994). Photoinduced toxicity of three polycyclic aromatic hydrocarbons (Fluoranthene, Pyrene, and Naphthalene) to the Duckweed *Lemna gibba* L. G-3, *Ecotoxicol. Environ. Safety* 28: 160-171.

Bioavailability of Polycyclic Aromatic Hydrocarbons Studied Through Single-Species Ecotoxicity Tests and Laboratory Microcosm Assays

133

Southworth, G.R.; Beauchamp, J.J. & Schmieders, P.K. (1978). Bioaccumulation potential of polycyclic aromatic hydrocarbons in *Daphnia pulex*, *Water Research*, 12: 973-977.

Swartz, R.C.; Ferraro, S.P., Lamberson, J.O., Cole, F.A., Ozretich, R.J., Boese, B.L., Schults, D.W., Behrenfeld, M. & Ankley, G.T. (1997). Photoactivation and toxicity of hydrocarbon compounds in marine sediment, *Environ. Toxicol. Chem.* 10: 2151-2157.

Swartz, R.C.; Schults, D.W., Ozretich, R.J., Lamberson, J.O., Cole, F.A., Dewitt, T.H., Redmond, M.S. & Ferraro, S.P. (1995). PAHs: a model to predict the toxicity of polynuclear aromatic hydrocarbon mixtures in field-collected sediments, *Environ. Toxicol. Chem.* 14:1977-1987.

Valette-Silver N., Hammed M.J., Efurd D.W., Robertson A. (1999). Status of the contamination in sediments and biota form the western Beaufort Sea (Alaska), *Mar. Poll. Bull.* 38: 702-722.

Verrhiest, G. (2001). Toxicité de sédiments d'eau douce contaminés par des HAPs. Influence de la nature des sédiments sur la biodisponibilité des HAPs, Thèse de Doctorat de l'Université de Savoie (spécialité: Biologie et Biochimie Appliquées).

Verrhiest, G.; Clément, B. & Blake, G. (2001). Single and combined effects of sediment-associated PAHs on three species of freshwater macroinvertebrates, *Ecotoxicology* 10: 363-372.

Verrhiest, G.; Clément, B. & Merlin,G. (2000). Influence of sediment organic matter and fluoranthene-spiked sediments on some bacterial parameters in laboratory freshwater / formulated sediment microcosms, *Aquat. Ecosystem Health Manag.* 3: 359-368.

Verrhiest, G.; Clément, B., Volat, B., Montuelle, B. & Perrodin, Y. (2002a). Interactions between a polycyclic aromatic hydrocarbons mixture and the microbial communities in a natural freshwater sediment. *Chemosphere* 46: 187-196.

Verrhiest, G.; Cortes, S., Clément, B. & Montuelle, B. (2002b). Chemical and bacterial changes during laboratory conditioning of formulated and natural sediments, *Chemosphere* 46: 961-974.

Vindimian, E.; Bisson, M., Dujardin, R., Flammarion, P., Garric, J., Babut, M., Lamy, M.-H., Porcher, J.-M. & Thybaud, E. (2000). Complément au SEQ-Eau: méthode de détermination des seuils de qualité pour les substances génotoxiques. Rapport final", 2000, 151 p, INERIS, Agence de l'eau Rhin-Meuse - Verneuil-en-Halatte - avril 2000.

Warshawsky, D.; Cody, T., Radike, M., Reilman, R, Schumann, B., LaDow, K. & Schneider, J. (1995). Biotransformation of benzo[a]pyrene and other polycyclic aromatic hydrocarbons and heterocyclic analogs by several green algae and other algal species under gold and white light, *Chemico-Biol. Interact.* 97: 131-148.

Wernersson, A.S.; Dave, G.R. & Nilsson, E. (1999). Combining sediment quality criteria and sediment bioassays with photoactivation for assessing sediment quality along the Swedish West Coast, *Aquat. Ecosystem Health Manag.* 2: 379-389.

Wernersson, A-S. & Dave, G.R. (1997). Phototoxicity identification by solid phase extraction and photoinduced toxicity to *Daphnia magna*, *Arch. Environ. Contam. Toxicol.* 32: 268-273.

Wilcoxen, S.E.; Meier, P.G. & Landrum, P.F. (2003). The toxicity of fluoranthene to *Hyalella azteca* in sediment and water-only exposures under varying light spectra, *Ecotoxicol. Environ. Safety* 54: 105 -117.

Witt, G. (1995). Polycyclic aromatic hydrocarbons in water and sediment of Baltic Sea, *Mar. Poll. Bull.* 31: 237-248.

6

Depositional History of Polycyclic Aromatic Hydrocarbons: Reconstruction of Petroleum Pollution Record in Peninsular Malaysia

Mahyar Sakari

Water Research Unit & School of Science and Technology,
Universiti Malaysia Sabah,
Malaysia

1. Introduction

In the last century, the world has experienced huge and various types of environmental threats. An important group of them is generated from the wide use of fossil fuel such as petroleum as the source of energy in industries, urban development and transportation. Hydrocarbons are the main constituents of fossil fuels thus petroleum hydrocarbons are possible and important source of pollution worldwide. Petroleum hydrocarbons enter the environment from accidental oil spill, natural leaks, industrial releases, vehicles or as by-products from commercial or domestic uses (Ou *et al.*, 2004). Hydrocarbons in petroleum include several types and categories of normal alkanes (saturated, n-alkane), unsaturated hydrocarbons, non-symmetric cyclic hydrocarbons (terpanes) and polycyclic aromatic hydrocarbons (PAHs). Predominance of these compounds in the environmental compartments or samples may indicate petroleum pollution. Petroleum hydrocarbon may disperse in the environment via atmospheric transportation and/or lateral transport. Petroleum contaminants are subject to several processes and changes after production or release such as degradation, photooxidation and decay. The trend over hydrocarbon changes in the environment depends on their chemical characteristics and depositional locations. Locations such as depository sediments under the sea bed surface usually keep hydrocarbon contents unchanged.

2. Polycyclic Aromatic Hydrocarbon (PAHs)

An important class of petroleum hydrocarbons is polycyclic aromatic hydrocarbons (PAHs). PAHs and their derivatives are ubiquitous in the environment such as air, water, soil, sediments and living organisms. PAHs are group of chemicals with more than 10,000 compounds that consist of two or more fused benzene rings (Fig. 1) in different arrangements (Blumer, 1976).

Among PAHs compounds, some have potential for being carcinogen, mutagen and disturbing human endocrine systems (Neff, 1979). Therefore they are categorized as environmental high priority contaminants. PAHs are lipophilic compounds consist of 2 to 7 benzene rings; the 2-4 rings are classified as Lower Molecular Weight (LMW) since 5-7 as Higher Molecular Weight (HMW). The LMW PAHs are more soluble in water and are acutely toxic to human and living organisms whereas HMW are highly soluble in lipid and more carcinogenic, mutagenic with more time period effects (Neff, 1979). The hydrophobic and lipophilic properties of some HMW PAHs make them relatively insoluble in water and tend to accumulate on surfaces or in non-polar matrices.

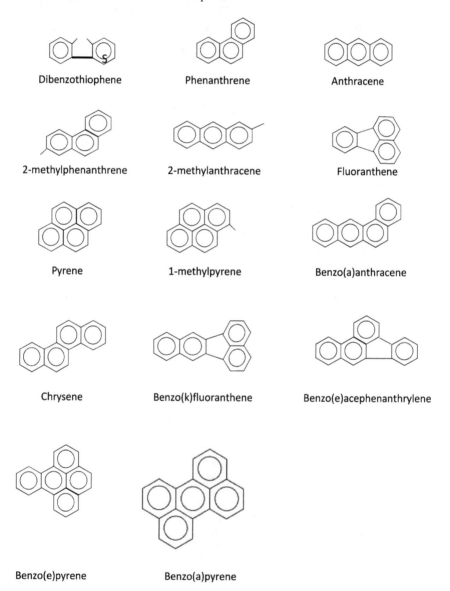

Fig. 1. Some of Polycyclic Aromatic Hydrocarbons Compounds.

Depositional History of Polycyclic Aromatic Hydrocarbons: Reconstruction of Petroleum Pollution Record in
Peninsular Malaysia

137

The sorptive properties of PAHs are largely controlled by the organic particulate fraction of suspended and deposited sediments (Baker *et al.*, 1986). Particle bound PAHs have a short residence time in the water column before they are settled to the bottom sediments where they may be re-suspended, degraded or subjected to long-term retention. The extent of any release back to the water column depends on the degree of bioturbation, physical re-suspension and the physico-chemical properties of the compound (Wong *et al.*, 1995).

Concerns over PAHs compounds in the environment arise since they are persistent in the environment for a long period of time. PAHs are generated from anthropogenic source as well as nature. In the environment, natural products of PAHs are limited to few types such as Perylene, Phenantherene and Retene where there is no health effect on human and the environment (Neff, 1979).

Anthropogenic PAHs are widespread in the environment as pollutants produced from incomplete combustion of fossil fuel and biomass burning. Anthropogenic PAHs enter marine environment from two primary sources of Pyrogenic and Petrogenic. Pyrogenic source PAHs come from pyrolytic processes such as combustion of fossil fuel, urban and industrial activities, natural fire and biomass burning that produce high molecular weight and less or non-alkylated PAHs. Combusted PAHs after production attach into soot particles, move in far distances and get deposited on soil, terrestrial plants or surface layers of sediments at sea bottom. Some of pyrogenic products of PAHs such as fine particles from charcoals also are washed out from the place of production via sewage plants or precipitations to the marine environment.

Petrogenic sources of PAHs are mainly derived from the release of crude oil and petroleum products such as lubricating oil, diesel fuel, gasoline, asphalt and kerosene. This class of PAHs enters the environment via oil spill, tanker accident, routine tanker operation such as ballast water discharge and discharge from vehicle workshops (NAS, 2002).

3. Pentacyclic triterpanes

Major class of pentacyclic triterpanes, hopanes, is derived from precursor in bacterial membrane (microbial origins) of bacteriohopentetrol. Hopanes are the constituents of crude oil and some petroleum products. Hopanes are believed to be synthesized in the nature by cyclization of squalene precursor during the diagenesis (Fig. 2).

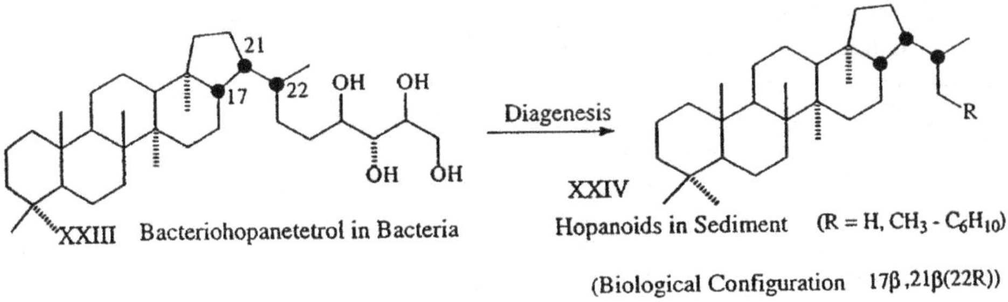

XXIII   Bacteriohopanetetrol in Bacteria

Diagenesis

XXIV   Hopanoids in Sediment   (R = H, CH$_3$ - C$_6$H$_{10}$)

(Biological Configuration   17β, 21β(22R))

Fig. 2. Diagenesis process that converts bacteriohopentetrol in bacteria to ββ (22R) stereochemistry of hopane. This unstable configuration changes to more stable αβ and βα hopane during the same process.

Classification of hopanes is based on the degree of oil maturation from specific source rock. Hopane itself is not categorized as pollution however they are distributed in the environment with petroleum hydrocarbon pollution and consequently are ubiquitous. They resist degradation processes thus are persistent component in crude oil and petroleum products.Due to this, they are widely used as the sources identifier of oil pollution in the environment. They are relatively existed in small amounts (usually <1% by weight) among other hydrocarbons. Hopanes are commonly found in C_{29}-C_{35}, together with two C_{27} species called regular steranes.

Homohopanes are the name of hopane series which the number of carbon arises by 30. The hopanes are composed of three sterioisomeric series, namely 17α(H),21β(H)- Hopanes, 17β(H),21β(H)- Hopanes and 17β(H),21α(H)- Hopanes. Hopanes with αβ configuration range from C_{27} to C_{35} are characteristics of petroleum because of their greater thermodynamic stability compare to other epimeric (ββ and βα) series. In geologically mature samples, αβ epimeric isomers are greater predominant over βα isomer (moretane). However ββ-isomers are commonly found in living organisms. Ts (18α(H)-22,29,30-trisnorneohopane) and Tm (17α(H)-22,29,30-trisnorhopane) can be a sensitive indicator of thermal maturity when capering oil or sediment samples from the same source. In addition, Hopanes with predominance of 17α(H), 21β(H)- stereochemistry indicates a substantial contribution from petroleum pollution.

Hopane distributions are usually recorded using the m/z 191 in mass chromatograms. An unusually high proportion of the C_{29} hopane is often associated with oil derived from carbonate source rock oil which includes most of those from the Middle East. Those oils also show a slightly enhanced abundance of the C_{35} extended hopane compare with the C_{34} homohopane. In C_{31}-C_{35} hopanes the biologically conferred 22R configuration is preserved during the initial stages of diagenesis. Subsequent isomerization results in a final equilibrium mixture containing approximately equal amount of 22R and 22S isomers. Oleanane is another triterpanes commonly associated with oil derived predominantly from higher plant sources. In conclusion, pentacyclic triterpanes are useful biomarker to identify plant and petroleum input of PAHs into the aquatic environment as well as sediments (Wakeham et al., 1980; Tan and Heit, 1981; Bouloubassi and Saliot, 1993; Yunker and McDonald, 1995; Chandru et al, 2008).

4. Source, distribution and fate of PAHs in aquatic environment

PAHs are released into the environment via natural and anthropogenic sources. Natural source includes oil seeps, volcanoes, grass fires, chlorophyllous and nonchlorophyllous (bacteria and fungi) plants. Anthropogenic sources of PAHs include discharge from routine oil transportation, oil spill, power plants based on fossil fuel consumption, biomass burning, pyrolysis of wood and internal combustion in industrial and vehicle engines.

PAHs enter into the marine environment usually by anthropogenic sources while natural sources have less contribution in this process. Among many possible sources for PAHs contamination in the marine environments; municipal and industrial wastes, city runoff, riverine discharge and atmospheric input have higher proportion.

Petrogenic and pyrogenic based PAHs usually show similar behaviors and fates after entering the environment. Pyrogenic PAHs that are produced via combustion processes have a high and strong affinity to airborne organic particles that may move in greater distances by wind and other atmospheric phenomena. PAHs associated with airborne particles reach to the top layer of the water column in the marine environment, moving to the water column and the bottom of the sea where settles in the sediment. Petroleum and petroleum products which are originated from concentrated hydrocarbon sources enter the marine environment and subjects to dispersion, evaporations, settlement in the bottom on the sediments, weathering, chemical changes, sunlight effects (photooxidation) and microbial degradation (bacteria, yeast and fungi) in short and long term period (Neff, 1979). Petrogenic sources of PAHs on the sediment stick into the particles and consequently is subjected to different chemical and biological changes. Heavier and more complex compounds of crude oil and its products are more resistant to microbial degradation. Regardless of the origin of PAHs, in the marine environment, they adhere to the particles (clay, silt, organisms, detritus and microbes) and settle on the sediments, where a variety of microbes metabolize it into some simple and light compound structure. Accumulation and bioaccumulation of PAHs in the marine environment and organisms are inversely correlated to the potential and ability of hydrocarbons to metabolize them either chemically or biologically.

Finding the source of hydrocarbon pollution is a great concern for many scientists all around the world. Although the first track in this line had been started in 1970s, many researchers are currently try to identify the source of hydrocarbon pollution in the marine environment. In Southeast Asia, the pioneering studies on specific compound analysis have been started by intensive survey in the Straits of Malacca (Zakaria et al., 1999, 2000, 2001, 2002, 2006) and followed laterally in a study in Gulf of Thailand (Boonyatumanoond et al., 2006 and 2007).

In order to identify the sources of hydrocarbon pollution in the environment, there are many techniques such as use of isomer pair ratios (Yunker, 2002), individual compound ratios (Hase and Hites, 1976; Laflamme and Hites, 1978; Baumard et al., 1998; Zakaria et al., 2000) and biomarkers (Volkman et al., 1997; Zakaria et al., 2002 and Wang and Fingas, 2005).

Some molecular ratios of specific hydrocarbons were developed to distinguish differences between PAHs originating from various origins and sources. Among those, the ratio of Phenantherene/anthracene (Ph/An) and flouranthene/pyrene (Fl/Py) were widely used by scientists (Steinhauer and Boehm, 1992; Budzinski et al., 1997; Baumard et al., 1998, 1999). The ratio of Fl/Py (fluoranthene/pyrene) has been used to identify fuel sources, showing values < 1.4 for coal combustion (Lee et al., 1977) and < 1.0 for wood (Lee et al., 1977; Knight et al., 1983). In sediments, value for this ratio was 1.3-1.7 at remote sites and < 1.0 near to urban centers (Gschwend and Hites, 1981; Helfrich and Armstrong, 1986). The phenanthrene/anthracene ratio also applies as an indicator for measuring the remoteness (>15) or vicinity (<10) of PAHs sources to urban areas (Zhang et al., 1993).

High-temperature processes such as combustion of organic matters generates PAHs characterized by low ratio of Ph/An (<10), whereas slow maturation during catagenesis, reach to higher values (Ph/An <15). Same trend observed in ratio of flouranthene/pyrene (Fl/Py), where values greater than 1 come from pyrogenic sources and less than unity is indicative of petroleum input. Another ratio which is summarized by Yunker et al., (2002) is Flu over Flu plus Pyr (Flu/(Flu+Pyr)) that is generally greater than 0.5 in grass, wood or coal combustion, and the petroleum boundary ratio appears closer to 0.40 than 0.50, whereas the Flu/(Flu+Pyr) ratios between 0.40 and 0.50 are more characteristic of liquid fossil fuel combustion. Above values were shown to be relatively less reliable in different geographical locations due to the various combustion material sources (Budzinski et al., 1997).

PAHs in the environment have definite behaviors which are controlled by several processes. Processes which can control the transport and degradation of PAHs in sediment include: 1) partition of the compounds between aqueous (pore-water) and particulate phase, 2) microbial degradation, 3) uptake, metabolisms and depuration of PAHs by the benthoses 4) photo-oxidation (surface sediment), chemical oxidation and 5) biosynthesis. Moreover the compounds specific selections for above mentioned processes are absolutely selective.

It is now well established that microbial degradation of PAH occurs primarily in the aerobic zone (Bauer and Capone, 1985) with highest rates occurring with low molecular weight homologues (Lee et al., 1977; Gardner et al., 1979; Readman et al., 1982).

Consequently, any degradation should result in selective losses of, anthracene relative to benzo[a]pyrene, and so affect the ratio of residual PAH. Readman et al., (1987) calculated that up to 80% of anthracene and 40% of benzo[a]pyrene could theoretically be degraded during the approximately 2 year particle/PAH passage through the aerobic layer at the laboratory condition (Table 1).

PAHs	Degradation Rate (10⁻³)	Half Life (years)	PAH Percentage Aerobically Degraded (2 years)
Anthracene	2.18	0.87	80
Fluoranthene	1.67	1.14	71
Benzo(a) Antheracene	1.17	1.63	57
Benzo(a)Pyrene	0.67	2.73	40

Table 1. Theorical impact of degradation on sedimentary PAHs (after Readman et al., 1987).

At the same time, the anthracene/benzo[a]pyrene ratio would be expected to decrease from surface sediment around one third in the anaerobic sediments. Jones et al., (1986), show that oil-derived aromatic hydrocarbons can be rapidly biodegraded in sediments, but combustion-derived aromatic hydrocarbons in the same sediments are relatively resistant to degradation. Similar anomalous behavior of PAH has also been reported by Farrington et al., (1983) where it was suggested that petroleum-derived PAH are more available for uptake by mussels than are pyrogenic PAH. Another process that controls the characteristics and concentration of PAHs is the phenomenon of photo oxidation. Photooxidation is a process which starts from the beginning stage of PAHs production in surface layers of soil, sediment, water or during the transportation in the air. There are selective photooxidation for specific PAHs such as Benz[a]anthracene which is more labile to photooxidation than chrysene + triphenylene (Kamens et al., 1986, 1988), therefore the benz[a]anthracene/(benz[a]anthracene + chrysene + triphenylene) ratio are supposed to be lower in summer than winter samples (Fernandez et al., 2002). Benzo[a]pyrene is photochemically less stable than benzo[e]pyrene where light exposure transforms unstable BaP to more stable BeP (Nielsen, 1988). The indeno[1,2,3-cd]pyrene/(indeno[1,2,3- d]pyrene + benzo[ghi] perylene) ratio is a prior and more stable to photooxidation than the ratios discussed above. Therefore there is expectation for Southeast Asian countries environment to show less concentrations of the low stable PAHs due to heavy and continued sunshine. Interesting to know that PAHs are penetrating in sediment layers after the deposition, where in upper layers of surface sediments PAHs are rich in 5-6 rings PAHs rather than 2-3 rings which migrate downward from upper to deeper layers due to integrity with fine particles and their physical migration with fine particles (Curtosi *et al.*, 2007).

Malaysia, which is located in Southeast Asia, has a unique tropical environment and climate. It is surrounded by the Straits of Malacca in the west and the South China Sea in the west of Peninsular Malaysia. The western part of peninsular Malaysia has been experiencing rapid development during the last half century. On the other hand, the strategic location of this country has made Malaysia as one of the busiest shipping route in the world due to huge petroleum demand from the Middle East to Japan and China (Fig. 3).

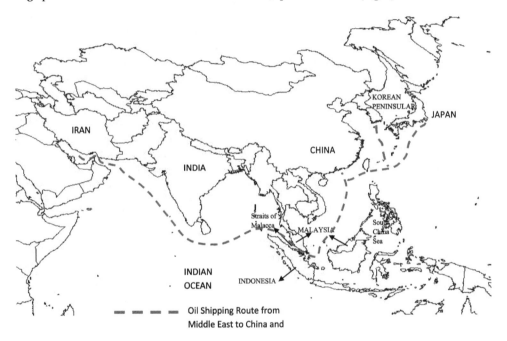

Fig. 3. Oil shipping route from the Middle East to the Far East via Straits of Malacca.

While Malaysia is experiencing extraordinary economic and population growth, it is also developing fast in industrialization, urbanization and motorization in last few decades. As a result of this development, the environment of this country is receiving more threats and hazards especially from the main source of energy which is petroleum. In Malaysia, the concentration and sources of hydrocarbon pollution vary according to locations. For instance, in western P. Malaysia, existence of rapid urban development and the establishment of several industrial areas, the hydrocarbon pollution is introduced throughout non-point and pointed sources. In the eastern P. Malaysia the pollution mostly comes from the urban area and less from industries, due to less industrial developments.

Besides that, Malaysian Marine Department reported 127 oil spill incidents since 1976 due to heavy oil tankers traffic in Straits of Malacca (Malaysian Marine Department, 2003). Zakaria and Takada (2003) believe that the Malaysian environment is under increasing threat of petroleum pollution; although this is not well-documented and recorded. To understand the petroleum hydrocarbon pollution levels in the environment, scientists usually study different environmental samples such as water, sediment, bio-monitoring agents, particles and aerosols. In Malaysia, few researchers have studied petroleum hydrocarbon pollution and were used one or more types of above mentioned samples to demonstrate the status of hydrocarbon pollution in the country.

5. A brief on global historical records of PAHs

Studies on hydrocarbon pollution in historical trend were started alongside studies on recent and modern sediment hydrocarbon pollution. Among the first records, Hites *et al.*, (1977) studied PAHs concentration in Buzzards Bay, Massachusetts for the period from 1900 to 1970. Later on, Wakeham *et al.*, (1979) reported PAHs from sedimentary records of several lakes in Switzerland and Washington, indicative of high levels of PAHs pollution in modern input. In 1984, Prahl and Carpenter published data on PAHs and aliphatic hydrocarbons from Washington coastal sediments indicated that naturally derived aliphatic hydrocarbons are very frequent in ancient sediments while anthropogenic PAHs show high concentrations in recent deposited sediments. In Lake Michigan, the PAHs from sedimentary record reported by Christensen and Zhang (1993) showing a constant trend with fossil fuel consumption. Zhang *et al.*, (1993) reported PAHs maxima (maximum concentration) for early 1950s and 1985 for cores from Green Bay and Lake Michigan. Taylor and Lester (1995) reported significant decrease in PAHs concentrations since 1966 because of less coal combustion. This study is confirmed by Gevao *et al.*, (1997) where they reported PAHs sub-surface maxima in late 1960s and early 1970s from a small rural lake of Cambria, UK. This sub-surface maxima reported by Pereira *et al.*, (1999) from San Francisco Bay in California during 1950s.

There are many more studies on investigation of hydrocarbon pollution in sedimentary record. For instance, Hostettler *et al.*, (1998) studied the trace of biomarker profiles in San Francisco Bay dated sediments, showing anthropogenic input of hydrocarbons in recent decades depositions. Okuda *et al.*, (2002) reported high PAHs concentrations from Chidorigafuchi Moat, Japan, in 1960s. Furthermore, Yim *et al.*, (2005) reported high flux of PAHs in 1950s and 1980 from Masan Bay in Korea. Liu *et al.*, (2005) reported the PAHs fluxes from Pearl River Estuary, South China shows first sharp peak of PAHs levels in 1950s and consequently 1990s. Hartmann *et al.*, (2005) were reported depositional history of organic contaminants from Narragansett Bay, Rhode Island in the United States of America, showing highest PAHs fluxes in modern sediments from Apponaug Cove and sub-surface maxima in Seekonk River core sdiment.

There is an unpublished report of historical record of aliphatic and aromatic hydrocarbons in the straits of Malacca during 1980's by Law in a curtsey communication. Unfortunately, no more details are available since the scientist is deceased. In South East Asia the first available and published study pioneered by Boonyatumanond *et al.*, (2007) from Gulf of Thailand. In this report PAHs fluxes show high levels in 1950s and 1970s because of rapid increase in numbers of vehicles and their usage in Thailand. Also the molecular marker of Hopane shows high contribution of Petroleum and anthropogenic hydrocarbons in dated sediments from the studied area.

Coastal sediment containing a mixture of natural and anthropogenic PAHs presents two important problems for assessing the fate and effect of PAHs in the environment (Wakeham *et al.* 1980). At first, anthropogenic PAHs should be evaluated by site-specific background of

Depositional History of Polycyclic Aromatic Hydrocarbons: Reconstruction of Petroleum Pollution Record in
Peninsular Malaysia

143

the PAHs derived naturally to the studied area. Secondly, parent PAHs data, solely, does not deliver reliable information due to overlap of source composition. This paper investigates the reconstruction of PAHs history and hopanes in one of the highly developing country in tropical Asia that experiences the rapid industrialization, motorization and urbanization.

PAHs derive to the environment via natural and anthropogenic processes. Natural processes of PAHs production occur during diagenesis and microbial activities as well as natural seeps of hydrocarbons and forest fires. Anthropogenic PAHs productions were consistent entering to the environment since mankind used fire for any purposes. The most recent PAHs derive to the environment so called modern input, have been increased since industrial revolution, when man used fossil fuel in industrial wheels. Beginning of the 20th century was in conjunction with rapid increase in PAHs flow to the environment when oil production contributed in fast development of the globe. PAHs come from oil origin enter the environment via petroleum and petroleum products such as gasoline and lubricating oils and their combustion. Combustion derived PAHs are dominated by the un-substituted moieties, whereas PAHs in petroleum are dominated by the alkylated homolog (LaFlamme and Hites, 1978). Lower formation temperatures such as in the formation of petroleum during the diagenesis preserve a higher degree of alkylated compounds (Youngblood and Blumer, 1975). Alkylated and non-alkylated PAHs are the basic knowledge of source identification of hydrocarbon and petroleum pollution in the environment. The degree of alkylation and alkyl homologs existed in environmental samples provides information on the sources of pollution. Youngblood and Blumer (1975) have proposed natural combustion such as natural fires in the forest as primary source of PAHs in deep layers of long core sediments. This idea is criticized by Wakeham et al., (1979) where forest fire might have constant input to the marine environment. In addition to the recent arguments, PAHs from natural fire does not occur often, since forest fire in the environment is not a predominant event. The concentrations of PAHs in the core sediments do not always correspond the unity of sources and input. This is due to various sources and concentrations that possibly interfere in a single layer along sedimentary intervals. Complex mixtures of different sources usually demonstrate irregular ratios that interfere with the results of source identification. Due to the complexity of different PAHs sources, PAHs compound specific ratios are still the most valuable tool for determination of pollution origin (Yunker et al., 1999).

Applications of different ratios are based on molecular structure of specific PAHs compounds in the environment. Lower Molecular Weight (LMW) PAHs is categorized by 2-3 benzene rings while 4-7 rings are known as Higher Molecular Weight (HMW). Pyrogenic PAHs (combusted) are characterized by high abundance of HMW compounds (4-6 benzene rings) and un-substituted (parent) compounds, whereas petrogenic PAHs are dominated by alkyl substituted and abundance of LMW (2-3 benzene rings) PAHs (Garrigues et al., 1995; Budzenski et al., 1997). The ratio of LMW over HMW PAHs is often applied for source discrimination of PAHs in environmental studies. The ratio of LMW/HMW PAHs for values lower than unity indicates pyrogenic source while 2 to 6 is an indicative for petrogenic input into the marine environment. The relative ratio of Methylphenanthrenes over Phenanthrene (MP/P) is also frequently applied for source identification of PAHs. The MP/P is another valuable ratio, as discussed earlier, based on un-substituted moieties and alkylated homolog frequencies in the sample. The MP/P fluctuates among values such as

0.5-1 for combustion derived PAHs in the sediments and 2-6 in sediment dominate by fossil fuel direct release (Prahl and Carpenter, 1983; Garrigues et al., 1995; Budzenski et al., 1997). The MP/P ratio around 4.0 is reported to be derived from crankcase oil (Pruel and Quinn, 1988), close to 1.0 for street and urban dust samples (Takada et al., 1990, 1991) and around 0.5 for atmospheric fallout (Takada et al., 1991). This ratio is higher for coal combustion sources than petroleum (Lee et al., 1977; Takada et al., 1990, 1991).

Among other ratios, scientists use other permanent ratios such as the relevant concentrations of sum of methylfluoranthenes and methylpyrenes over fluoranthene (Youngblood and Blumer, 1975; Laflamme and Hites, 1978; Gustafson et al., 1997). In this ratio, values above the unity (>1) indicate the petroleum sources of pollutions. In addition, the results of these two recent ratios (MP/P and (MFl+MPy)/Fl) are not necessarily same but in some studies deliver similar trends (Pereira et al., 1999).

Some specific compounds are well known in their characteristics. Among those compounds, Benzo(a)pyrene are proven carcinogens material to living organisms (Neff, 1979). Characteristics of specific compound PAHs are derived in the environment are strongly associated with the origins. One of these ratios is benzo(ghi)perylene to indeno(1,2,3-cd)pyrene (BghiP/IPy) where the high values come from the automotive exhaust particles (Marr et al., 1999; Nielsen et al., 1996 and Tuominen et al., 1987).

Okuda et al., (2002) reported specific compound PAHs from the core collected from Chidorigafuchi Moat in Japan. They showed that in surface sediments (0-8 cm) there are significant and high values of ratio of benzo(ghi)perylene to indeno(1,2,3-cd)pyrene than lower sections of the core, whereas the ratio showed a constant value around unity for depth below 20 cm and increasing for higher levels up to the sediments from the surface. This strongly suggests that since 1990 the PAHs are more influenced by automotive exhausts. This ratio shows a relatively high value in automotive exhaust particles as it is constant with the socio-economic condition of the study area.

Yim et al., (2005) reported specific hydrocarbon compounds for source identification of the PAHs in a study conducted in Masan Bay, Korea. The ratio that used including Phenantherene/anthracene (Ph/An) and fluorenthene/pyrene (Fl/Py) indicates pyrolitic origins. In the ratio of Fl/Py, scientists reported lower values than those reported from same source of American and European coal. Although Budzinski et al., (1997) showed this low value from coal sources of Australia (Fl/Py: 0.3-0.7).

The amounts of HMW and LMW PAHs in environmental samples are possibly indicative of pollution sources. For instance, coal usually produces high amounts of PAHs than other fuel materials such as petroleum and natural gas. The highest concentration of PAHs (maxima) in core samples collected from Masan Bay, Korea, indicates this issue, where the sharpest peak appeared from layers corresponds 1950s-1980s. During the above era the country was widely used coal for various energy purposes (Yim et al., 2005). Although previous studies found PAHs maximum concentration (maxima) in the sedimentary environment during 1940-1950, as an indication of fuel type changes from coal to oil and gas. The fuel type changes usually reveal increase or decrease in the total concentration of PAHs in the environment (Gschwend and Hites, 1981; Bates et al., 1984; Barrick and Prahl, 1987). Later on, in Masan Bay, Korea this pollution input trends have been decreasing due to Pollution Prevention Act established in 1963 and strong environmental control and monitoring conducted by local and national authorities (Yim, 2005).

Beside the PAHs analysis to determine the concentration and sources of pollution in the environment, there are other tools such as measurements of magnetic susceptibility. This is a fast and cost effective method based on the presence of magnetic-rich spherules that forms during the combustion processes by oxidation of pyrite to magnetic. This method successfully applied by Morris *et al.*, 1994 in Hamilton Harbour, Western Lake Ontario, Canada to compare the method efficiency with PAHs analysis, approve the pyrolytic sources of PAHs pollution in the studied area.

6. PAHs in depositional record, Malaysia

The depositional record of PAHs in peninsular Malaysia is studied during a 4 years scientific investigation. Eight sedimentary core samples were collected from 4 identical coastal areas. Each area represented a historical background of development and socio-economic events of peninsular Malaysia. They are consisted of Klang (Port and Offshore), Malacca (Near and Offshore), Johor (Near and Offshore) and Tebrau (No. 1 and 2). Thus there were 2 core samples taken from each location mainly from near shore and offshore locations to evaluate the distance factor effect of distribution and concentration of PAHs compounds in the environment (Fig. 4).

Fig. 4. Map of sampling locations.

The study elucidated the "Distribution, Sources and Depositional History of PAHs and Hopanes in Selected Locations in Peninsular Malaysia" using chemical molecular markers such as PAHs and Pentacyclic Triterpanes (Hopane) in deposited sediments. The cores lengths were ranged from 21 to 56 cm. The ^{210}Pb was used to reconstruct the pollution history of collected cores, revealed a time period of 60 to 280 years in different cores. Table 2 shows the concentration of PAHs in the sedimentary core of the study areas.

The core that was collected from the Klang City station showed that since 1945, there was an increasing trend in total deposited PAHs (Sakari *et al.*, 2010a). The highest concentration of total PAHs was observed during the era of 1990 to 1998 (2442 ng/g d. w.) as a sub-surface maxima which is interestingly followed by minimum PAHs level of 33 (ng/g d. w.) for the era of 1999 to 2007. Although in lower layers, the total PAHs of 161 (ng/g d. w.) was reported from the period of 1954-1962. In all sedimentary layers and intervals, specific compounds such as BkF, BeP and BaP were the leading PAH among others. This trend of PAHs increase is highly correlated to population increase of surrounded area, increase in registered cars and economic data of the study area. The rapid and sudden drop of total PAHs was interpreted as a joint function of physical phenomena, as well as weather condition, improvements of vehicle engine performance and enforcement of law and legislations.

In Offshore Klang station, the results showed very much depleted concentrations compared to Klang City core. Except for the recent decades deposited sediments, that showed lower concentration usually the core revealed homogenized concentration of PAHs fluctuating from 20 to 32 (ng/g d. w.). It is reported that the highest level of 32 ng/g d.w. happened at the beginning of the 20th century. The PAHs input in this core was not correlated to any of above mentioned socio-economic data, indicative of constant input via atmospheric fallout, where the BkF and BeP were the leading compounds throughout the core intervals. In Offshore Klang core, again the signature of pyrogenic input of PAHs was observed as those MP/P and L/H ratios. For both cores of Klang area, it is found that pyrogenic input from vehicle's emission and asphalt are the main contributor of PAHs into the marine environment of this area, although none of hopane signatures showed definite sign of any specific oil sources mainly due to combustion effect of pyrogenic sources on molecular structure of hopane.

In Malacca, the first core was collected from the near shore location showed the highest concentration of total PAHs in the entire study areas. The highest concentration of total PAHs (4195 ng/g d. w.) was reported from 1977 to 1983 while the lowest were observed at the beginning of the 20th century (Sakari *et al.*, 2011). Interesting to see that very severe depletion of HMW exists in this core and in all layers. The predominant of compound in this core was shown to be Phenanthrene and most of the sediment intervals revealed pyrogenic sources with MP/P ratio below the unity.

Offshore Malacca core was shown lower concentration than near shore but still Phenanthrene and its derivatives are the main PAHs contributor to the total PAHs. The highest and lowest concentrations of PAHs were revealed during 1963 to 1969 and 1914 to 1920, respectively. The signature of PAHs likewise near shore station was shown pyrogenic. This is reconfirmed by MP/P values. Identification of the PAHs origin using hopane marker showed street and urban dusts of Malacca City as the main contributor as observed in near shore Malacca.

In Johor, the first core collected from a location near the city where the connecting bridge (causeway) commutes Malaysia to Singapore. In this location, along the core, the PAHs concentration has ranged from the minimum of 44 (ng/g d. w.) to the maximum concentration of 1129 (ng/g d. w.). The highest concentration of PAHs was observed during the era from 1922 to 1969. Moreover, the lowest concentration observed in recent deposited sediments (Sakari *et al.*, 2011b). The PAHs signature showed a mixture of pyrogenic and petrogenic input where most modern input showed more combusted materials than old sediments with pyrogenic signature. This statement is evidenced by MP/P and LMW/HMW ratios. The leading PAHs compounds along core intervals were BeP, BkF and Phe and its alkyl substitutes.

Depositional History of Polycyclic Aromatic Hydrocarbons: Reconstruction of Petroleum Pollution Record in Peninsular Malaysia

147

Klang City Core

	Concentration (ng/g d.w.)						
Sediment age (year)*	1945-1953	1954-1962	1963-1971	1972-1980	1981-1989	1990-1998	1999-2007
aTotal PAHs (ng/g d.w.)	488.99	161.35	238.11	1032.52	1572.97	2422.93	33.85
bL/H PAHs	0.55	0.60	0.56	0.75	0.35	0.45	0.92
cMP/P	0.63	1.11	0.92	0.66	0.83	0.83	1.11
dTOC mg/g	61.23	51.40	51.37	55.93	59.45	71.04	64.96

Offshore Klang Core

	Concentration (ng/g d.w.)													
Sediment age (year)	1910-1916	1917-1923	1924-1930	1931-1937	1938-1944	1945-1951	1952-1958	1959-1965	1966-1972	1973-1979	1980-1986	1987-1993	1994-2000	2001-2007
aTotal PAHs (ng/g d.w.)	29.92	32.05	10.34	24.66	23.29	24.05	20.04	11.21	12.67	7.31	20.10	9.54	15.42	11.91
bL/H PAHs	0.55	0.82	0.49	0.49	0.36	0.96	0.41	0.34	0.38	0.34	1.26	0.34	0.99	0.40
cMP/P	0.23	0.24	0.22	0.22	0.34	0.46	0.32	0.59	0.37	0.52	0.51	0.54	0.64	0.83
dTOC mg/g	69.09	85.67	85.15	65.21	77.23	72.13	45.60	61.23	51.40	51.37	55.98	59.45	71.04	64.96

Near Shore Malacca

	Concentration (ng/g d.w.)																	
Sediment age (year)	1879-1885	1886-1892	1893-1899	1900-1906	1907-1913	1914-1920	1921-1927	1928-1934	1935-1941	1942-1948	1949-1955	1956-1962	1963-1969	1970-1976	1977-1983	1984-1990	1991-1997	1998-2005
aTotal PAHs (ng/g d.w.)	378.02	358.08	402.13	275.10	508.48	431.91	603.70	266.29	967.16	910.25	176.61	3111.47	4447.14	999.16	4195.07	2660.95	335.36	451.87
bL/H PAHs	96.84	23.31	126.59	39.66	22.75	12.93	22.87	8.98	29.36	26.70	174.55	144.23	195.50	204.02	122.70	505.00	24.16	20.22
cMP/P	0.78	1.26	3.17	0.77	0.84	0.76	0.88	1.09	2.39	1.03	0.72	0.80	0.71	1.31	0.93	0.86	3.19	8.58
dTOC mg/g	124.20	117.00	126.90	117.30	113.40	135.30	129.80	149.70	143.50	145.00	137.00	143.90	148.90	143.40	134.40	147.60	133.70	137.20

Offshore Malacca Core

	Concentration (ng/g d.w.)																
Sediment age (year)	1886-1892	1893-1899	1900-1906	1907-1913	1914-1920	1921-1927	1928-1934	1935-1941	1942-1948	1949-1955	1959-1962	1963-1969	1970-1976	1977-1983	1984-1990	1991-1997	1998-2005
aTotal PAHs (ng/g d.w.)	48.00	2.63	61.03	2.92	1.71	325.84	60.42	73.60	10.19	107.11	71.70	714.37	15.26	70.64	19.28	126.74	98.37
bL/H PAHs	3.96	NA	3.90	0.32	0.32	0.61	2.13	3.64	0.78	3.79	7.28	0.25	0.62	5.79	0.83	8.60	7.06
cMP/P	0.77	0.92	0.79	1.29	2.06	0.90	22.84	0.91	4.95	0.85	0.74	0.80	5.11	0.90	9.92	0.83	0.98
dTOC mg/g	143.30	144.60	139.80	137.90	154.20	149.90	141.00	66.90	138.50	136.10	105.60	141.30	129.20	159.60	172.60	156.10	184.10

Johor City Core

Sediment age (year)	1874-1885	1886-1897	1898-1909	1910-1921	1922-1933	1934-1945	1946-1957	1958-1969	1970-1981	1982-1993	1994-2005
				Concentration (ng/g d.w.)							
[a]Total PAHs (ng/g d.w.)	580.51	1005.40	648.67	478.73	1129.52	935.25	725.76	920.98	89.11	44.64	171.95
[b]L/H PAHs	0.95	0.15	1.93	0.60	4.27	0.60	0.44	0.54	2.03	1.02	3.05
[c]MP/P	2.45	2.39	2.41	1.93	1.87	1.66	1.61	1.04	0.72	1.59	0.74
[d]TOC mg/g	227.30	218.60	221.20	202.60	207.60	221.20	108.20	256.00	178.40	167.20	206.10

Offshore Johor Core

Sediment age (year)	1886-1895	1896-1905	1906-1915	1916-1925	1926-1935	1936-1945	1946-1955	1956-1965	1966-1975	1976-1985	1986-1995	1996-2005
					Concentration (ng/g d.w.)							
[a]Total PAHs (ng/g d.w.)	304.64	321.69	92.06	68.58	411.00	363.70	521.15	215.28	286.63	221.57	126.65	99.60
[b]L/H PAHs	107.31	108.39	73.57	47.71	291.92	33.05	9.01	8.60	3.36	9.62	12.20	10.24
[c]MP/P	1.38	0.96	1.87	1.55	0.90	1.00	1.00	2.43	3.08	0.90	1.41	1.70
[d]TOC mg/g	77.90	82.10	153.50	53.40	50.30	55.40	44.80	51.70	45.40	52.70	52.20	51.20

Tebrau Core I

Sediment age (year)	1728-1847	1748-1767	1768-1787	1788-1807	1808-1827	1828-1847	1848-1867	1868-1887	1888-1907	1908-1927	1928-1947	1948-1967	1968-1987	1988-2007
							Concentration (ng/g d.w.)							
[a]Total PAHs (ng/g d.w.)	6.86	11.83	25.42	24.87	23.99	3.51	11.50	21.82	18.02	20.41	31.02	67.23	48.13	310.92
[b]L/H PAHs	NA	NA	NA	NA	NA	NA	NA	NA	NA	NA	8.26	0.67	1.78	0.40
[c]MP/P	0.29	2.00	0.95	0.95	0.99	0.00	0.72	0.90	0.76	0.87	0.93	0.85	0.74	0.97
[d]TOC mg/g	125.30	126.70	156.70	162.80	109.80	160.50	183.00	103.00	213.80	140.00	177.90	146.70	126.80	126.50

Tebrau Core II

Sediment age (year)	1862-1874	1875-1884	1886-1897	1898-1909	1910-1921	1922-1933	1934-1945	1946-1957	1958-1969	1970-1981	1982-1994	1995-2006
					Concentration (ng/g d.w.)							
[a]Total PAHs (ng/g d.w.)	10.50	9.65	7.68	12.36	9.53	4.63	13.56	14.59	8.03	10.25	12.64	38.72
[b]L/H PAHs	6.16	23.23	NA	8.05	NA	NA	17.88	2.22	NA	34.33	4.05	8.10
[c]MP/P	0.86	0.92	1.00	0.80	0.51	0.93	0.89	0.59	0.85	0.86	0.92	0.13
[d]TOC mg/g	95.80	96.10	104.80	95.20	95.30	92.20	112.60	110.70	102.40	98.80	103.60	75.40

aTotal PAHs: sum of 18 PAHs ranging from Dibenzothiophene to Dibenzo (a,h) antharacene; bL/H PAHs: ratio of LMW over HMW PAHs; cMP/P: ratio of sum of 3-Methylphenanthrene, 2-Methylphenanthrene, 9-Methylphenanthrene and 1-Methylphenanthrene to Phenanthrene; dTOC: Total Organic Carbon.

Table 2. The Concentrations of Polycyclic Aromatic Hydrocarbon (PAHs) and TOC and ratios of L/H and MP/P in cores collected from the study area.

The second core was collected from Johor strait. This core showed generally lower concentration than those observed in Johor City core. The highest concentration of PAHs (521 ng/g d. w.) observed soon after the WWII and during the independency. The main source of PAHs in this core showed petrogenic signature using MP/P and hopane ratio.

Tebrau Strait is the main gateway connecting Singapore and Malaysia to the waters of South China Sea. The cores from Tebrau Strait were collected from eastern part of Johor-Singapore waterway. The first core revealed the highest concentration during the modern era (1988-2005; 311 ng/g d. w.) and the lowest concentration during the ancient time (1827-1847; 3.51 ng/g d. w.). Since the study was not revealed significant HMW PAHs, the ratio of L/H was not mathematically available in this core however MP/P ratio showed pyrogenic input to the marine environment of the study area (Sakari, 2009).

The second core in Tebrau Strait likewise showed same increasing trend where the highest concentration observed in the recent deposited sediments. In general, the concentration in this core is lower than the first Tebrau core. The sources of PAHs again indicate that there is pyrogenic input received in this location where MP/P was shown values below the unity. The hopane ratio showed that mostly Southeast Asia Crude Oil is the main contributor of PAHs in these cores. This statement is confirmed by ratios such as C_{29}/C_{30} from the hopane compounds.

In conclusion, the concentration of PAHs and hopane in all cores showed that the increase in populations, number of cars, socio-economic indicators such as GDP and GNP, industries, urbanizations, oil production and transportation accelerate the pollution trend. The overall view of PAHs concentration showed that near shore locations demonstrate higher PAHs contribution than offshore stations.

The total concentration of PAHs in this study ranged from 1.7 to 4447 (ng/g d. w.) with a mean value of 381 (n=105). The results of all source identification tools have been shown that a range of highly pyrogenic to extremely petrogenic PAHs are existed in the study area where a zero value of other PAHs is observed in conjunction with a minimum Methyl Phenantherene concentration that possibly indicates negligible nature derived compounds. Total organic carbon (TOC) in this study were fluctuated from 44 to 256 (mg/g) with an average of 117 mg/g (n=105) that statistically showed to be in a very low to negative correlations with total PAHs. The source identification parameters that has applied in this research were ratios such as Ph/(Ph+An), Ph/An, Fluo/Pyr, Fluo/(Fluo+Pyr), BeP/(BeP+BaP), Phe/(Phe+An) and BaA/(BaA+Chry). The application of these ratios revealed vicinity of sources such as adjacent cities, vehicles and industries to the study areas. This study has concluded that these sources emit gasses and particle based materials that transfers via lateral movements by daily rain wash and flushing into the marine environment thru canals, rivers, and drainage and finally settle down to the estuaries and straits. It is also emphasized that shipping and oil transportation play an important role in releasing PAHs into the study areas where the daily heavy ocean going vessels transport goods and oil.

7. Sources and origins of PAHs in deposited environment

Several studies around the world were conducted to understand and determine the sources of hydrocarbon pollution in sedimentary cores. One of the most pioneering studies is conducted by Hites et al., (1977), where three stages of hydrocarbon deposition were reported from 1850 till 1970 from the Buzzards Bay, Massachusetts. This report indicates the sources that were almost constant from 1850 till 1900. The constant source of PAH pollution has been determined as combustion processes, regardless to its origin from natural or anthropogenic sides.

In the UK, Readman et al., (1987) reported PAHs from Tamar Estuary, showing predominant of parent compound rather than alkyl homologues, a clear indication of pyrogenic input correlate with increased motor vehicle activity and road runoff. This is remarkable that compositional uniformity of PAHs throughout the polluted sedimentary core characterize biogeochemical transformation and exchange processes (sorption/leaching; microbial breakdown; photodegradation; etc). Thus it has been concluded that the majority of un-substitute PAHs comes from combusted fossil fuel and/or street dust. Rapid reduction in PAHs concentrations since 1940s may come due to fuel consumption changes from coal to petroleum (Gschwend and Hites, 1981; Bates et al., 1984; Barrick and Prahl, 1987).

Industries are one of the most important contributors of PAHs input into the environment. Appearances of pollutions such as PAHs depends on the time and location of production and deposition. Martel et al., (1987) reported considerable increase of PAHs concentration since 1930 from Saguenay Fjord, Quebec in Canada where two aluminum reduction plants increased the PAHs concentrations in the studied area. The above statement was approved after a couple of years by Cranwell and Koul, (1989) where anthropogenic PAH input that peaked 1900-1920 in Windermere North Basin is tentatively attributed to local industrial input. The decline in post-1975 flux values may result from replacement of coal as the source of energy by oil or gas however flux values remain ten times higher than in the pre-industrial age.

The sources of pollution are always not unique or with a same pattern. It can be a contribution of different sources such as natural and anthropogenic. Christensen and Zhang, (1993) identified a combination of sources including coal, petroleum and wood from four sediment cores collected in Lake Michigan for Source identification. In this study, the sedimentary record of PAHs high flux is reported with petroleum origins (oil and gas during 1985) but the high PAHs flux for 1950s was clarified when coal was used. As the background data, the concentration of PAHs was zero during 1900 for petroleum derived PAHs.

In another study Su et al., (1998) analyzed 6 cores from Green Bay, Wisconsin in order to identify the PAHs concentration and sources. This study showed the same trend in source combinations for PAHs in the studied area. The total concentrations were reported from 0.46 to 8.04 ppm with combination of combustion sources from coal, wood and petroleum hydrocarbon.

Based on the regulations and the availability of different sources, in the energy markets, some of those are decreased in consumption or fully stopped. For instance, Taylor and Lester, (1995) showed that since 1966 that coal combustion had been banned; the coal derived PAHs has decreased and shifted to the oil and gas sources. Although the usage of coal is limited in many countries all around the world, there are still footprints of its application in many countries. For example, Liu et al., 2005 reported 30% of coal combusted PAHs from air particles collected from Guangzhou, China atmospheric environment, due to wide use as energy source.

The historical profile of PAHs from the sedimentary cores collected from Lake Michigan, USA, showed the Wisconsin coal profile exhibit similar trends with peaks for 1946-1951 and 1968-1973, indicative of coal combustion source material (>36%) and petroleum sources (>76%) in various samples using Factor Analysis (FA) model (Rachdawong et al., 1998).

Although the atmospheric environment distributes the PAHs in a homogenized concentrations, lateral transportation such as movements via rivers demonstrate irregular and high amounts of PAHs concentration in the environment. Witt and Trost, (1999) indicate significant contribution of river discharge of the petrogenic hydrocarbon to the sediment with predominant of higher molecular weight PAHs due to its stability in German coastal waters. The highest concentration of the PAHs occurred in recent sediments presented from 1 to 8 cm of surface, indicative of modern input. Petrogenic PAHs pollution which are mainly enter into the marine environment via lateral transport contribute to the pollution history of the world since past centuries. This petrogenic PAHs are abundant in riverine systems due to wash out phenomenon from the city run off. Liu et al., (2000) reported the sources of PAHs from core collected in Yangtze Estuary, China; that mainly was petrogenic origin. However, PAHs concentration in sedimentary records may be affected due to physico-chemical conditions during sedimentation, the nature of inputs, biodegradation, and bioturbation (surface sediments).

As petrogenic PAHs affect the marine environment in short distances, the pyrogenic PAHs are subject to long range transportation via atmospheric movements (Prahl and Carpenter, 1983). This model of transport is able to influence remote and pristine areas. Rose and Rippey (2002) were reported low concentration in recent PAHs deposition via atmospheric movement into a remote lake in the north-west England. Specific compounds analysis for the definition of ratios tested for this study (phenantherene:anthracene and fluorenthene:pyrene) do not identify and clarify any specific reason while shows less urban discharges to the lake comparing to the era of pre-1830.

Natural disasters affect the concentration of the PAHs deposited in the marine environment. Flood as a natural disaster contributes in PAHs irregular concentrations where it washes out city surface to water bodies such as rivers and streams. Ikenaka et al., (2005) reported the highest PAHs of core layers with multiple pyrogenic sources from Lake Suwa, Japan when the heavy rain and consequently flood had been historically consistent. Discharges from natural disasters and local input are characterized by irregular distributions of PAHs. Since PAHs enter locally in mass amounts distribute according to the physical and chemical properties of the destination points. In a study (Moriwaki et al, 2005) on historical trend of PAHs in reservoir sediment core of Osaka, Japan; however the sources of PAHs in the sedimentary record is found a combination of grass, wood and coal for pre-industrial era, for early 20th century and petroleum and its combusted derived materials in recent and mostly modern input.

As discussed earlier in previous sections, there are possibilities for natural inputs of PAHs entering into the enviroment. PAHs naturally derive from higher plant detritus and degradations products (Simoneit and Mazurek, 1982; Yunker and McDonald, 1995). Four and 5 benzene rings PAHs can be produced from microbial breakdown of plant wax and woody tissue. Prahl and Carpenter (1983) were reported natural sources of PAHs in Washington coastal sediments with a constant input of clay samples represents the era of Pleistocene. Quiroz et al., (2005) reported 50 years of PAHs depositions into the Laja Lake from south central Chile, were showed relatively low PAHs concentration (226 to 620 ng/g d. w.) with mostly natural origins. One of the most important and mostly natural PAHs is Perylene. Perylene is 5 benzene rings PAHs which there are doubts in its origin. High temperature combustion of the internal engines produces perylene however other sources originate. This is strongly believed among scientist that perylene can produce naturally in the deep sediments via diagenesis. Interestingly, there are several reports that indicate the high perylene concentration from surface sediments (Zakaria *et al.*, 2002; Tolosa *et al.*, 2004) and throughout the core (Barra *et al.*, 2006).

PAHs studies in core samples are subject to scientific debates. Core samples have different properties than surface sediments samples thus the fate of contaminants are various from surface. In surficial sediment, there are chemical, physical and biological properties which affect the targeted compounds in analytical analysis and interpretation of their data. In sediments collected from a core, there is no expectation of aerobic condition whereas the anoxic characteristics are notable for any possible chemical and biological changes. The profile of individual PAH concentrations with depth in sediment often reflect changes in source input over time rather than significant *in situ* biological degradation (Hites et el, 1977; Prahl and Carpenter, 1979). Although some organisms were capable to biosynthesize naturally the hydrocarbons (Graef and Diehl, 1966; Hancock *et al.*, 1970), other studies reveal the bioaccumulation effects of organisms in the core sediments (Hase and Hites, 1976). Moreover, some PAHs are generated by post-deposition transformations of biogenic precursors over a short period of time. This subject was confirmed for Perylene in research conducted by scientists (Aizenshtat, 1973; LaFlamme and Hites, 1978).

While aliphatic hydrocarbons are subjected to dissolution and microbial degradation, PAHs remain less or unchanged (Yunker *et al.*, 1999). Although Wakeham *et al.*, (1979) believe that lower molecular weight PAHs contribute in lower concentration in the total PAHs comparing high molecular weight in the core sediments. This interprets as a faster degradation of the PAHs that enter into the marine sediments. PAHs are always associated with organic carbon in sedimentary environment and are integrated with those values, but sometimes greater values of PAHs are not associated with TOC, indicative of soot particle existence in the sediments. For example Richardson Bay studies in the United States showed high values of PAHs with low values of TOC. This is an indicative of soot particle associated in the sediments come from the atmospheric transportations of combusted fossil fuel and organic contents that are less available in partitioned PAHs in the organic carbons (Pereira *et al.*, 1999). PAHs associated with soot particles are less biologically available for uptake than the PAHs derived from the petroleum and oil spill (Farrington *et al.*, 1983; Gustafson *et al.*, 1997). PAHs studies through the core samples are usually consistent with gradual and little changes either in concentration or in ratios in a normal condition but environmental disasters such as oil spills show significant changes. Therefore, core sections always do not show sudden changes in PAHs ratio values.

8. Application of biomarkers in petroleum pollution studies

The forensic chemistry techniques fingerprint pollutants in environmental samples. This technique is based on existence and abundance of the biomarkers. Terpanes are a group of biomarkers that are ubiquitous in the environment together with hydrocarbons. Pentacyclic triterpanes (hopanes) are the group of biomarkers that existed in crude oil and some petroleum products. As discussed earlier, hopane as a fingerprinting biomarker delivers from precursor of bacterial membrane (microbial origins) of bacteriohopentetrol (Fig. 2). The production of hopane in the nature is due to cyclization of bacteriohopanetetrol during the diagenesis. Basically, the stereochemistry of hopane makes them thermodynamically unstable (Peters and Moldowan, 1993). Hopanes in their biological origins present 17β(H), 21β(H) compounds that is known as biological stereochemistry.

Depositional History of Polycyclic Aromatic Hydrocarbons: Reconstruction of Petroleum Pollution Record in Peninsular Malaysia

153

Instability against temperature increase is due to their polar and non-polar ends. Upon change, they might convert from $\beta\beta$ to more stable configuration of $\alpha\beta$ and $\beta\alpha$ structures. The $\alpha\beta$ that is called hopane is predominantly available in crude oil and some petroleum products. Hopanes are relatively involatile, resist biodegradation, geologically mature and relatively stable in the environment (Simoneit et al., 1988) however there are chemical characteristics and properties among hopanes that compounds with higher number of carbon (*e.g.* C_{35}) shows bigger resistibility against biodegradation than lower numbered such as C_{31} (Frontera et al, 2002) (Fig. 5).

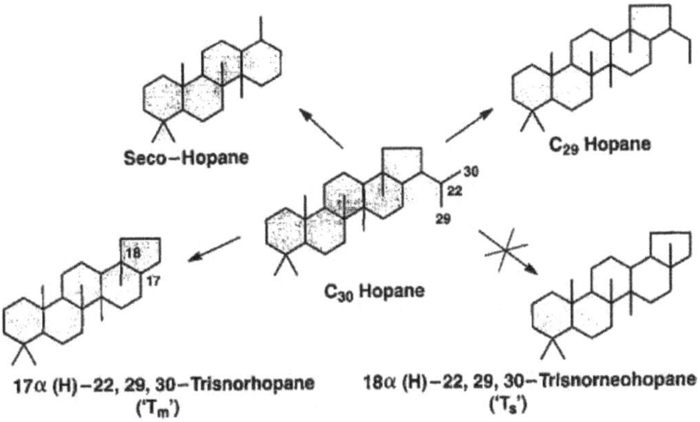

Fig. 5. Hopane chemical structures.

Homohopanes are the name of hopane series in which the number of carbon arises by thirty (n=30). The relatively more abundant homohopanes (C_{31}-C_{35}) are showed in less oil contaminated sites with significant loss of C_{30} (Colombo et al, 2005). In environmental samples from the biomass burning, hopane appears together with moretanes with abundance of C_{27} and C_{31} (Standley and Semoneit, 1987) that overlaps in common peaks (Omar et al., 2006).

Hopane is found in mineral oil and coal based fuel and lubricants (Kapalan et al, 2001). An unusually high proportion of the C_{29} hopane is often associated with oil derived from carbonate source rock oil which includes most of those from the Middle East however existence of oleanane is indicative of Southeast Asian oil.

9. New dimension of biomarkers; A reliability test over hopane

PAHs and hopane compounds in the study area were identified by comparing chromatograms of samples and standard solutions in the GC-MS. Chromatograms were compared in their retention time, surface area and mass spectra. Some of samples from the core intervals in various locations presented an absence or irregularity of appearances of peaks (representing compounds) in the chromatograms. Since some of Hopane compounds were missing, the identification of sources was difficult. We noticed that hopanes with carbon numbers from 27 to 30 (single peaks) are the most missing or depleting compounds among others. However hopanes with carbon numbers from 31 to 35 that appears in twin peaks (S and R stereochemistry) are dominant showing an unchanged structure. In this report, samples mostly from offshore locations such as Malacca and Klang showed this phenomenon. These samples are observed to be identical in their source and origin of production. They are highly pyrogenic appearing in extremely depleted MP/P and L/H ratios. Meantime, it is observed that hopane chromatograms appearing shorten or fade up in single peaks (C_{27}-C_{30}). Twin peaks however showed more resistance than single peaks, the depletion was also observed among them.

The scenarios are different among samples with petrogenic sources such as crankcase oil, spilled oil and lubricating oil. They show taller and sharper single peaks together with complete twin peaks representing C_{31} to C_{35}. We believe that the stereochemistry of twin peaks provides resistibility against temperature increase rather than single peaks. Thus, it is criticized that high temperature especially in combustion process of petroleum in internal engines may cause destruction on compound structure appear as demolished or depleted peaks in chromatograms (Peters et al 1992; Colombo et al, 2005). Hence, application of some compound and isomer pair ratios of hopane are failed to assist source identification of PAHs.

The correlation was applied for statistical comparison between MP/P and other ratios such as C_{31}-C_{35}/C_{30}, C_{29}/C_{30}, Tm/Ts and C_{31}-C_{35}/C_{29}. Increasing of hopane indices like C_{31}-C_{35}/C_{30} and C_{31}-C_{35}/C_{29} with depletion of MP/P (combusted) may suggest demolishing of chemical structures in C_{30}, C_{29} and MP compounds during the high temperature combustion (Peters et al, 1992).

Therefore in an environmental sample such as sediment, theoretically a decrease in MP/P values renders high temperature in combustion process. Here, the theory criticizes the possible changes on chemical structure of single peak hopane compounds with carbon numbers ranging from C_{27} to C_{30}.

A positive correlation were observed between Offshore locations such as Malacca and Klang together with near shore station such as Klang City demonstrate combustion of petroleum where several and average MP/P appear to be lower than unity. Here there are negative correlation between the MP/P and hopane indices of C_{31}-C_{35}/C_{30} and C_{31}-C_{35}/C_{29}. These correlation values indicate that combustion results lower values of either C_{29} or C_{30}. (Takada et al., 1990; Prahl and Carpenter, 1983; Pruel and Guinn, 1987; Garrigues et al., 1995).

10. Natural vs. anthropogenic PAHs

PAHs are known as anthropogenic and/or natural compounds, based on their sources of production (Simoneit and Mazurek, 1982; National Academy of Science, 2002). Natural process is called a procedure that bacterial and algae are involved. This process results in-situ production of PAHs that produce limited concentrations (Hites et al., 1977; Prahl and Carpenter, 1979). Anthropogenic processes usually produce greater concentration of PAH in the environment. They include combustion of organic matter such as plant and/or oil and direct release of oil and its derivatives into the environment (Neff, 1979). Thus natural in-situ PAH generation in sedimentary environment is negligible in the total concentration of anthropogenic amount. There are limited locations around the world that produce natural based PAHs. These reports are mostly from Brazilian tropical forest where scientists report appearance of Phenanthrene, Naphthalene and Perylene in remote and virgin locations (Wilcke et al, 2003, 2004).

Depositional History of Polycyclic Aromatic Hydrocarbons: Reconstruction of Petroleum Pollution Record in Peninsular Malaysia

155

Likewise reports from tropical forest of Brazil, we expect to have a natural contribution of some specific polycyclic aromatic hydrocarbons such as Phenanthrene and Perylene in Peninsular Malaysia. As it has been discussed earlier, it is too hard to differentiate specific compounds as background level from nature from those as anthropogenic input using ordinary instrumentations. The application of hopane assisted us to technically differentiate those natural from anthropogenic individual PAH. Dated sedimentary intervals along the core have shown deposited PAH in which represents an era before the oil exploration and usage contain limited but detectable concentrations of Phenanthrene. The same signature has been frequently found in dated ancient sediment from Malacca and Tebrau in which intervals represent an era of 17th century. This research expects a constant input of natural hydrocarbons into the study area.

11. High fluxes; Climate contribution to distribution of PAHs

Malaysia is located near the equator where the weather is characterized as hot and humid with constant daylight time of around 12 hours and heavy rainfall. This cause abundance of plants and thus increase in available organic matter in the environment. Daily heavy rainfall basically washes away organic material such as total organic carbon that is associated with PAHs from the city and land surface into the water. Since the media for PAHs transport is always available in the environment, it is expected to record the highest existed PAHs in the sedimentary environment.

Malaysia has been experiencing a rapid development in modernization, transport, urbanization and industries starting 1950's. Hence, it is hypothesized that due to the massive land development of the post-independence, the marine environment of Malaysia such as estuaries and coastal water should receive a considerable amount of TOC via rainfall and drainage runoff. Organic compounds including PAHs adhere to organic contents and are therefore able to travel over distances. In almost all stations of the study area, the concentrations of TOC were found to be very high as compared to other areas in the world.

There are basically four phases in cores from near shore stations namely pre-war (Pre-WWII), war-independency era, rapid development and finally modern input. The first phase belongs to era represents sediments with natural PAHs input or minimum anthropogenic input from pyrolysis of organic matters. The second phase intervals represent deposition during WWII and pre-independence. The third phase represents post-independence and rapid development (1956-1990) that shows the highest PAHs with oil signature. The last phase represents mostly sudden drop in PAHs in concentration however the sources are remained same as phase two and three (Fig. 6). The samples from the offshore unlikely have shown different results from the near shore cores. The Offshore cores are more erratic and expected to be derived presumably by the input from atmospheric movement. The results from the Offshore core show that MP/P ratios of 1 (in average) suggesting that the source of the PAHs were pyrogenic originated from street and urban dust and transported with atmospheric movements (Takada et al., 1990; 1991).

The near shore stations, the MP/P values indicated highly matched identified source comparing offshore cores. Since offshore PAHs are mainly derived by atmosphere, near shore locations receive via street run off, canals and drainage systems due to climate condition and rainfall.

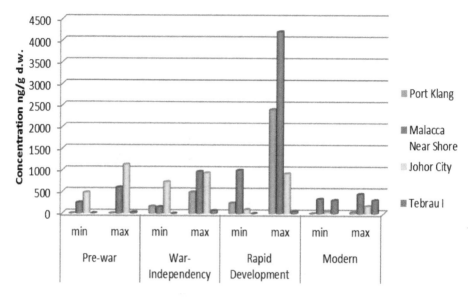

Historical Time Intervals

Fig. 6. The concentration of Polycyclic Aromatic Hydrocarbon (ng/g d.w.) during four identical time period in Malaysian history in selected study area.

12. References

Aizenshtat, Z., 1973. Perylene and its geochemical significance. *Geochimica et Cosmochimica Acta* 37: 559-567.

Baker, J. E., Capel P.D. and Eisenreich S.J., 1986. Influence of colloids on sediment-water partition coefficients of polychlorobiphenyl congeners in natural waters. *Environmental Science and Technology*, 20: 1136-1143.

Barra, R., P. Popp, R. Quiroz, H. Treutler, A. Araneda, C. Bauer, R. Urrutia, 2006. Polycyclic aromatic hydrocarbons fluxes during the past 50 years observed in dated sediment cores from Andean mountain lakes in central south Chile. *Ecotoxicology and Environmental Safety.* 63: 52-60.

Barrick, R. C. and F. G. Prahl, 1987. Hydrocarbon Geochemistry of the Puget Sound Region - III. Polycyclic Aromatic Hydrocarbons in Sediments. *Estuarine, Coastal and Shelf Science* 25: 175-191.

Bates, T. S., Hamilton S. E. and Cline J.D., 1984. Vertical transport and sedimentation of hydrocarbons in the central main basin of Puget Sound, Washington. *Environmental Science and Technology*, 18: 299-305.

Bauer, J.E. and D.G. Capone, 1985. Degradation and mineralization of the polycyclic aromatic hydrocarbons anthracene and naphthalene in intertidal marine sediments. *Applied Environmental Microbiology*, 50: 81-90.

Baumard, P., H. Budzinski, P. Garrigues, T. Burgeot, X. Michel, J. Bellocq (1999). Polycyclic aromatic hydrocarbons (PAH) burden of mussels (*Mytilus* sp.) in different marine environments in relation with sediment PAH contamination, and bioavailability. *Marine Environmental Research* 47: 415-439.

Baumard, P., Budzinski, H., Garrigues, P., Sorbe, J. C., Burgeot, T. and Bellocq, J., 1998. Concentrations of PAHs (Polycyclic Aromatic Hydrocarbons) in various marine organisms in relation to those in sediments and to trophic level. *Marine Pollution Bulletin*, 36: 951-960

Blumer M. 1976. Polycyclic aromatic compounds in nature. *Science*, 234: 34-45.

Boonyatumanond, R., G. Wattayakorn, A. Amano, Y. Inouchi, H. Takada, 2007. Reconstruction of pollution history of organic contaminants in the upper Gulf of Thailand by using sediment cores: First report from Tropical Asia Core (TACO) project. *Marine pollution bulletin* 54: 554-565.

Boonyatumanond R., Wattayakom, G., Togo A. and Takada, H., 2006. Distribution and origins of Polycyclic Aromatic Hydrocarbons in estuarine, rivers and marine sediments in Thailand. *Marine Pollution Bulletin*, 52: 942-956

Bouloubassi, I. and A. Saliot (1993). Dissolved, Particulate and Sedimentary Naturally Derived Polycyclic Aromatic-hydrocarbons in a Coastal Environment - Geochemical Significance. *Marine Chemistry* 42: 127-143.

Budzinski, H., I. Jones, J. Bellocq, C. Pierrard, P. Garrigues, 1997. Evaluation of sediment contamination by polycyclic aromatic hydrocarbons in the sediment of the Georges River estuary. *Marine chemistry*, 58: 85-97.

Chandru, K., M. P. Zakaria, S. Anita, A. Shahbazi, M. Sakari, P.S. Bahry, C. A. R. Mohamed (2008). Characterization of alkanes, hopanes, and polycyclic aromatic hydrocarbons (PAHs) in tar-balls collected from the east coast of peninsular Malaysia. *Marine Pollution Bulletin*, 56: 950-962.

Christensen, E. R., X. Zhang, 1993. Sources of polycyclic aromatic hydrocarbons to Lake Michigan determined from sedimentary records. *Environmental Science and Technology*, 27: 139-146.

Colombo, J.C., Cappelletti, N., Lasci, J., Migoya, M.C., Speranza, E. and Skorupka, C.N. (2005). Sources, Vertical Fluxes and Accumulation of Aliphatic Hydrocarbons in Coastal Sediments of the Rio de la Plata Estuary, Argentina. *Environmental Science and Technology*, 39, 8227-8234.

Cranwell P. A. and V. K. Koul, 1989. Sedimentary Record of Polycyclic Aromatic and aliphatic Hydrocarbons in the Windermere Catchment. *Water Research*, 23: 275-283

Curtosi, A. E. Pelletier, C. L. Vodopivez, W. P. MacCormack, 2007. Polycyclic aromatic hydrocarbons in soil and surface marine sediment near Jubany Station Antarctica. Role of permafrost as a low-permeability barrier. *Science of the Total Environment* 383: 193-204

Farrington, J. W., E. D. Goldberg, R. W. Risebrough, J. H. Martin, V. T. Bowen 1983. "Mussel Watch" 1976-1978: An overview of the trace-metal, DDE, PCB, Hydrocarbons, and artificial radionuclide data. *Environmental Science and Technology*. 17: 490-496.

Fernandez, P., Rose, N.L., Vilanova, R.M., Grimalt, J.O., 2002. Spatial and temporal comparison of polycyclic aromatic hydrocarbons and spheroidal carbonaceous particles in remote European lakes. *Water, Air and Soil Pollution: Focus* 2: 261-274.

Frontera-Suana, R., Bost, F. D., McDonald, T. J. and Morris, P. J., (2002). Aerobic Biodegradation of Hopanes and other Biomarkers by Crude Oil-Degrading Enrichment Culture. *Environmental Science and Technology*, 36, 4585-4592.

Gardner, W.S., R.F. Lee, K.R. Tenore and L.W. Smith, 1979. Degradation of selected polycyclic aromatic hydrocarbons in coastal sediments: importance of microbes and polychaete worms. *Water, Air, Soil Pollution*, 11: 339-347.

Garrigues, P., Budzinski, H., Manitz, M. P. and Wise, S. A. 1995. Pyrolitic and petrogenic input in recent sediments: A definitive signature through Phenantherene and Chrysene compounds distribution. *Journal of Polycyclic Aromatic Compounds* 7: 175-184.

Gevao, B., J. Hamilton-Taylor, C. Murdoch, K. C. Jones, M. Kelly, B. J. Tabner, 1997. Depositional time trends and remobilization of PCBs in lake sediments. *Environmental Science and Technology*, 31: 3274-3280.

Graef, W. and H. Diehl, 1966. The natural Normal Levels of Carcinogenic PCAH and the Reasons therefore. *Arch. Hyg. Bakteriol.* 150: 49-59

Gschwend, P. M. and R. A. Hites, R. A., 1981. Fluxes of polycyclic aromatic hydrocarbons to marine lacustrine sediments in the northeastern United States. *Geochimica et Cosmochimica Acta*, 45: 2359-2367.

Gustafsson, O., F. Haghseta, C. Chan, J. Macfarlane, P. Gschwend (1997). Quantification of The Dilute Sedimentary Soot Phase: Implications for PAH Speciation and Bioavailability. *Environmental Science and Technology* 31(1): 203-209.

Hancock J. L., H. G. Applegate and J. D. Dodd, 1970. Polynuclear Aromatic Hydrocarbons on Leaves. *Atmospheric Environment*. 4: 363-370.

Hartmann P. C., J. G. Quinn, R. W. Carins, J. W. King, 2005. Depositional History of Organic Contaminants in Narragansett Bay, Rhode Island, USA. *Marine Pollution Bulletin* 50 (4): 388-395.

Hase A. and Hites R.A. 1976. Identification and Analysis of Organic Pollutants in Water. *Geochimica et Cosmochimia Acta* 40: 1141.

Helfrich, J. and D. E. Armstrong, 1986. Polycyclic aromatic hydrocarbons in sediment of Lake Michigan. *Journal of Great Lake Research*, 12: 192-199.

Hites, R. A., R. E. Laflamme, J. W. Farrington, 1977. Sedimentary polycyclic aromatic hydrocarbons: the historical record. *Science*, 198: 829-831.

Hostettler, F. D., W. E. Pereira, K. A. Kvenvolden, A. Geen, S. N. Luoma, C. C. Fuller, R. Anima, 1999. A record of hydrocarbon input to San Francisco Bay as traced by biomarker profiles in surface sediment and sediment cores. *Marine Chemistry* 64:115–127

Ikenaka, Y., H. Eun, E. Watanabe, F. Kumon, Y. Miyabara, 2005. Estimation of sources and inflow of dioxin and polycyclic aromatic hydrocarbons from the sediment core of Lake Suwa, Japan. *Environmental Pollution*, 138: 529-537.

Jones, D.M., S.J. Rowland, A.G. Douglas and S. Howells, 1986. An examination of the fate of Nigerian crude oil in surface sediments of the Humber Estuary by Gas Chromatography and Gas Chromatography-Mass Spectrometry. *International Journal of Environmental Analytical Chemistry*, 24: 227 247.

Kamens, R.M., Guo, Z., Fulcher, J.N., Bell, D.A., 1988. Influence of humidity, sunlight, and temperature on the daytime decay of polycyclic aromatic hydrocarbons on atmospheric soot particles. *Environmental Science and Technology*, 22: 103-108.

Kamens, R.M., Fulcher, J.N., Guo, Z., 1986. Effects of temperature on wood soot PAH decay in atmospheres with sunlight and low NOx. *Atmosphere Environment*, 20: 1579-1587.

Kapalan, I. R.; Lu, S.-T.; Alimi, H. M.; MacMurphey, J. (2001). Fingerprinting of high boiling hydrocarbon fuels, asphalts and lubricants. *Environ. Forensics*, 2, 231-248.

Knight C. V., Graham M. S. and Neal B. S., 1983. Polynuclear Aromatic Hydrocarbons: Formation, Metabolism and Measurement *In Polynuclear Aromatic Hydrocarbons and associated organic emissions for catalytic and noncatalytic wood heaters.* (Edited by Cooke M. and Dennis A. J.), 689-708. Battelle Press, Columbus, Ohio.

Laflamme, R.E. and R.A. Hites, 1978. The global distribution of polycyclic aromatic hydrocarbons in recent sediments. *Geochimica Cosmochima Acta*, 42: 289-303.

Lee, M. L., G. P. Prado, J. B. Howard, R. A. Hites, 1977. Source identification of urban airborne polycyclic aromatic hydrocarbons by gas chromatography mass spectrometry and high resolution mass spectrometry. *Biomedical Mass Spectrometry*, 4: 182-186.

Liu, G. Q., G. Zhang, X. D. Li, J. Li, X. Z. Peng, S. H. Qi, 2005. Sedimentary record of polycyclic aromatic hydrocarbons in a sediment core from the Pearl River estuary, South China. *Marine Pollution Bulettin*, 51: 912-921.

Malaysian Marine Department, 2003. Annual Report.

Marr, L.C., Kirchstetter, T.W., Harley, R.A., Miguel, A.H., Hering, S.V., Hammond, S.K., 1999. *Environmental Science and Technology* 33:3091-3099.

Martel, L., M. J. Gagnon, R. Masse and A. Leclerc, 1987. The spatio-temporal variations and fluxes of polycyclic aromatic hydrocarbons in the sediment of the Saguenay Fjord, Quebec, Canada. *Water Research*, 21: 699-707.

Moriwaki, H., K. Katahiraa, O. Yamamotoa, J. Fukuyamaa, T. Kamiuraa, H. Yamazakib, S. Yoshikawac, 2005. Historical trends of polycyclic aromatic hydrocarbons in the reservoir sediment core at Osaka. *Atmospheric Environment*, 39: 1019–1025

Morris W. A., J.K. Versteeg a, C.H. Marvin b, B.E. McCarry b, N.A., Rukavina, 1994. Preliminary comparisons between magnetic susceptibility and polycyclic aromatic hydrocarbon content in sediments from Hamilton Harbour, western Lake Ontario. *The Science of the Total Environment* 152: 153-160

National Academy of Science (2002). *Oil in the Sea; input, fates and effects.* National Academy Press, Washington D.C. Press.

Neff, J. M. 1979. *Polycyclic Aromatic Hydrocarbon in the Aquatic Environment: Sources, Fates and Biological Effects.* Applied Science Publishers, London.

Nielsen, T., Jorgensen, H.E., Larsen, J.C., Poulsen, M. 1996. City air pollution of polycyclic aromatic hydrocarbons and other mutagens: occurrence, sources and health effects. *Science of the Total Environment* 189(190): 41-49.

Nielsen, T., 1988. The decay of benzo(a)pyrene and cyclopenteno(cd)pyrene in the atmosphere. *Atmospheric Environment* 22: 2249–2254.

Okuda, T., Kumata, H., Zakaria, M.P., Naroaka, H., Ishiwatari, R., and Takada, H. 2002. Source identification of Malaysian atmospheric polycyclic aromatic hydrocarbons nearby forest fires using molecular and isotopic compositions. *Atmospheric Environment* 36: 611-618.

Ou, S. M., J.H. Zheng, J.S. Zheng, B.J. Richardson and P.K.S. Lam (2004). Petroleum hydrocarbons and polycyclic aromatic hydrocarbons in the surficial sediments of Xiamen Harbour and Yuan Dan Lake, China, *Chemosphere* 56: 107–112.

Pereira, W. E., Hostettler, F. D., Luoma, S. N. Van Geen, A., Fuller, C. C., Anima, R. G., 1999. Sedimentary record of anthropogenic and biogenic Polycyclic Aromatic Hydrocarbons (PAHs) in San Francisco Bay, California. *Marine Chemistry* 64: 99-113.

Peters, K. E. and Moldowan, J. M. 1993. *The Biomarker Guide: Interpreting Molecular Fossils in Petroleum and Ancient Sediments*. Prentice Hall, Englewood Cliff, N.J. 363 p.

Peters, K.E., Scheuerman, G.L., Lee, C.Y., Moldowan, J.M., Reynolds, R.N. and Pena, M.M. (1992). Effects of refinery processes on biological markers. *Energy and Fuels*, 6, 560-577.

Prahl, F.G. and Carpenter, R., 1984. Hydrocarbons in Washington coastal sediments. *Estuarine, Coastal Shelf Science* 18: 703-720.

Prahl, F. G. and Carpenter, R. 1983. Polycyclic aromatic hydrocarbon (PAH)-phase associations in Washington coastal sediment. *Geochimica et Cosmochimica Acta*, 47(6): 1013-1023

Prahl F.G. and Carpenter R. 1979. The role of zooplankton fecal pellets in the sedimentation of PAHs in Dabob Bay, Washington. *Geochim Cosmochemica Acta,* 43: 1959-1968.

Pruel, R.J. and J. G. Quinn, 1988. Accumulation of polycyclic aromatic hydrocarbons in crankcase oil. *Environmental pollution*, 49: 89–97.

Omar, N.Y.M.J., T. C. Mon, N. A. Rahman, M. R. Abas. 2006. Distribution and health risk of polycyclic aromatic hydrocarbons (PAHs) in atmospheric aerosols of Kuala Lumpur, Malaysia. *Science of the Total Environment* 369: 76-81

Quiroz R., P. Popp, R. Urrutia, C. Bauer, A. Araneda, H. C. Treutler, R. Barra, 2005. PAH fluxes in the Laja Lake of south central Chile Andes over the last 50 years: Evidence from a dated sediment core. *Science of the Total Environment*, 349: 150–160

Rachdawong, P., E. R. Christensen and J. F. Carls, 1998. Historical PAH Fluxes to Lake Michigan Sediments Determined by Factor Analysis. *Water Research*, 32 (8): 2422-2430.

Readman, J. W., R.F.C. Mantoura and M. M. Rhead, 1987. A record of the polycyclic aromatic hydrocarbons (PAHs) pollution obtained from accreting sediments of the Tamar Estuary, UK; Evidences for non-equilibrium behavior of PAH. *The Science of the Total Environment,* 66: 73-94

Readman, J.W., R.F.C. Mantoura, M.M. Rhead and L. Brown, 1982. Aquatic distribution and heterotrophic degradation of polycyclic aromatic hydrocarbons in the Tamar Estuary. *Estuarine Coastal Shelf Science*, 14: 36-38.

Rose, N. L. and B. Rippey, 2002. The historical record of PAH, PCB, trace metal and fly-ash particle deposition at the remote lake in north-west Scotland. *Environmental Pollution*, 117: 121-132.

Sakari, M., Zakaria, M. P., Lajis, N., Mohamed, C. A. R., Chandru, K., Shahpoury. 2011. Polycyclic Aromatic Hydrocarbons and Hopane in Malacca Coastal Water: 130 Years of Evidence for their Land-based Sources. *Environ. Forensic.* 12: 63-78.

Sakari, M., Zakaria, M. P., Lajis, N., Mohamed, C. A. R., Chandru, K., Shahpoury, P., Shahbazi, A. and Anita, S. 2010a. Historical profiles of Polycyclic Aromatic Hydrocarbons (PAHs), sources and origins in dated sediment cores from Port Klang, Straits of Malacca, Malaysia. *Coast. Mar. Sci.* 34: (1): 140-155.

Sakari, M., Zakaria, M. P., Lajis, N., Mohamed, C. A. R., Chandru, K., Shahpoury, P., Mokhtar, M. and Shahbazi, A. 2010b. Urban vs. Marine Based Oil Pollution in the Strait of Johor, Malaysia: A Century Record. *Soil & Sed. Cont.* 19: 644-666.

Sakari, M 2009. Characterization, Concentration and Depositional History of Polycyclic Aromatic Hydrocarbons and Hopane in Selected Locations of Peninsular Malaysia. PhD dissertation. Universiti Putra Malaysia, Sri Serdang, Selangor, Malaysia.

Simoneit, B. R. T., Cox, R. E. and Standley, L. J. (1988). Organic Matter of the Troposphere-IV. Lipids in Harmattan aerosols of Nigeria. *Atmospheric Environment*, 22, 983-1004.

Simoneit, B.R.T. and Mazurek, M.A., 1982. Organic matter of the troposphere - II. Natural background of biogenic lipid matter in aerosols over the rural western United States. *Atmospheric Environment*.16: 2139-2159.

Standley, L. J. and Semoneit, B. R. T. (1987). Characterization of extractable plant wax, resin, and thermally matured components in smoke particles from prescribed burns. *Environmental Science and Technology* 21, 163-169.

Steinhauer, S. S. and P. D. Boehm, 1992. The composition and distribution of saturated and aromatic hydrocarbons in near shore sediments, river sediments and coastal peat of the Alaskan Beaufort Sea: Implications for detecting anthropogenic hydrocarbon inputs. *Marine Environmental Research* 33: 223-253.

Su, M., E. R. Christensen, J. F. Karls, 1998. Determination of PAHs sources in dated sediments from Green Bay, Wisconsin, by a chemical mass balance model. *Environmental Pollution*, 98: 411-419.

Takada, H., Onda, T., Ogura, N. 1990. Determination of polycyclic aromatic hydrocarbons in urban street dusts and their source materials by capillary gas chromatography. *Environmental Science and Technology* 24: 1179-1186.

Takada, H., Onda, T., Harada, M., Ogura, N. 1991. Distribution and sources of polycyclic aromatic hydrocarbons (PAHs) in street dust from Tokyo metropolitan area. *Science of the Total Environment* 107: 45-69.

Tan, Y. L. & Heit, M. 1981 Biogenic and abiogenic polynuclear aromatic hydrocarbons in sediments from two remote Adirondack Lakes. *Geochimica et Cosmochimica Acta* 45: 2267-2279.

Taylor, P. N. and J. N. Lester, 1995. Polynuclear aromatic hydrocarbons in a River Thames sediment core. *Environmental Technology*, 16: 1155-1163.

Tolosa, I., de Mora, S., Sheikholeslami, M.R., Villeneuve, J.P., Bartocci, J., Cattini, C., 2004. Aliphatic and aromatic hydrocarbons in coastal Caspian Sea sediments. *Marine Pollution Bulletin* 48: 44-60.

Tuominen, J., H. Pyysalo, J. Laurikko, T. Nurmela, 1987. Application of GLC-selected ion monitoring (SIM)-technique in analyzing Polycyclic Organic Compounds in Vehicle Emissions. *The Science of Total Environment* 59: 207-210.

Volkman, J. K., A. T. Revill, A. P. Murray (1997). Applications of Biomarkers for Identifying Sources of Natural and Pollutant Hydrocarbons in Aquatic Environments. In *Molecular Markers in Environmental Geochemistry*, ed. R. P. Eganhouse, American Chemical Society Press, Washington, D.C. pp. 279-313.

Wakeham, S. G., C. Schaffner, W. Giger (1980). Polycyclic aromatic hydrocarbons in recent lake sediments - I. Compounds having anthropogenic origin. *Geochimica et Cosmochimica Acta* 44: 403-413.

Wakeham, S. G., C. Schaffner, W. Giger, 1979. Polycyclic aromatic hydrocarbons in recent lake sediments- I. compounds having anthropogenic origins. *Environmental Science and Techonology*, 403-413.

Wang, Z., Fingas, M., 2005. Oil and petroleum product fingerprinting analysis by gas chromatographic techniques. In: Nollet, L. (Ed.), *Chromatographic Analysis of the Environment*, third ed. CRC Press, New York.

Wilcke W., Amelung W, Krauss M., Mrtius C, Bandeira A and Garcia MVB. 2003. Polycyclic aromatic hydrocarbon (PAH) patterns in climatically different ecological zones of Brazil. Org Geochem 34: 1405–1417.

Wilcke W., Krauss M., Lilienfein J. and Amelung W. 2004. Polycyclic aromatic hydrocarbon storage in a typical Cerrado of the Brazilian savanna. J Environ Qual 33: 946–955

Witt, G. and E. Trost, 1999. Polycyclic aromatic hydrocarbons (PAHs) in sediments of the Baltic Sea and of the German coastal waters. *Chemosphere*, 38: 1603-1614.

Wong CS., G. Snders, DR. Engstrom, DT. Long, DL. Swackhamer, SJ. Eisenreich, 1995. Accumulation, Inventory and Diagenesis of Chlorinated Hydrocarbons in Lake Ontario Sediments. *Environmental Science and Technology* 29: 2661-2672.

Yim, U. H., Hong, S. H., Shim, W. J., Oh, J. R., Chang, M., 2005. Spatio-temporal distribution and characteristics of PAHs in sediments from Masan Bay, Korea. *Marine Pollution Bulletin* 50: 319-326.

Youngblood, W. W. and Blumer, M. 1975. Polycyclic aromatic hydrocarbons in the environment: homologous series in soils and recent marine sediments. *Geochimica et Cosmochimica Acta*, 39: 1303-1314.

Yunker, M.B., Macdonald, R.W., Vingarzan, R., Mitchell, R.H., Goyette, D., Sylvestre, S., 2002. PAHs in the Fraser River basin: a critical appraisal of PAH ratios as indicators of PAH source and composition. *Organic Geochemistry*, 33: 489-515.

Yunker, M. B., R. W. MacDonald, D. Goyette, D. W. Paton, B. R. Fowler, D. Sullivan, J. Boyd (1999). Natural and Anthropogenic Inputs of Hydrocarbons to the Strait of Georgia. *The Science of the Total Environment* 225(3): 181-209.

Yunker, M. B. and Macdonald R. W., 1995. Composition and Origin of Polycyclic Aromatic Hydrocarbons in the Mackenzie River and on the Beaufort Sea Shelf. *Arctic*, 48(2): 118-129.

Zakaria, M. P., Takada, H., Tsutsumi, S., 1999. American Chemical Society (ACS) national meeting. Division of Environmental Chemistry (J). 39(2): 6.

Zakaria M.P., Horinouchi A., Tsutsumi S., Takada H., Tanabe S. and Ismail A. 2000. Oil Pollution in the Strait of Malacca, Malaysia: Application of Molecular Markers for Source Identification. *Environmental Science and Technology* 34: 1189-1196

Zakaria, M. P., Okuda, T., Takada, H., 2001. Polycyclic Aromatic Hydrocarbons (PAHs) and Hopanes in stranded tar-balls on the coasts of Peninsular Malaysia: applications of biomarkers for identifying sources of oil pollution. *Marine Pollution Bulletin* 42(12): 1357-1366.

Zakaria M. P., Takada H., Kumata H., Nakada N., Ohno K., Mihoko Y. 2002. Distribution of Polycyclic Aromatic Hydrocarbons (PAHs) in rivers and estuaries in Malaysia: widespread Input of petrogenic hydrocarbons. *Environmental Science and Technology* 36:1907-1918.

Zakaria, M. P., Takada H. 2003. *Petroleum hydrocarbon pollution: A closer look at the Malaysian legislations on marine environment*. Paper presented at First joint seminar on oceanography, NRCT-JSPS. Chien Mai, Thailand. December 2003.

Zakaria M.P. and Mahat A. A. (2006). Distribution of Polycyclic Aromatic Hydrocarbon (PAHs) in Sediments in the Langat Estuary. *Coastal Marine Science Journal*. 30(1).

Zhang, X., E. R. Christensen, M. F. Gin, 1993. Polycyclic aromatic hydrocarbons in dated sediments from Green Bay and Lake Michigan. *Estuaries*, 16: 638-652.

7

Global Distillation in an Era of Climate Change

Ross Sadler and Des Connell
Griffith University
Australia

1. Introduction

When a chemical enters the natural environment it undergoes change by several processes. It can be transported by the movement of the sector of the environment it enters, for example ocean currents and atmospheric movements, to a different geographical location. In addition as it is transported it is usually diluted so that its concentration is reduced. Chemical degradation processes take place which result in the production of more polar and water soluble products and leaving residual initial chemical. As all of these processes occur there is a distribution of the chemical between the phases in the environment. These processes have been described in a series of papers and books by Mackay as reported in Mackay et al (2009).

The processes shown in Figure 1 illustrate the basic processes involved in the partitioning of a chemical into phases in the environment. All of the processes shown involve two phases and movement backwards and forwards of the chemical and thus can be characterised by a partition coefficient. The partition coefficient is the ratio of the chemical in the two phases at equilibrium and isoften represented by the symbol, K. Thus the air – water partition coefficient is represented by K_{AW} and is better known as the dimensionless Henrys Law constant (Shiu and Mackay, 1986), the fish – water partition coefficient is K_B, (Connell, 1990) the sediment – water coefficient is K_D (Gobas and Maclean, 2003) and the other partition processes can be represented in a similar way. It is of interest to note that as a result of these partition processes a chemical can occur in very low concentrations in the atmosphere, low concentration in water but relatively high concentrations in fish and other aquatic biota.

Most of these partition coefficients can be calculated from partition values arrived at by laboratory measurements. For example the Henrys Law constant and the octanol – water partition coefficient, K_{OW} , can be measured in the laboratory. The K_{OW} can be used to calculate the fish – water partition coefficient, K_B value and also the K_D value and is extensively used to model chemical partitions in the environment. The octanol – air partition coefficient, K_{OA}, can be calculated or measured in the laboratory and can be used to evaluate the partitioning of a chemical into organisms as a result of concentrations in the atmosphere. The Persistent Organic Pollutants (POPs) are a group of mainly chlorinated hydrocarbon insecticides and dioxins which are often reported to undergo global distillation (Fernandez and Grimalt, 2003). These substances undergo the partition processes illustrated diagrammatically in Figure 1. However if they are discharged to the environment in the warmer zones of the planet a proportion will partition into the atmospheric phase according to the partition coefficient at that temperature. Movement in the atmosphere through winds results in transport to different geographical regions which, with the polar zones, have a colder temperature. The partition coefficient values are displaced to favour the solid or liquid phase and not the atmospheric phase. Thus a transfer of chemical can occur from warmer to the polar zones of the planet by a process somewhat similar to distillation – global distillation. This can occur in a series of steps or hops (Ma, 2010).

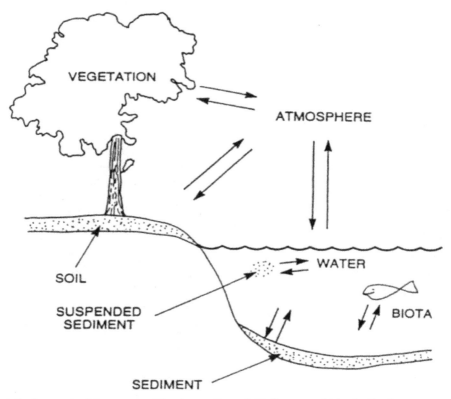

Fig. 1. Pathways for Movement of Persistent Organic Pollutants within the Environment

1.1 Physicochemical properties of POPS

There is no direct evidence that all of the compounds which are currently classified as POPs (see Table 1) exhibit the property of global distillation. However the POPs which have been identified as exhibiting global distillation share a set of properties which lead to their long range transport to the polar zones in the Arctic and are presumed (rather than demonstrated to also do the same in the Antarctic). Firstly all of these compounds are chlorohydrocarbons which means there is a limited range of covalent bonds present in them. These bonds are mainly C-C, C=C aromatic, C=C, C-H, C-Cl which are resistant to oxidation and hydrolysis – the main degradation processes in the environment. This leads to environmental persistence with a half life of between 2 to 10 years in soil and 0.25 to 2.5 years in air. Aqueous solubility is low at 1 to 7 g/m^3 with a correspondingly high solubility in lipid reflected in the K_{OW} values from about 10^2 to 10^6.

1.2 Climate change and long range transport of POPs

The effects of climate change on the global distillation process have been evaluated by several authors (e.g. Macdonald et al., (2003), Sadler et al., (2011)). Macdonald et al., (2003) pointed out that while the partition process can be successfully modelled, the impact of other environmental processes such as altered rainfall and changes to the particulate content of the atmosphere are difficult to assess. In the years since this review, a great deal more has been learned regarding the processes that underlie global distillation and also the effects and severity of climate change. For this reason, an updated review of the literature has been undertaken with a view to providing a prediction of long range transport of POPs for both the Northern and Southern Hemispheres that is relevant to the current state of knowledge.

2. Evidence for global distillation

2.1 Northern Hemisphere

Since the process was first proposed by Wania and Mackay (1995), numerous studies (notably in the Northern Hemisphere) have provided evidence that would be supportive of a global distillation model in respect of global transport of persistent organic pollutants. Polychlorinated biphenyls (PCB) have been amongst the most heavily studied of pollutants in this regard and reports of PCB migration include those in Northern Hemisphere soils (Meijer, et al., 2002), European high mountain lakes (Carera et al, 2002) levels observed in Norwegian mosses (Lead et al., 1996), deposition patterns in Canadian lakes (Muir et al., 1996) and Russian Lakes (McConnell *et al.*, 1996). Studies universally point to enhanced retention of the more chlorinated PCB congeners at lower latitudes, probably reflecting a more facile volatilization of the simpler PCBs and their subsequent migration to the poles. It is clear that although overall distribution patterns are consistent with the theory of global distribution and transport, a number of other local factors have to be taken into consideration when ascribing differences to global transport. Regression plots generally show considerable scatter with regard to the trend line, particularly with octa-PCBs (Meijer et al., 2002). This may account for the fact that other studies have found only partial evidence in support of the expected global transport process.

Reports of similar distribution patterns have also been produced in respect of global distribution of PCBs and organochlorine pesticides in other media. For example Kalantzi et al., (2001), noted that the distribution of PCB congeners was highest in European and North American butter and lowest in butter samples from the southern hemisphere. But the authors rightly point out that there are a number of compounding factors, such as relative efficiency of uptake from air/foliage, history of use in the area, etc. A number of other POPs included in the study showed maximal levels in areas where they are still in use. Trapping of PCBs and organochlorine pesticides in pine needles in Canada has also been ascribed to their delivery via global transport processes (Davidson et al., 2003).

Additional evidence of POP migration in accordance with the global transport/distillation model has come from measurement of levels in air over an extended period of time. There are now a number of networks dedicated to this type of surveillance. The underlying premise is that with a decrease in evolution of POPs following withdrawal of their use in many countries, levels in air should decrease with increasing time. For example, Sweetman and Jones (2000) reported a decrease of this kind with respect to PCBs. Their study, which centered around Hazelrigg, UK, noted a significant downward trend in PCB congener levels with time. Similar results were also obtained as regards levels of PCB congeners around the

Great Lakes region (Slmck et al., 1999). In contrast, the situation in the Arctic is more complex, as there will be an ongoing input of POPs from the continuing use in certain countries and also in terms of completion of global migration which commenced in other areas when the pesticides were routinely employed. The levels observed in Arctic air will principally result from global transport and also to a lesser extent, from revolatilization of POPs already in the Arctic environment. Hung et al., 2001 observed a rather mixed response in terms of trends for PCB congeners in Arctic air with time. There was generally a lack in decline of temporal trend, particularly as regards the more chlorinated congeners. The authors considered this to be basically supportive of a global transport model. Nevertheless, there were some exceptions, notably PCB 180, which whilst polychlorinated, showed a distinctive downward temporal trend. It was postulated that the heavier congeners may be more subject to differential removal processes by snow and particulate scavenging, whereas the lighter PBCs may be more prone to attack by hydroxyl radicals.

There have also been a number of reports of global transport in respect of a variety of pollutants apart from PCB congeners. Several of these have been included in the discussion above, but pollutants for which similar patterns have been observed include endosulfan, HCB and HCH (Carrera et al., 2002), lindane (Zhang et al., 2008) and phthalates (Xie et al., 2007). Significant attention has been paid to α- and γ-HCH (lindane) as well as endosulfan, because of their relatively recent use. In 2002, γ-HCH was shown to have reached phase equilibrium in the North Atlantic as has α-HCH, but the surface waters of the tropical and southern Atlantic were strongly undersaturated with γ-HCH, especially between 30°N and 20 °S. It must be noted that the state of the air-sea equilibria of α-HCH and γ-HCH are very different. The sporadic occurrence of γ-HCH in air makes it difficult to obtain representative results from transect cruises (Lakaschus et al., 2002). Weber et al., (2006) collated data for these pesticides from a number of sampling expeditions over the past decade. Although all could be detected in the Arctic environment, different patterns of distribution were observed for each substance with α-endosulfan predominating in the western Arctic and γ-HCH in the central Arctic. It was concluded that coastal sources may be important as regards γ-HCH, whereas air exchange is the major pathway of input as regards α-endosulfan.

Hargrave et al., 1997 examined the Arctic air-seawater fluxes of a number of POPs in the Canadian Arctic Archipelago during 1993. All of the pesticides showed a bimodal seasonal distribution of concentrations with maxima in February-May and July-August separated by minimum values in June. Organochlorine levels in air increased sharply during April and May and decreased during June, coincident with the onset of the open water period. Volatilization loses for HCH isomers during the open water period were estimated to have been small, whilst those for HCB and dieldrin were significant. These workers obtained evidence for significant deposition of toxaphenes, chlordanes and α-endosulfan during the open water period. The former USSR and Eastern Europe were probable emission sources for atmospheric contaminants.

An alternative approach to demonstrating the operation of a global distillation process would be the demonstration of summer vs. winter differences as regards deposition of POPs. In theory, higher levels of POPs would be expected to be transported during the summer, as a result of increased volatiliziation. Although data of this kind have been demonstrated for a number of European sites (cf. Carrera et al., 2002), it is impossible to separate the processes of revolatiliziation resulting from summer temperatures and increased use patterns at this same period. Diel variations in polybrominated diphenyl ethers and chlordanes in air have also been noted (Moeckel et al., 2008). The use of isotope tracers or more probably isotope ratios has also been advanced as a possible means of confirming the processes involved in long range transport of POPs (Dickhut et al., 2004).

There is also evidence that apart from global migration towards the poles, distillation phenomena can be invoked to explain the increasing concentrations of these pollutants with altitude. For example, Gallego et al., 2007 studied the distribution of PBDEs and PCBs in fish from high European mountain lakes (Pyranees and Tatra Mountains respectively). PCB levels in fish muscle were found to be significant in both study areas, with higher levels being recorded in the Tatra Mountains. This would be expected as the latter region experiences lower minima than the former. As regards the recorded levels of PBDEs, there was a good correlation between their levels and those of the PCBs in samples from the lakes in the Pyranees. The authors considered that this pointed to a predominance of temperature effects in this case, as would be predicted from the global distillation model. This relationship did not apply in the Tatra Mountains, probably reflecting the activity of other (unspecified) processes and also the fact that PBDEs are of later introduction than PCBs. One possibility would be the more recent and probably less restrained use of PCDEs during and immediately after the existence of the Eastern Bloc regimes (cf. Daly and Wania, 2005). Similar conclusions were reached by Demers et al., (2007), in a study of PCB and other organochlorine levels in trout from lakes in British Columbia and Alberta. Although the results obtained were in keeping with overall global distillation, this process alone could not explain the levels of contaminants found in the trout. The authors considered that both the feeding behavior of the fish and the octanol-water partition coefficient of the contaminant in question were also important factors in determining the observed distribution.

However, Wania and Westgate (2008) have pointed to several important differences between so-called mountain cold trapping and polar cold trapping. Although the same families of pollutants, in particular the PCBs, have been shown to concentrate in both high latitude and high altitude locations, a detailed comparison of studies shows that different chemicals concentrate in high latitudes than at high elevations. The chemicals that become enriched in mountains tend to be less volatile than those preferentially accumulating in polar regions by about two orders of magnitude. Wania and Westgate (2008) hypothesized that the temperature dependence of the precipitation scavenging efficiency of organic chemicals underlies mountain cold-trapping. In both polar and mountain cold trapping, temperature gradients and their impact on gas phase/condensed phase partitioning play a crucial role. Nevertheless, these processes are controlled by different mechanisms and affect different chemicals. In the case of polar cold-trapping the temperature dependence of partitioning between the Earth's surface and atmosphere is at the basis of the grass-hopper effect. In the case of mountain cold trapping the temperature dependence of partitioning between the various atmospheric components (gas phase vs. particles, rain droplets and snow flakes) is important. Franz and Eisenreich (1998) have pointed to the relative efficiency of snow as compared to rain for scavenging of PCBs and PAHs from the atmosphere, citing porosity differences as a major factor.

2.2 Southern Hemisphere

Evidence for operation of the process of global distillation in the Southern Hemisphere is generally more limited, albeit with at least some grounds for supporting the hypothesis (Corsolini et al., 2002; Noël et al., 2009). A major factor to be considered when comparing the two processes is the relative differences in landforms between the two hemispheres. Whereas an extensive landmass extends towards the poles in the northern hemisphere, the Southern Hemisphere consists of large expanses of open ocean, with only South America extending close to the Antarctic. Consequently, it is not surprising that the few studies of POPs distribution in the Southern Hemisphere have not provided much evidence of global migration. Studies of seawater are subject to many compounding effects and thus a North-South transect that included a large number of sampling sites off the coast of Africa, failed to show any evidence of increased PCB content at southern latitudes (Nizzetto *et al.*, 2008). The concentrations of HCHs in air and surface water of the Arctic have been shown to exceed those of the Antarctic by one to two orders of magnitude (Lakaschus *et al.*, 2002). Soil and sediment samples from James Ross Island were shown to contain low levels of PCBs, PAHs, p,p'-DDT, DDE, and DDD, with generally lower levels being detected in the sediments. A prevalence of low-mass PAHs, less chlorinated PCBs, and more volatile chemicals was taken to indicate that the long-range atmospheric transport from populated areas of Africa, South America, and Australia as the most probable contamination source for the solid matrices in James Ross Island (Klánová *et al.*, 2008).

Low levels of various organochlorine samples have been recorded in moss samples from high altitude southern hemisphere locations (Grimalt *et al.*, 2004) and also from the Antarctic (Focardi *et al.*, 1991). (Mosses are favoured for this type of observation because of their rather simple physiology compared to higher plants.). When compared to the levels of these pollutants found in similar locations within the northern hemisphere, the levels are small, with the Antarctic values being particularly low. As has already been explained, there are differences in the underlying principles of high altitude vs. polar deposition, even though there is some overall resemblance in the processes.

The general lack of supporting observations for operation of global distillation in the Southern Hemisphere is probably a reflection of a number of factors. As mentioned above, the Southern Hemisphere contains far larger expanses of ocean than does the Northern Hemisphere. There has to date been little evidence obtained to support the operation of global distillation in oceans, probably because of the susceptibility of deposited POPs to scavenging by water-borne particulates (notably those of phytoplankton origin), with subsequent benthic deposition (Dachs *et al.*, 2002).

Moreover, it must be conceded that the overall shape of continents in the Southern Hemisphere differs markedly from that of their Northern Hemisphere counterparts. The two land masses that would provide a terrestrial pathway for POPs from the equator towards the poles (viz. South America and Africa) are both triangular in shape. Given that global distillation works best in terrestrial environments (Dachs *et al.*, 2002), a persistent organic pollutant moving towards the Arctic in the Northern Hemisphere would have a relatively good chance of progressing via land. In contrast, movement southwards in Africa or South America would significantly increase the chance of a persistent organic pollutants' entering the marine environment and hence being lost to the global distillation process. And although the southern tip of South America does lie close to Antarctica, southward migration would still require that the pollutant traverse a significant tract of the Southern Ocean. In the case of Africa, the migration distance across water would be so significant, as to probably relegate the global distillation process to a minor role. Interestingly, Gioia *et al.*, 2008 noted that PCB levels in the North Atlantic appeared to be governed primarily by transport from source regions, whereas corresponding levels in the South Atlantic seemed to be driven by temperature changes via air–water exchange with the ocean. This would be in keeping with the overall observations above.

Finally, evidence of pollutant migration throughout the African and South American continents would probably be blurred by the ongoing/recent use of organochlorine pesticides in some of the countries.

The relative differences in pollutant transport by global distillation in the two hemispheres are highlighted in the accompanying figure (Figure 2).

2.3 Alternative explanations

Reference has already been made to the fact that global distillation may, in certain circumstances provide only a partial explanation for the observed long-range transport of pollutants (Dachs *et al.*, 2002). von Waldow *et al.*, 2010 proposed an alternative differential removal hypothesis, which proposes that fractionation results from different loss rates from the atmosphere, acting along a gradient of remoteness from emission sources. They successfully applied the associated model to explain the observed differences in PCB concentrations in European air, using data from an ongoing study in which transects from England to Norway are monitored. From their data analysis, the observed concentrations were better correlated with distance from the source than with temperature.

It must be emphasized that these data pertain only to a relatively limited transect within the Northern Hemisphere. Moreover, the authors point to possible anomalies in data collection using semipermeable membrane devices as a source of error in some calculations. The failure of this study to find evidence in support of global distillation does not negate the concept overall. Various other examples have been given above in which global distillation was shown to be of secondary importance to other processes. The two processes (viz. global distillation and differential removal) may be seen as competing and further discussion of their relative importance in terms of climate change is provided below.

Northern Hemisphere

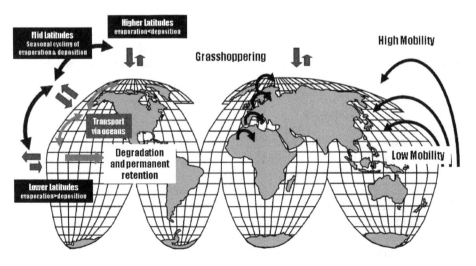

Southern Hemisphere

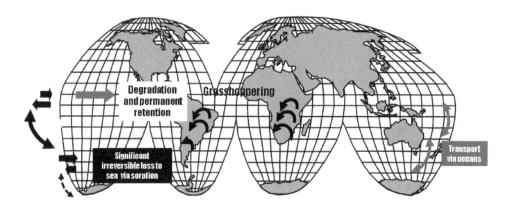

[1]Although multiple hops are involved in the global distillation process of POPs (Wania, 2003), only one is shown in these diagrams for simplicity.

Fig. 2. Global Distillation Processes in Northern and Southern Hemispheres[1]

3. Processes involved in global distillation

Wania (2003) divided pollutants into:

- 'Fliers' which are very volatile pollutants and unlikely to deposit even at the poles.
- 'Single hoppers' which are volatile enough to be carried all the way to the poles, where they will ultimately deposit by condensation.
- 'Multiple hoppers' which will be carried toward the poles by repeated evaporation and condensation cycles.
- 'Swimmers', which are non-volatile and will be transported via the oceans.

Of these four groups, it is the multiple hoppers that will be subject to transport by global distillation. This group includes the persistent organic pollutants, which are enumerated in the Table below (Table 1).

3.1 Basic underlying processes

The ability of a substance to undergo long range transport via global distillation will be governed by three factors:

1. Its ability to enter the air compartment by volatilization, if it is not emitted directly to air.
2. Its ability to remain stable during the transport process.
3. Its ability to deposit in the Arctic region.

Thus the main physicochemical parameters that govern global distillation will be the Henry's Law coefficient (H), the octanol-air partition coefficient (K_{OA}) and the air-water partition coefficient (K_{AW}).

Chemical	Log K_{OA}	Log k_{AW}	H
Aldrin	4.44	-2.23[11]	7.0669 x 10[-3] atm-m[3]/mole
Chlordane	9.21 (cis)[4] 9.16 (trans)[4]	2.31[12]	4.105 x 10[-3] atm-m[3]/mole
p,p'-Dichlorodiphenyltrichloroethane (DDT)	10.08[4]	-2.48[11]	2.232 x 10[-3] atm-m[3]/mole
Dieldrin	8.54	-3.56[11]	1.1099 x 10[-4] atm-m[3]/mole
Endrin	10.1[5]	-4.66[12]	3.08 x 10[-4] atm-m[3]/mole
Heptachlor	6.12[4]	-0.02	4.47 x 10[-2] atm-m[3]/mole
Hexachlorobenzene (HCB)	7.563[2]	-1.64[11]	3.61 x 10[-2]
Mirex	10.7[12]	-5.42[12]	8.28 x 10[-3] atm-m[3]/mole
Polychlorinated biphenyls	7.01 -10.75[2]	0.349- (-5.08)[12]	3.3x10[-4] - 5x10[-5] atm-m[3]/mole
Polychlorinated dibenzo-p-dioxins	8.564 – 11.66[2]	-5.02 (TCDD[12])	1.47 x 10[-3] (TCDD)
Polychlorinated dibenzofurans	10.281[2]	(-2.511) – (-5.47)[12]	8.1 x 10[-5] atm-m[3]/mole
Toxaphene	7.6[12]	-1.16[12]	6.0 x 10[-6] atm-m[3]/mole
Alpha hexachlorocyclohexane	7.26[5]	-3.51[11]	6.9 x 10[-6] atm-m[3]/mole
Beta hexachlorocyclohexane	8.13[5]	-5.32[11]	2.35 x 10[-9]
Chlordecone	8.92[12]	-6.69[6]	2.53 x 10[-3]
Technical endosulfan and its related isomers	6.00 (α-endosulfan[5])	-2.72 (α-endosulfan[11])	1.06
Hexabromobiphenyl	13.6[12]	–4.62[6]	1.38 x 10[-6] to 5.7 x 10[-3]
Hexabromodiphenyl ether and heptabromodiphenyl ether (commercial octabromodiphenyl ether)	12.113 – 12.273[2]	-4.29[12]	1.88 x 10[-7] – 7.48 x 10[-8] atm-m[3]/mole
Lindane	8.04[4]	–3.96[6]	6.1 x 10[-5]
Pentachlorobenzene	6.539[2, 3]		7.03 x 10[-4]
Perfluorooctane sulfonic acid, its salts and perfluorooctane sulfonyl fluoride	6.5 – 7.5[9]	<2 x 10[-6 10]	3.09 x 10[-9] atm-m[3]/mole
Tetrabromodiphenyl ether and pentabromodiphenyl ether (commercial pentabromodiphenyl ether)	10.04 (TBDE)[8] 11.2 (PBDE)	–3.67[6, 7]	1.18 x 10[-6] – 3.54 x 10[-6] atm-m[3]/mole

[1]As defined by the Stockholm Convention (http://chm.pops.int/Home/tabid/2121/Default.aspx);
[2]Data from Li et al., 2006, all values for 20°C unless otherwise stated. [3]Data for 18.7°C.
[4]Data from Harner and Bidleman, 1998 (all data for 20°C). [5]Data from Shoeib and Harner, 2002 (all data for 25°C).[6]Data from Scheringer et al., 2006. [7]Data for BDE-99. [8]Data from Wania et al., 2002 (all data for 20°C). [9]Data from Shoeib et al., 2002 (all data for 20°C). [10]Data from POPRC, 2007. [11]Data for 25°C
[12]Calculated according to the method of Meylan and Howard (2005)

Table 1. Persistent Organic Pollutants[1]

Using the above modelling parameters, Wania (2003) developed an averaged global distribution model (the Globo-POP model) to predict both intermediate term (1 year ongoing emissions) and long term (10 year ongoing emissions) Arctic Contamination Potential (ACP). The ACP of most chemicals is sensitive to the temperature dependence of the partition coefficients, temperature, atmospheric mixing coefficients, and sea ice cover. The substances with significant ACP were in the range elevated ACP overlap in the range $6.5 < \log K_{OA} < 10$ and $-0.5 > \log K_{AW} > -3$, corresponding to chemicals that have significant ability to bioaccumulate. From the modelling, a parameter mACP, representing the potential for relative enrichment in the Arctic was developed.

$$mACP = [m_{T1} - m_{A1}]/m_{TG} \times 100\% \tag{1}$$

where m_{T1} and m_{A1} are the mass of chemical in all compartments and in the four atmospheric compartments of zone 1 (N-Polar) of the Globo-POP model, respectively. m_{TG} is the mass of chemicals in all model compartments. Wania (2006) later added a parameter for estimating absolute contamination of the Arctic environment (eACP):

$$eACP = [m_{T1} - m_{A1}]/e_{TG} \times 100\% \tag{2}$$

where e_{TG} is the mass of chemical emitted cumulatively to the global environment. m_{TG} will always be smaller than e_{TG} because of chemical losses by degradation and loss processes such as transfer to the deep sea, with subsequent loss to deeper sediment layers.

3.2 Atmospheric sorption

During transport, a persistent organic pollutant will always be subject to loss processes. These could include sorption onto particles with subsequent deposition, either by wet or dry deposition processes. Little is known of the relative importance of wet vs. dry deposition fluxes for persistent organic pollutants, although in a study of a number of sites in Canada, Yao et al. (2008) demonstrated that wet deposition was more important than dry deposition for most organochlorine pesticides. Clearly, there is a need for similar studies in other geographic regions. Moreover, it cannot be assumed that all scavenging will take place via rain, particularly in higher latitudes, where a significant amount of precipitation takes place via snowfall. Snow has been demonstrated to be a more efficient scavenger of airborne particulates than normal rainfall, probably largely because of the larger size and surface area of snowflakes. Franz and Eisenreich (1998) studied the scavenging of PCBs and PAHs by snow in Minnesota. Although anomalous, gas scavenging of the two pollutant classes was a relatively minor contributor to the overall snow removal flux. Particle scavenging ratios ranged from 5×10^4 to 5×10^7 for the snow events, as compared to 10^3 to 10^5 for rainfall events. Snow particle scavenging was slightly greater for the more volatile PCBs and PAHs. The authors hypothesized that lower molecular weight semi-volatile organic carbon compounds are associated with a different particle size spectrum than their less volatile counterparts.

The characteristics of air-particulate sorption for POPs are somewhat obscure and it is almost certain that the sorption characteristics of particulates from different sources will not be the same (cf. Gustafason and Dickhut, 1997). Once sorbed to a particle and carried to the earth by some form of deposition, a pollutant can be considered to have been removed from the global distillation process, at least temporarily (cf. Scheringer et al., 2000). Such removal is particularly significant in the case of persistent organic pollutants, as a result of their ability to sorb to particulates. Removal of this kind does not preclude the pollutant from undergoing desorption and revolatilization at some later stage although it will cause a temporary disruption to the global transport process.

3.3 Atmospheric destruction
Atmospheric destruction of POPs is obviously another competing process and one with the potential to permanently remove the molecules from global transport processes. The mechanisms are generally less well understood than are those of destruction of volatile organics in air, but hydroxyl radicals, nitrate radicals, and ozone are believed to be involved (Boethling et al., 2009). The potential for POPs to associate with atmospheric particles provides a further complication, although there is some evidence to suggest that sorbed POPs may still be subject to oxidation processes. In the case of atmospheric ozone and hydroxyl radical destruction, the half life ($t_{1/2}$)of a persistent organic pollutant is given by:

$$t_{1/2} = 0.693 k_{Oxidant}[Oxidant] \tag{3}$$

where $k_{Oxidant}$ is the rate constant in units of cm^3 molecule^{-1} s^{-1}, and [Oxidant] is the oxidant concentration in units of molecules (or radicals) cm^{-3}. On a global basis, the situation is complicated by the fact that the concentration of these oxidants vary from place to place and also exhibit seasonal as well as diel variation.

3.4 Aquatic sorption
When considering atmospheric volatilization and deposition processes, it is important to appreciate that the situation over water will differ from that over land. Over the oceans, the operative processes will be diffusive air-water exchange, as well as exchange involving wet and dry particulates. In terms of deposition, the latter terms will equate to wet and dry deposition. These processes will also be operative in the case of terrestrial systems but the situation will be complicated by air-soil exchange and air-vegetation exchange. Under normal circumstances in the oceans, air-water exchange is likely to be the dominant process (Dachs et al., 2002). The air-water partition coefficient is related to the Henry's Law Coefficient by the following equation:

$$\frac{1}{k_{aw}} = \frac{1}{k_a H'} + \frac{1}{k_w} \tag{4}$$

where k_{AW} is the air-water partition coefficient, k_A and k_W are the POP mass transfer coefficients (m d^{-1}) in the air and water films, respectively and H′ is the temperature corrected and dimensionless Henry's law Constant. k_{AW} is known to be very sensitive to temperature, because of its dependence on the Henry's Law Constant, but it can also be affected by a number of other environmental variables such as wind speed. Therefore, the latitudinal variation of k_{AW} will also be a function of prevailing winds. The air-water flux for a persistent organic pollutant will be described by the equation:

$$F_{aw} = k_{aw} \left\{ \frac{C_a}{H'} - C_w \right\} \tag{5}$$

where C_A and C_W are the concentrations in the air and water phases respectively.

Once a persistent organic pollutant lands on the water surface, it is unlikely to dissolve because of its low inherent solubility. The pollutant will however be subject to processes of aquatic sorption by suspended particles, notably plankton. As with atmospheric deposition (considered above), the pollutant could theoretically later desorb and revolatilize. This is however less likely in the case of phytoplankron sorption, as the phytoplankton cell will eventually senesce and deposit in the benthic layer. Dachs *et al.*, (2002) modelled the phytoplankton uptake flux (F_{WP}) according to the equation:

$$F_{wp} = k_{wp} \left\{ C_w - \frac{k_{dep}}{k_{up}} C_p \right\}$$ (6)

where k_{WP} is the water-phytoplankton mass transfer rate constant, k_{dep} and k_{up} are the depuration and uptake rate constants respectively and C_P is the concentration in the phytoplankton. The overall sinking flux for a POP, associated with phytoplankton (F_{sink}) could be calculated from:

$$\log F_{sink} = 1.8 \times F_{OC} \times C_p$$

$$\text{where } \log F_{OC} = 2.09 + 0.81 \log [\text{Chlorophyll}]$$ (7)

The authors suggested that the process would be of major significance in regions of ocean upwelling, where nutrient rich waters are brought to the surface, resulting in high primary productivity. In the overall transfer of POPs from the atmosphere to benthic sediment, the rate limiting processes will either be air-water transfer rate or the sinking rate. With POPs of $\log K_{OW} < 6.0$, the sinking process will provide the rate determinant step in most cases, whereas for more lipophilic compounds, sinking fluxes are probably rate limiting at lower latitudes and air-water exchange at higher latitudes. These biogeochemical processes have the ability to disrupt global distillation processes over the oceans and as has already been mentioned, are probably important in terms of explaining the differences in global distillation between the northern and southern hemisphere. Supporting evidence in respect of the role of these plankton mediated biogeochemical controls in the Mediterranean Sea has recently been produced by Berrojalbiz *et al.*, (2011).

4. Climate change scenarios

Since 1990, considerable attention has been paid to climate change on an international basis, largely through the efforts of the Intergovernmental Panel on Climate Change (IPCC). To date, four assessment reports have been published, the most recent being in 2007 (IPCC, 2007a; IPCC 2007b; IPCC 2007c) and a fifth assessment report is expected to be published in due course. However, the topic of pollutant behaviour in response to climate change remains under-investigated. By far the most detailed studies of pollutant behaviour in response to climate change are those of MacDonald (MacDonald et al., 2002, MacDonald et al., 2003) which pertain to the Canadian Arctic. Although the majority of attention has been devoted to effects in the Northern Hemisphere, a recent review has considered possible climate change effects on pollutant behaviour in the Southern Hemisphere (Sadler et al., 2011). Further studies of the effects of selected climate change phenomena on behaviour of certain persistent organic pollutants have been carried out by some authors (Dalla Vale et al., 2007; Lamon et al., 2009, Ma and Cao, 2010).

When considering effects of a phenomenon such as climate change upon a complex process such as global transport, it is necessary to conduct the assessment with reference to as many variables s possible if a true picture is to be obtained. In a number of studies that have been conducted to date, there has been a tendency to consider temperature as the major driver associated with climate change. This is clearly not the case (cf. IPCC 2007a, IPCC 2007b). Climate change can be considered to consist of the following manifestations:
* Increases in land, water and air temperatures
* Increases in the intensity of extreme weather events
* Changes in salinization patterns within the oceans
* Rises in sea levels

In respect of temperature changes, it must be noted that although it is the most frequently discussed aspect of climate change, current predictions of global temperature rise are well below 10°C in 100 years. Most of these projections constitute relatively small changes and are unlikely on their own to have a major effect on pollutant behaviour. It is our view that more significant effects on pollutant behaviour (including ones mediated by temperature) will be occasioned by the increased intensity of extreme weather events. IPCC predictions point to significant changes in the intensity of heatwaves, tropical storms (whose range may well extend into areas currently free of such events). Such events have the potential to cause far greater changes in environmental parameters, notably temperature albeit over a shorter period. Hence, the modelling of climate change mediated effects becomes a far more complex process.

Another frequently neglected aspect of climate change is its potential to alter physical conditions leading to changes in both production environments and also receiving environments. This may be reflected in a number of ways, including patterns of use. Persistent organic pollutants such as DDT have been banned in many countries and hence their evolution rates declined significantly in the recent past. But the effects of warming have seen the occurrence of diseases such as Malaria increase in geographical scope in some countries (notably Africa). This has occasioned a reintroduction of these pesticides and hence a potential for increased evolution rates (cf. WHO 2009).

Equally, the warming effects associated with retreat of the polar ice caps and glaciers have major potential to alter the scenarios as regards persistent organic pollutant sorption. The air-ice interface has been recognized as a major site of sorption for these pollutants and loss of this interface will almost certainly lead to a remobilization of pollutants both from polar and high altitude regions. Thus, even if global/mountain transport processes continue to operate at their present or some altered rate, persistent organic pollutants delivered to these sites cannot be expected to remain in their traditional niches.

Intense extreme weather events also have the potential to cause a significant export of nutrients from land and offering the potential for increased primary productivity in the oceans (Sadler et al., 2011). As has already been pointed out above and will be discussed more fully below, this too has significant ability to affect global transport phenomena. Glacial melt can also make a significant similar contribution in this type of area, with the potential to also transport persistent organic pollutants released from the ice.

It follows from the foregoing remarks that a detailed consideration of the processes involved in global transport of pollutants is required if a meaningful assessment of climate change effects is to be made. Whilst temperature will be an important driver for all these processes, it or any other factor cannot be considered in isolation.

4.1 Effects on modelling parameters

As has already been outlined, the most simple parameters associated with global distillation modelling are the octanol-air partition coefficient, the air-water partition coefficient and the Henry's Law Constant. The octanol-air partition coefficient is known to vary log-linearly with temperature. Shoeib and Harner (2002) measured octanol-air partition coefficients for a range of persistent organic pollutants, over the range 278-308°K and plotted the response as log K_{OA} vs 1000/T (°K). From their data, they concluded that it was possible to calculate the variation in log K_{OA} from a simple regression equation. Using this equation, the variation in log K_{OA} for a 5°K and a 10°K temperature change as associated with global warming scenarios has been calculated for a number of persistent organic pollutants and is shown in Table 2.

POP	293°K (20°C)	298°K (25°C)	303°K (30°C)
HCB	7.55	7.38	7.23
α-HCH	7.80	7.61	7.43
γ-HCH	8.04	7.85	7.66
Heptachlor	7.84	7.64	7.45
Aldrin	8.29	8.08	7.87
trans-Chlordane	9.16	8.87	8.59
cis-Chlordane	9.21	8.92	8.63
α-Endosulfan	8.88	8.64	8.40
Dieldrin	9.11	8.90	8.69
Endrin	8.38	8.13	7.89
p, p-DDT	10.08	9.82	9.56

[1]Data calculated using the regression equation of Shoeib and Harner 2002.

Table 2. Variation in log K_{OA} values for Temperature Rises associated with Global Warming[1]

It is readily seen that the temperature rises associated with climate change (and the values given in the above table are on the extreme side of predictions) cause relatively small changes in the octanol-air partition coefficient. In general, rising temperatures will tend to decrease sorption of persistent organic pollutants on particulates. But it must be remembered that the changes referred to above are also well within the range of uncertainty associated with determinations of octanol-air partition coefficients and temperature.

It must also be remembered that the majority of observations, linking K_{OA} to particle sorption pertain to urban situations (cf. Radonić et al., 2011). The nature of the particles with which persistent organic pollutants will interact in non-urban situations may be entirely different, particularly in a climate change scenario and the subject may warrant some further specific consideration. Predictions of climate change scenarios generally include an increased frequency of intense dust storms and although currently, the majority of pollutants appear to exist in the gaseous phase, there have been contrary reports from some agricultural areas (cf. Yao et al., 2008). There is a need to obtain more information on the partitioning behaviour of pesticides with dust derived from agricultural soils, which will almost certainly form the majority of climate change associated dust.

Although it is possible to make theoretical estimates of log K_{AW} (cf. Table 1, Meylan and Howard, 2005), in the actual environment a number of parameters are known to affect the actual value of this parameter. Temperature plays an important role in affecting K_{AW}, but the parameter is also subject to other influences, notably wind speed and various diffusion/transport variables (cf. Dachs et al., 2002). Prediction of future wind speed changes with current state of computer power remains a difficult task and thus a subject of active research. While definite trend changes in severe cyclonic wind intensity and frequency have not yet established, preliminary research results suggest that significant alteration in cyclonic wind intensity and frequency are possible. In terms of non-cyclonic wind intensity, there appears good evidence to suggest significant increases, based upon modelling using the IPCC A1F1 scenario (temperature change +2.4-6.4°C, best estimate +4°C, over the final decade of the 21st century vs. the final decade of the 20th century) (CSIRO 2007).

In terms of equation 2, it can be shown that both k_A and k_w are sensitive to wind speed. The equation is clearly sensitive to changes in Henry's Law Coefficient, which is known to respond to temperature (see below). Obviously, climate change scenarios would be expected to involve changes of temperature. But even in the absence of any temperature effect, it can be shown that a 5% increase in wind speed (considered an upper best estimate by CSIRO (2007) for the next 50 years) can have a noticeable effect on the value of k_{AW}. Later in the 21st century, there may be greater increases in wind speed, depending on the forcings operative at the time.

That temperature has a major effect on the value of the Henry's Law Coefficient has been demonstrated both experimentally and by calculation (Staudinger and Roberts, 2001). The variation of Henry's Law Coefficient with temperature is given by:

$$H_{t'} = \left\{ H_{293°K} \right\} \left\{ 10^{-\frac{\Delta H^0}{R} \left[\frac{1}{T} - \frac{1}{293} \right]} \right\} \tag{8}$$

where $H_{T'}$ is the value of H' at a given temperature, T, ΔH^0 is the enthalpy of phase change (J mol^{-1}) and R is the universal gas constant (8.314 J mol K^{-1}).

The variation of H' with temperature is not uniform for all environmental contaminants. On the basis of an extensive literature survey, Staudinger and Roberts (2001) estimated the average % rise in Henry's Law Constant/10°C to be 60% in the case of hydrocarbons, 90% in the case of miscellaneous substances and 140% in the case of chlorinated organic pesticides and PCBs. Unfortunately, insufficient data of this kind are currently available in the case of the more recent additions to the list of persistent organic pollutants. Kűhne et al., (2005) suggested that temperature variation of Henry's Law Coefficient is maximal for compounds that are polar or have significant hydrogen bond interaction capacity. In this case, it could be expected that the temperature variation of the more recent additions to the POPs register would be significant.

The effect of a 5°C temperature rise on the temperature corrected Henry's Law constant for a number of persistent organic pollutants is shown in Table 3. These values were derived using the USEPA's on-line calculator (USEPA 1996).

Because of its effects on modelling parameters, notably k_{AW}, the temperature effect on the Henry's Law Constant is probably the largest single effect that climate change is likely to exert. From the data presented above and Equation 4, it can be seen that the difference caused by even a 5°C rise of temperature would have a significant effect on the Henry's Law constant and also on dependent parameters such as k_{AW}. Although the temperature effect is the most prominent influence of all on the Henry's Law Constant, a number of other climate change factors have the ability to affect this parameter as well. These have recently been discussed (Sadler et al., 2011) and although their effects will probably be secondary to the temperature perturbations associated with climate change, they may be of significant in some instances. Of special relevance to climate change scenarios is the effect of suspended particles, which would tend to offset the temperature increases in the Henry's Law Constant. These will be further discussed below.

POP	H' at 20°C	H' at 25°C	Ratio of H' at 25°C/H' at 20°C (%)
HCB	0.0310	0.0540	174
α-HCH	0.000241	0.000433	237
γ-HCH	0.000319	0.000572	179
Heptachlor	0.0268	0.0446	166
Aldrin	0.00384	0.00695	181
α-Endosulfan	0.000261	0.000458	198
Dieldrin	0.000312	0.000617	198
Endrin	0.000165	0.000307	186
p, p-DDT	0.000139	0.000331	238

[1]Data obtained using on-line calculator (USEPA, 1996)

Table 3. Variation in H' values for a 5°C Temperature Rise associated with Global Warming[1]

4.2 Modelling studies relating to effects of climate change on global transport

To date, only a few specific studies relating climate change to effects on global transport have been published, although it is possible to make extrapolations from other published work. Lamon et al., (2009) examined the effects of climate change on global levels of PCBs. Using two IPCC scenarios, they were able to demonstrate that increased temperature would probably be the major driver as regards climate change effects on PCB transport. Higher temperatures were considered to drive increased primary and secondary volatilization emissions of PCBs and enhance transport from temperate regions to the Arctic. The largest relative increase in concentrations of both PCBs in air was predicted to occur in the high Arctic and the remote Pacific Ocean. Higher wind speeds were predicted to result in more efficient intercontinental transport of PCB congeners.

Ma and Cao (2010) developed a perturbed air-surface coupled to simulate and predict perturbations of POPs concentrations in various environmental media under given climate change scenarios. Their studies pertained to α- and γ-HCHs, HCB, PCB-28 and -153. The HCHs, HCB and OCB 153, showed strong perturbations as regards emissions under climate change scenarios. The largest perturbation was found in the soil-air system for all these chemicals with the exception of α-HCH, which showed a stronger perturbation in the water-air system. The study also included terms for degradation, as would be operative under climate change scenarios. As would be expected, climate change also has a significant effect on the degradative pathways for pollutants, potentially undergoing global transport. The scenarios investigated resulted in a maximum 12% increase of HCB and a 10% increase of α-HCH levels in air during the first year of disturbed conditions. Following the increases during the first several years, the perturbed atmospheric concentrations tend to result in decreased air levels due to degradation processes. It is clear from this study that simple modelling of global transport cannot be undertaken, without a proper consideration of all factors that might be affected by climate forcings. As has already been noted in this review, the so-called global distillation is only one pathway available for transport and fate of persistent organic pollutants. Its relative importance may well change under some climate change scenarios.

4.3 A possible role for phytoplankton

All of the above studies pertain to Northern Hemisphere situations. As has already been pointed out above, the role of deposition on the oceans provides an important sink for persistent organic pollutants and is likely to be operative particularly in the Southern Hemisphere (cf. Dachs et. al. 2002). Jurado et al., (2004) noted the importance of phytoplankton, especially in areas of high primary productivity as regards controlling levels of PCBs in the oceans. They stressed the need for further research, particularly as regards seasonal influences on phytoplankton populations and the associated biogeochemical cycles. The effect of climate change on phytoplankton levels has been the subject of a number of papers in recent years (Hayes et al., 2005; Hallegraeff et al., 2009; Sadler et al., 2011). All of these studies have identified the need for intense ongoing monitoring. There is a general belief that the effects of climate change will see an increase in the frequency and intensity of blooms, an expansion of the geographical range and seasonal window for warm water species. The general concern expressed by these studies has been a possible increase in toxic algae and the attendant public health problems. But it is equally true that these projected increases could translate into increased sinking fluxes of persistent organic pollutants, deposited on the ocean surface.

In order to appreciate the magnitude of what phytoplankton blooms could achieve in terms of removal of persistent organic pollutants, it is necessary to consider the magnitude of the F_{sink} term (Equation 7). Dachs et al.,(2002) calculated sinking fluxes on the basis of chlorophyll levels in oceans, as recorded by satellite imaging. Although chlorophyll concentrations are a less than satisfactory indicator of algal biomass, under the circumstances, it must be conceded that this was probably the best available indicator.

Seasonal growth of algae represents what is probably the normal situation at present, but given the fact that blooms will become more frequent with climate change, it is important to examine the relative ability of such occurrences in terms of capturing persistent organic pollutants. For comparison, therefore, data on two *Trichodesmium* bloom events are included. This organism has been chosen because of its adaptive abilities (cf. Bell et al., 2005) which make it an ideal candidate to successfully proliferate under climate change conditions. As is shown in the table below, bloom situations could effectively remove four to six times as much persistent organic pollutant than the relatively static situation reported by Dachs et al., (2002). This means that the phytoplankton blooms which are expected to result from climate change, particularly in the Southern Hemisphere provide an important and hitherto neglected contribution to the overall observed global transport of pollutants.

The possibility also exists that in the Arctic Region, with the retreat of the Arctic ice cover, phytoplankton blooms will occur in the resulting ocean, providing a sink for released pollutants (see next section). There is already evidence of earlier algal blooms in the Arctic Region (Kahru et al., 2010) but although considerable effort has been devoted to effects of climate change on higher vertebrates in the Arctic, there remains a significant gap of knowledge with respect to dynamics of phytoplankton (Wassmann et al., 2011).

Scenario	Reference	No of Phytoplankton Cellsx 10^3 L^{-1}	[Chlorophyll] (mg m^{-3})	F_{OC}[1]
Ocean at 60-75 S	Dachs et al (2002)	-	0.1-18	21-1068
Trichodesmium bloom, Arabian Sea	Desa et al., 2005	690.9	126.75	6214
Trichodesmium bloom Moreton Bay, Queensland	Queensland Department of Environment and Resource Management, unpublished Data	500 (calculated from number of trichomes, assuming 100 cells per trichome)	87	4581

[1]log FOC = 2.09 + 0.81 log [Chlorophyll]

Table 4. Phytoplankton and Organic Carbon Sinking Fluxes

The end product of this type of pollutant sink will almost certainly be an increase in the levels of persistent organic pollutants in benthic communities. Results such as those obtained by Berrojalbiz et al., (2011) tend to highlight the importance of phytoplankton processes as sinks for persistent organic pollutants. With regard to climate change phenomena, a full discussion of causal factors for algal blooms is given elsewhere (Hayes et al., 2005; Hallegraeff et al., 2009; Sadler et al., 2011). It will be particularly important to pinpoint sites of upwellings. Upwellings are of two kinds:
1.   Upwellings caused by trade winds.
2.   Upwellings caused by extreme weather events such as cyclones (hurricanes, typhoons).
Both types have the potential to be affected by climate change. Upwellings caused by trade winds are known to decrease during El Niño events and increase during La Niña events. With the expected accentuation of these events as a result of climate change (IPCC 2007a) it could be expected that changes in the intensity and patterns of these upwellings would be observed. Similarly, the expected increase in intensity of cyclones (hurricanes, typhoons) with climate change will have an influence on the second type of upwelling. It is significant that some surveys of POPs in ocean transects have shown increased levels in waters off the west coast of Africa – a major site for ocean upwellings (Gioia et al., 2008).

4.4 Fate of transported pollutants

It is established that both Arctic cold trapping and mountain cold trapping lead to an accumulation of pollutants in the coldest regions, although the actual pattern of pollutant retention is different (Daly and Wania, 2005). To date however, there seems to have been a tacit assumption that climate change has the potential to alter these transport processes, but that any pollutants reaching the usual site of accumulation will be retained in the normal manner. Such a view is clearly untenable, particularly in respect of the Arctic, where significant ice melt has already taken place. It is known that the air-ice interface provides a particularly strong site for sorption of persistent organic pollutants (Hoff et al., 1995) and hence some examination of the fate of cold trapped pollutants is warranted. In fact, regions of the Arctic are particularly sensitive to changes in temperature as the Greenland Ice Sheet is normally only very close to freezing point (Archer and Rahmstorf, 2010).

Daly and Wania (2005) noted that concentrations of persistent organic pollutants in air and water increased during periods of melt activity. More recently, Stocker et al., (2007) modelled the effect of snow and ice on distribution of semivolatile organic compounds, with particular reference to long range transport. It was concluded that polar ice receives most chemicals from snow precipitation, although HCH and PBDEs are largely brought to the Arctic region in a particle-bound form. The polar ice caps serve to prevent air-water exchange of gaseous pollutants and the major removal pathway of pollutants in surface ice is via transport to lower layers of the ice cap. Because of its open nature, snow offers a greater potential for both diffusion and revolatilization than does ice in the polar regions. Therefore, the loss of ice from both polar and mountain regions has the potential to bring about a redistribution of pollutants already in this layer. It can be expected that the pollutants will largely move to the aquatic phase in the first instance, although more severe warming could lead to increases in the air phase as well. In the case of the Arctic, the pollutants released to the aquatic compartment will have a significantly higher potential to enter food webs. Pollutants entering mountain meltwater will be carried downstream and it is noteworthy that a number of major river-systems throughout the world have a glacial origin.

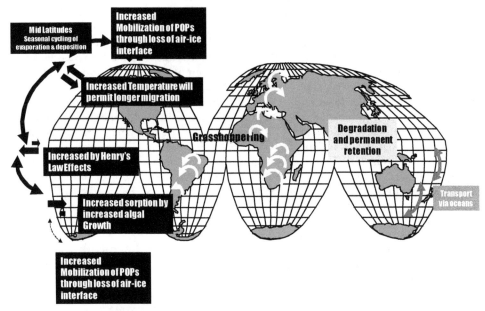

Fig. 3. Global Transport Processes for POPs as modified by Climate Change

5. Overall conclusions

The topic of global distillation has been reviewed with respect to both Northern and Southern Hemispheres. It appears that the global distillation phenomenon is best suited for operation in the Northern Hemisphere, although even here, it may be subject to significant competition from other processes. It is hypothesized that climate change will bring about significant interference with the operation of global distillation in both hemispheres. Although the increased temperatures associated with climate change have the potential to result in increased volatilization of persistent organic pollutants, the altered conditions may lead to increased losses of pollutants from the atmosphere through increased availability of hydroxyl radicals and other species active in the destruction of these substances in the atmosphere. There will be decreases in snow and rain scavenging in many areas, as a result of changes in precipitation patterns and a higher probability of particulate sorption in the atmosphere. The increase in phytoplankton blooms that is predicted to accompany climate change will provide a significant sink for persistent organic pollutants. Finally, the loss of ice at the poles will result in a redistribution of pollutants that have already been transport there, or those which subsequently arrive as a result of long range transport. The accompanying figure (Figure 3) provides a summary of the expected effects of climate change on global transport of persistent organic pollutants.

6. References

Archer, D., and Rahmstorf, S., 2010. *The Climate Crisis*. Cambridge University Press, ISBN 978-0-521-73255-0 UK.

Bell, P. R. F., Unwins, P. J. R., Elnmetri, I., Phillips, J. A., Fu, F-X. & Yago, A. J. E. (2005). Laboratory Culture Studies of *Trichodesmium* isolated from the Great Barrier Reef Lagoon, Australia. *Hydrobiologia*, Vol. 532, pp. 9-21, ISSN: 0018-8158

Berrojalbiz, N., Dachs J., Del Vento, S., Ojeda, M. J., Valle, M. C., Castro-Jiminéz, J., Mariani, G., Wollgast, J. & Hanke, G. (2011). Persistent Organic Pollutants in Mediterranean Seawater and Processes Affecting Their Accumulation in Plankton. *Environmental Science and Technology*, Vol. 45, pp. 4315-4222, ISSN 0013-936X.

Boethling, R., Fenner, K., Howard, P., Klečka, G., Madsen, T., Snape J. R. and Whelhan, M. J. (2009). Environmental Persistence of Organic Pollutants: Guidance for Development and Review of POP Risk Profiles. *Integrated Environmental Assessment and Management*, Vol. 5, pp. 539-556, ISSN 1551-3793.

Carrera, G., P. Fernandez, Grimalt, J.O., Ventura, M., Camarero, L., Catalan, J. Nickus, U., Thies H. & Psenner, R., 2002. Atmospheric Deposition of Organochlorine Compounds to Remote High Mountain Lakes of Europe. *Environmental Science and Technology*, Vol 36, pp. 2581-2588, ISSN 0013-936X.

Connell, D.W. (1990). *Bioaccumulation of Xenobiotic Compounds*. CRC Press, ISBN: 0849348102, Boca Raton, FL, USA.

Corsolini, S., Kannan, K. Imagawa, T. Focardi, S. & Geisy, J.P. (2002). Polychloronaphthalenes and Other Dioxin-like Compounds in Arctic and AntArctic Marine Food Webs. *Environmental Science and Technology*, Vol. 36, pp. 4229-4237, ISSN 0013-936X.

CSIRO (2007). Climate Change in Australia. ISBN 9781921232947. CSIRO, Canberra.

Dachs, J., Lohmann,R., Ockenden, W., Méjanelle, L., Eisenreich, S.J & Jones, K.C. (2002). Oceanic Biogeochemical Controls on Global Dynamics of Persistent Organic Pollutants. *Environmental Science and Technology*, Vol. 36, pp. 2851-2858, ISSN 0013-936X.

Dalla Vale, M., Codato, E. & Marcomini. A.(2007). Climate change influence on POPs distribution and fate: A case study. *Chemosphere*, 67, 1287-1295, ISSN: 0045-6535.

Daly, G. L. & Wania, F. (2005). Organic Contaminants in Mountains. *Environmental Science and Technology*, Vol. 39, pp. 385-398, ISSN 0013-936X.

Davidson D., Wilkinson, A.C., Blais, J.M., Kempe, L., McDonald, K.M. & Schindler, D.W. (2003). Orographic Cold-Trapping of Persistent Organic Pollutants by Vegetation in Mountains of Western Canada. *Environmental Science and Technology*, Vol. 37, pp. 209-215, ISSN 0013-936X.

Demers, M.J. Kelly, E.N. Blais, J.M. Pick, F.R. St Louis, V.L. & Schindler, D.W. (2007). Organochlorine Compounds in Trout from Lakes over a 1600 Meter Elevation Gradient in the Canadian Rocky Mountains. *Environmental Science and Technology*, Vol. 41, pp. 2723-2729, ISSN 0013-936X.

Desa, E., Suresh, T., Matondkar, S .G. P, Desa, E., Goes, J., Mascarenhas, A., Parab, S.G., Shaikh, N. & Fernandes, C. E. G. (2005). Detection of *Trichodesmium* Bloom Patches along the Eastern Arabian Sea by IRS/P4 OCM Ocean Colour Sensor and by in-site Measurements. *Indian Journal of Marine Sciences*, Vol. 34, pp. 374-386 ISSN: 0379-5136.

Dickhut, R.M., Padma, T.V. & Cincinelli, A. (2004). Fractionation of Stable Isotope-Labeled Organic Pollutants as a Potential Tracer of Atmospheric Transport Processes. *Environmental Science and Technology*, Vol.38, pp. 3871-3876, ISSN 0013-936X.

Fernandez, P. & Grimalt, J.O. (2003). On the global distribution of persistent organic pollutants. *Chimea*, Vol. 57, pp. 514 – 521, ISSN: 1080-6059

Focardi, S., Gaggi, C., Chemello, G. & Bacci, E. (1991). Organochlorine residues in moss and lichen samples from two AntArctic areas. *Polar Record,* Vol. 27, pp. 241-244, ISSN: 0032-2474.

Franz, T.P. & Eisenreich, S.J. (1998). Snow Scavenging of Polychlorinated Biphenyls and Polycyclic Aromatic Hydrocarbons in Minnesota. *Environmental Science and Technology,* Vol. 32, pp. 1771-1778, ISSN 0013-936X.

Gallego, E., Grimalt, J.O., Bartrons, M.,. Lopez, J.F., Camarero, L., Catalan, J., Stuchlik, E. & Battarbee, R. (2007). Altitudinal Gradients of PBDEs and PCBs in Fish from European High Mountain Lakes. *Environmental Science and Technology*, Vol. 41 pp., 2196-2201, ISSN 0013-936X.

Gobas, F. & Maclean, L.G. (2003). Sediment- water distribution of organic contaminants in aquatic ecosystems – the role of organic carbon mineralization. *Environmental. Science and Technology*, Vol. 37, pp, 735- 741, ISSN 0013-936X.

Grimalt, J.O., Borghini, F., Sanchez-Hernandez, J.C., Barra, R., Torres Garcia, C. J. & Focardi, S. (2004). Temperature Dependence of the Distribution of Organochlorine Compounds in the Mosses of the Andean Mountains. *Environmental Science and Technology*,Vol. 42, pp. 1416-1422, ISSN 0013-936X.

Gioia, R., Nizzetta, L., Lohmann, R., Dachs, J., Temme, C. & Jones, K.C. (2008). Polychlorinated Biphenyls (PCBs) in Air and Seawater of the Atlantic Ocean: Sources, Trends and Processes. *Environmental Science and Technology* Vol. 38, pp. 5386-5392, ISSN 0013-936X.

Gustafason, K.E. & Dickhut, R.M. (1997). Response to Comment on "Particle/Gas Concentrations and Distributions of PAHs in the Atmosphere of Southern Chesapeake Bay". *Environmental Science and Technology*, Vol. 31, pp. 3738-3739, ISSN 0013-936X.

Hallegraeff G., Beardall, J., Brett, S., Doblin M., Hosja, W., de Salas, M. and Thompson, P. Marine Climate Change in Australia. Impacts and Adaptation Responses. 2009 Report Card. Phytoplankton. In A Marine Climate Change Impacts and Adaptation Report Card for Australia 2009 (Eds. E.S. Poloczanska, A.J. Hobday and A.J. Richardson), NCCARF Publication 05/09, ISBN 978-1-921609-03-9, Gold Coast, Australia.

Hargrave, B. T., Barrie, L.A., Bidleman, T.F. & Welch, H.E. (1997). Seasonality in Exchange of Organochlorines between Arctic Air and Seawater. *Environmental Science and Technology*, Vol. 31, pp. 3258-3266, ISSN 0013-936X.

Harner, T & Bidleman, T.F. (1998). Measurement of octanol-air partition coefficients for polycyclic aromatic hydrocarbons and polychlorinated naphthalenes. *Journal of Chemical and Engineering Data*, Vol. 43, pp. 40-46, ISSN 0021-9568.

Hayes G. C., Richardson, A. J. & Robinson, C (2005). Climate Change and Marine Phytoplankton. *Trends in Ecology and Evolution*, Vol. 20, pp. 337-344, ISSN: 0169-5347.

Hoff, J. T., Wania, F., Mackay, D. And Gilham, R. (1995). Sorption of Nonpolar Organic Vapors by Ice and Snow. *Environmental Science and Technology*, Vol. 29, pp. 1982-1989, ISSN 0013-936X.

Hung, H., Halsall, C.J., Blanchard, P., Li, H.H., Fellin, P., Stern, G. & D.B. Rosenberg, D.B. (2001). Are PCBs in the Canadian Arctic Atmosphere Declining? Evidence from 5 Years of Monitoring. *Environmental Science and Technology*, Vol. 35, pp. 1303-1311, ISSN 0013-936X.

IPCC 2007a. Intergovernmental Panel on Climate Change (IPCC). 2007. Fourth assessment report. Climate Change 2007 – The Physical Science Basis. Cambridge, ISBN 978 0521 88009-1, Cambridge University Press.

IPCC 2007b. Intergovernmental Panel on Climate Change (IPCC). 2007. Fourth assessment report. Climate Change 2007 –Impacts, Adaptation and Vulnerability. Cambridge, ISBN 978 0521 88010-7, Cambridge University Press.

IPCC 2007c. Intergovernmental Panel on Climate Change (IPCC). 2007. Fourth assessment report. Climate Change 2007 – Mitigation of Climate Change. Cambridge, 978 0521 88011-4, Cambridge University Press.

Kahru, M., Brotas, V., Manzano-Sarabia, M., and Mitchell, B. G. (2010). Are Phytoplankton Blooms occurring earlier in the Arctic? *Global Change Biology*, Vol. 17, pp. 1733-1739, ISSN 1365-2486.

Kalantzi, O.I., Alcock, R.E., Johnston, P.A., Santillo, D., Stringer, R.L.,. Thomas, G.O. & Jones, K.C. (2001). The Global Distribution of PCBs and Organochlorine Pesticides in Butter. *Environmental Science and Technology*, Vol. 35, pp. 1013-1018, ISSN 0013-936X.

Klánová, J., Matykiewiczová, N., Máčka, Z., Prošek, P., Láska, K. & Klán, P. (2008). Persistent Organic Pollutants in Soils and Sediments from James Ross Island, Antarctica. *Environmental Pollution*, Vol. 152, pp. 416-423, ISSN 0269-7491.

Kűhne, R., Ebert, R-U. & Schűűrmann, G. (2005). Prediction of the Temperature Dependency of Henry's Law Constant from Chemical Structure. *Environmental Science and Technology*, Vol. 39, pp. 6705-6711, ISSN 0013-936X.

Jurado, E., Lohmann, R.,. Meijer, S., Jones, K.C. and Dachs, J. (2004). Latitudinal and Seasonal Capacity of the Surface Oceans as a Reservoir of Polychlorinated Biphenyls. *Environmental Pollution* 128, 149-162. ISSN 0269-7491.

Lakaschus, S., Weber, K., Wania, F., Bruhn, R. & Schrems, O. (2002). The Air-Sea Equilibrium and TimeTrend of Hexachlorocyclohexanes in the Atlantic Ocean between theArctic and Antarctica. Environmental Science and Technology 36: 138-145, ISSN 0013-936X.

Lamon, L., von Waldow, H., Macleod, M., Scheringer, M., Marcomini, A. & Hungerbühler, K. (2009). Modeling the Global Levels and Distribution of Polychlorinated Biphenyls in Air under a Climate Change Scenario. *Environmental Science and Technology*, Vol. 43, pp. 5818-5824, ISSN 0013-936X.

Lead, W.A., Steinnes, E. & Jones, K.C. (2002). Atmospheric Deposition of PCBs to Moss (*Hylocomium splendens*) in Norway between 1977 and 1990. *Environmental Science and Technology* 30: 524-530, ISSN 0013-936X.

Li, X., Chen, J., Zhang, L., Quiao X. & Huang, L. (2006). The Fragment Constant Method for Predicting Octanol–Air PartitionCoefficients of Persistent Organic Pollutants at Different Temperatures. *Journal of Physical Chemistry* (Reference Data), Vol. 35, pp. 1366 – 1384, ISSN: 1520-6106.

Ma, J. (2010). Atmospheric transport of persistent semivolatile organic chemicals to the Arctic and cold condensation in the mid-troposphere – Part 1: 2-D modelling in mean atmosphere. *Atmospheric Chemistry and Physics*, Vol. 10, pp. 7303- 7314, ISSN ISSN 1680-7316.

Ma, J. & Cao, Z. (2010). Quantifying the Perturbations of Persistent Organic Pollutants Induced by Climate Change. *Environmental Science and Technology*, Vol. 44, pp. 8657-8573, ISSN 0013-936X.

Macdonald, R.W., Mackay, D. & Hickie, B. (2002). Contaminant amplification in the environment: Revealing the fundamental mechanisms. *Environmental Science and Technology*, Vol. 36, pp.: 457A–462A, ISSN 0013-936X.

Macdonald, R.W., Mackay, D., Li, Y-F & Hickie, B. (2003). How will global climate change affect risks from long range transport of persistent organic pollutants? *Ecological Risk Assessment*, Vol. 9, pp. 643-660, ISSN 1811-0231.

Mackay, D., Arnot, J.A., Webster, E. and Reid, L. (2009). The Evolution and Future of Environmental Fugacity Models. In *Ecotoxicology Modelling* . Devillers. J. Ed. Springer, pp 355 – 375, ISBN: 1441901965, Berlin, Germany.

McConnell, L., Kucklick, J.R., Bidleman, T.F., Ivanov, G.P. & Chernyak, S.M. (1996). Air-Water Gas Exchange of Organochlorine Compounds in Lake Baikal, Russia. *Environmental Science and Technology*, Vol. 30, pp. 2975-2983, ISSN 0013-936X.

Meijer, S.N., Steinnes, E., Ockenden, W.A. & Jones, K.C. (2002). Influence of Environmental Variables on the Spatial Distribution of PCBs in Norwegian and U.K. Soils: Implications for Global Cycling. *Environmental Science and Technology*, Vol. 36, pp. 2146-2153, ISSN 0013-936X.

Meylan, W.M & Howard, P.H. (2005). Estimating Octanol–Air Partition Coefficients with Octanol–Water Partition Coefficients and Henry's Law Constants. *Chemosphere*, Vol. 61, pp. 640-644, ISSN: 0045-6535.

Moeckel, C., MacLeod, M., Hungerbühler, K & Jones, K.C. (2008). Measurement and Modeling of Diel Variability of Polybrominated Diphenyl Ethers and Chlordanes in Air. *Environmental Science and Technology*, Vol. 42, pp. 3219–3225, ISSN 0013-936X.

Muir, D.C.G., Omelchenko, A., Grift, N.P., Savoie, D.A., Lockhardt, W.L., Wilkinson, P. & Brunskill, G.J. (1996). Spatial Trends and Historical Deposition of Polychlorinated Biphenyls in Canadian Midlatitude and Arctic Lake Sediments. *Environmental Science and Technology*, Vol. 30, pp. 3609–3617, ISSN 0013-936X.

Nizzetto, L., Lohmann, R, Gioia, G., Jahnke, A., Temme, C., Dachs, J., Herckes, P., Diguardo, A. & Jones, K.C. (2008). PAHs in Seawater along a North-South Atlantic Transect: Trends, Processes and Possible Sources. *Environmental Science and Technology*, Vol. 42, pp. 1580 – 1585, ISSN 0013-936X.

Noël, M., Barrett-Lennard, L., Guinet, C. Dangerfield, N. & Ross, P. S. (2009). Persistent organic pollutants (POPs) in killer whales (*Orcinus orca*) from the Crozet Archipelago, Southern Indian Ocean. *Marine Environmental Research*, Vol. 68, pp 196-202, ISSN: 0141-1136.

Radonić, J., Miloradov, M. V., Sekulić, M. T., Kiruski, J., Djogjo, M. & Milanovanović, D. (2011). The Octanol–Air Partition Coefficient, K_{OA}, as a Predictor of Gas–Particle Partitioning of Polycyclic Aromatic Hydrocarbons and Polychlorinated Biphenyls at Industrial and Urban Sites. *Journal of the Serbian Chemical Society*, Vol. 76, pp. 1-12, ISSN 0352-5139.

Sadler R., Gabric, A., Shaw, G., Shaw, E. & Connell, D. (2011). An opinion on the distribution and behaviour of chemicals in response to climate change, with particular reference to the Asia-Pacific Region. *Toxicological and Environmental Chemistry*, Vol. 93, pp. 1-29, ISSN: 0277-2248.

Scheringer, M., Wegman, F., Fenner, K. & Hungerbühler, K. (2000). Investigation of the Cold Condensation of Persistent Organic Pollutants with a Global Multimedia Fate Model. *Environmental Science and Technology*, Vol. 34, pp. 1842-1850, ISSN 0013-936X.

Scheringer, M., M. McLeod, and F. Wegmann, 2006. Analysis of Four Current POP Candidates with the OECD Pov and LRTP Screening Tool. http://www.sust-chem.ethz.ch/docs/POP_Candidates_OECD_Tool.pdf

Shiu, W.Y. & Mackay, D. (1986). A critical review of aqueous solubilities, vapour pressures. Henrys Law constants and octanol – water partition coefficients of polychlorinated biphenyls. *Journal of Physical Chemistry. Reference Data*, Vol. 15, pp. 911 – 929, ISSN 1520-6106.

Shoeib, M. & Harner,T. (2002). Using Measured Octanol-Air Partition Coefficients to Explain Environmental Partitioning of Organochlorine Pesticides. *Environmental Toxicology and Chemistry*, Vol. 21, pp. 984-990, ISSN 1552-8618.

Shoeib, M., T. Harner and K. Kannan, 2002. Octanol-Air Partition Coefficients and Indoor Air Measurements of PFOS and Precursor Compounds. SETAC Abstracts P212. http://abstracts.co.allenpress.com/pweb/setac2002/document/20575

Simcik, M., Basu, I., Sweet, C. & Hites, R. A., (1999). Temperature Dependence and Temporal Trends of Polychlorinated Biphenyl Congeners in the Great Lakes Atmosphere. *Environmental Science and Technology*, Vol. 33, pp. 1991-1995, ISSN 0013-936X.

Staudinger, J. & Roberts, P.V. (2001). A Critical Compilation of Henry's Law Constant Temperature Dependence Relations for Organic Compounds in Dilute Aqueous Solutions. *Chemosphere*, Vol. 44, pp. 561-576, ISSN: 0045-6535.

Stocker, J., Scheringer, M., Wegman, F., & Hungerbühler K. (2007). Modeling the Effect of Snow and Ice on the Global Environmental Fate and Long-Range Transport Potential of Semivolatile Organic Compounds. *Environmental Science and Technology*, Vol. 41, pp. 6192-6198, ISSN 0013-936X.

Sweetman, A.J. & Jones, K.C. (2000). Modeling Historical Emissions and Environmental Fate of PCBs in the United Kingdom. *Environmental Science and Technology*, Vol. 34, pp. 863-869, ISSN 0013-936X.

United States Environmental Protection Agency (USEPA) (1996). On-line Tools for Site Assessment Calculation. Estimated Henry's Law Constants. http://www.epa.gov/athens/learn2model/part-two/onsite/esthenry.html

Von Waldow, H., MacLeod, M., Jones, K. Scheringer, M. & Hungerbühler, K. (2010). Remoteness from Emission Sources Explains the Fractionation Pattern of Polychlorinated Biphenyls in the Northern Hemisphere. *Environmental Science and Technology*, Vol. 44, pp. 6183–6188, ISSN 0013-936X.

Wania, F., (2003). Assessing the Potential of Persistent Organic Chemicals for Long-Range Transport and Accumulation in Polar Regions. *Environmental Science and Technology*, Vol. 37, pp. 1344-1351, ISSN 0013-936X.

Wania F. (2006). Potential of degradable organic chemicals for absolute and relative enrichment in the Arctic. *Environmental Science and Technology*, Vol 40, pp. 569–577, ISSN 0013-936X.

Wania, F., & Mackay, D. (1995). A Global Distribution Model for Persistent Organic Chemicals. *Science of the Total Environment*, Vol. 160/161, pp. 211-232, ISSN: 0048-9697

Wania, F., & Mackay. (1996). Tracking the distribution of persistent organic pollutants. *Environmental Science and Technology* Vol. 30, pp. 390A-396A, ISSN 0013-936X.

Wania F. & Westgate, J.N. (2008). On the Mechanism of Mountain Cold-Trapping of Organic Chemicals. *Environmental Science and Technology*, Vol. 42, pp. 9092-9098, ISSN 0013-936X.

Wania, F., Lei, Y.D. & T. Harner, T. (2002). Estimating Octanol-Air Partition Coefficients of Nonpolar Semivolatile Organic Compounds from Gas Chromatographic Retention Times. *Analytical Chemistry* Vol. 74, pp. 3476-3483, ISSN 0003-2700.

Wassmann, P., Duarte, C. M., Agusti, S. & Sejr, M. K. (2011). Footprints of Climate Change in the Arctic Marine Ecosystem. *Global Change Biology*, Vol. 17, pp. 1325-1249, ISSN 1365-2486.

Weber, I., Halsall, C.J., Muir, D.C.G., Texiera, C., Burniston, D.A., Strachan, W.M.J., Hung, H., Mackay, N., Arnold, D. & Kylin, H. (2006). Endosulfan and γ-HCH in the Arctic: An Assessment of Surface Seawater Concentrations and Air-Sea Exchange. *Environmental Science and Technology*, Vol. 40, pp. 7570-7576, ISSN 0013-936X.

WHO (World Health Organization) (2009). Protecting health from climate change. World Health Organization, ISBN: 978 92 4 159888 0, Geneva, Switzerland.

Xie, Z., Erbinghaus, R., Temme, C., Lohmann, R., Caba, A. & Ruck, W.W. (2007). Occurrence and Air-Sea Exchange of Phthalates in the Arctic. *Environmental Science and Technology*, Vol. 41, pp. 4555-4560, ISSN 0013-936X.

Yao, Y., Harner, T., Blanchard, P., Tuduri, L., Waite, D., Poissant, L., Murphy, C., Belzer, W., Aulagnier, F. &. Sverko, E.(2008). Pesticides in the Atmosphere Across Canadian Agricultural Regions. *Environmental Science and Technology*, Vol. 42, pp. 5931-5937, ISSN 0013-936X.

Zhang, L., Ma, J., Venkatesh, S., Li, J-F. & Cheung. P. (2008). Modelling of Episodic Intercontinental Long-Range Transport of Lindane. *Environmental Science and Technology*, Vol. 42, pp. 8791-8797, ISSN 0013-936X.

8

The Mass Distribution of Particle-Bound PAH Among Aerosol Fractions: A Case-Study of an Urban Area in Poland

Wioletta Rogula-Kozłowska[1], Barbara Kozielska[2],
Barbara Błaszczak[1] and Krzysztof Klejnowski[1]
[1]Institute of Environmental Engineering of the Polish Academy of Sciences, Zabrze,
[2]Faculty of Energy and Environmental Engineering,
Department of Air Protection, Silesian University of Technology, Gliwice,
Poland

1. Introduction

Zabrze is one of the fourteen Silesian cities that form together the Silesian Agglomeration (Fig. 1-4). The Silesian Agglomeration lies in the center of the Silesia Province, occupies 1230 km², its population is about 2.1 million (1691 inhabitants per one square kilometer). It is one of the most urbanized and industrialized regions of Central Europe. Such a dense concentration of people on such a heavily urbanized and industrialized area is unique in Europe. About 50% of the Silesia Province gross product and 7% of the gross domestic product come from the Silesian Agglomeration. Six European capital cities, Berlin, Prague, Wien, Bratislava, Budapest and Warsaw, lie within 600 km from Katowice, the capital city of the Agglomeration. The main transport routes linking Poland with Western Europe run through it in all directions.

From the air protection point of view, the Silesian Agglomeration is one of the most interesting regions both in Poland and in Europe. Hard coal of better quality, mined in the western part of the Agglomeration since the 18th century, has been processed into coke and gas in five cities of the Agglomeration. Poorer quality coal, from the eastern part of the Agglomeration, is burnt in several great power stations, smaller power and heating plants, local heating plants and in domestic stoves. Almost all branches of industry, such as electrical, chemical, glass-making, textile, clothing, and ceramic industries, ferrous and non-ferrous metallurgy, machine-building, hard coal mining and coking very actively have been deteriorating the natural environment for about 200 years.

Nevertheless, three recent decades of economical changes forced in the Silesian Agglomeration the greatest in Poland drop of industrial air pollution (in Zabrze, yearly dust fall exceeded 2100 g m⁻² in the 70s, oscillated between 700 and 800 g m⁻² in the 80s, was less than 350 g m⁻² after 1995). The concentration of the ambient particulate matter (PM) dropped significantly (Fig. 1). The greatest drop of the PM concentrations occurred between 1985 and 2000, when the total industrial emission of air pollutants drastically decreased in the effect of the transformation of the whole national industry and closing or restructuring of many great plants in the 80s. In 1990, the list of 80 Polish plants exerting the greatest impact on the environment was announced. More than 20 of them were located in the Katowice Region whose the Silesian Agglomeration was a sub-region. The actions to lower the emission from these plants and limiting the output of heavy industry decreased the emission of dust and its gaseous precursors (Central Statistical Office of Poland [CSO], 1976-2001) and halved the ambient concentrations of dust at measuring points (at the beginning of the 80s, in Dąbrowa Górnicza the PM concentration was almost 0.5 mg m⁻³). During the 90s, the PM concentrations continued to drop and since 2000 the yearly PM concentrations have remained constant at the approximate level between 65-110 µg m⁻³ depending on a measuring site. The industrial emission of PM and its precursors (SO_2, NO_x) has also remained almost stable (Central Statistical Office of Poland [CSO], 2001-2011).

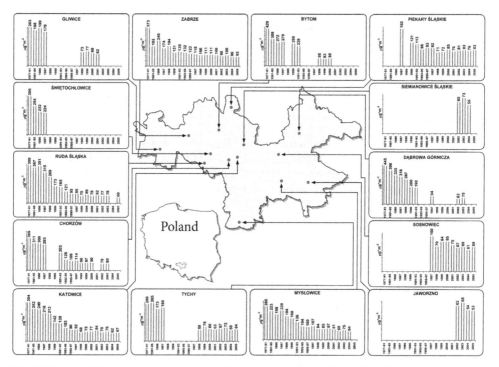

Fig. 1. The total PM concentrations (µg m⁻³) in 14 cities of the Silesian Agglomeration in 1977-2005 (SSI, 1980-2006)

The concentrations of PM_{10}[1], although slightly lower than in the 90s, are still high in the Silesian Agglomeration and have also been stabilized in each of the cities since 2000 (Fig. 2).

---

[1]For a number $d>0$, PM_d is the fraction of the particles that have the aerodynamic diameter not greater than d. For $0<c<d$, PM_{c-d} denotes the fraction of particles with the diameters between c and d. We have:
$PM_{2.5}$–particles with the aerodynamic diameter not greater than 2.5 µm (fine particles)
$PM_{2.5-10}$–particles with the aerodynamic diameter between 2.5 and 10 µm (coarse particles)
PM_{10}–particles with the aerodynamic diameter not greater than 10 µm ($PM_{2.5}$ and $PM_{2.5-10}$ together)

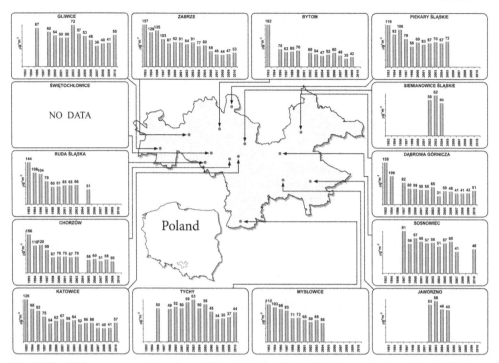

Fig. 2. The PM_{10} concentrations ($\mu g\ m^{-3}$) in 14 cities of the Silesian Agglomeration in 1993-2010 (SSI, 1980-2006; The Provincial Inspector for Environmental Protection in Katowice [PIEP], 2002-2011)

There are hundreds of organic compounds to be found in PM, among them more than one hundred polycyclic aromatic hydrocarbons (PAH). Since PAH tend to have low vapor pressures, they are usually adsorbed onto PM in the atmosphere. The vapor pressure of a PAH is inversely proportional to the number of rings it contains. As a result, the larger molecular weight PAH (≥ 4) are mostly adsorbed onto PM in atmospheric samples, while the lower molecular weight PAH can be found both free in the atmosphere and bound to particles. In some conditions, from 72% to even > 98% of four-, six-, and seven-ring ambient PAH are bound onto particles of PM_3 and PM_7 (Sheu et al., 1997). Seventeen PAH: acenaphthene (Acy), acenaphthylene (Ace), anthracene (An), benzo(a)anthracene (BaA), benzo(a)pyrene (BaP), benzo(e)pyrene (BeP), benzo(b)fluoranthene (BbF), benzo(j)fluoranthene (BjF), benzo(k)fluoranthene (BkF), benzo(g,h,i)perylene (BghiP), chrysene (Ch), dibenzo(a,h)anthracene (DBA), fluoranthene (Fl), fluorene (F), phenanthrene (Ph), pyrene (Py) and indeno(1,2,3-cd)pyrene (IP) are most often investigated. The most hazardous property of air pollutants is their carcinogenicity. Seven PAH: BaA, BaP, BbF, BkF, Ch, DBA and IP the USEPA classified as probable human carcinogens (http://www.epa.gov/).

The most important natural sources of PAH are forest fires and eruptions of volcanoes. However, amounts of the natural ambient PAH are small compared to the amounts of the anthropogenic PAH. The majority of anthropogenic PAH come from incomplete combustion of fossil fuels or organic matter (Sienra et al., 2005; Zou et al., 2003). In urban air this includes PAH combustion in car engines and residential heating (Harrison et al., 1996; Manoli et al., 2004; Kristensson et al., 2004).

Since 1977, in all cities of the Silesian Agglomeration, the State Sanitary Inspection (SSI, Department in Katowice) has measured ambient concentrations of PM-bound BaP continuously and 8 other PAH periodically (Fl, BaA, Ch, BbF, BkF, DBA, BghiP, IP; several years in each city). The total ambient concentration of these nine PAH drastically decreased in the period 1977–2005 (Fig. 3). The concentrations of PM-bound BaP dropped rapidly in all cities of the Agglomeration in 1980–1990 (Fig. 4). In each city the vast yearly concentrations of BaP from 1977–1981, reaching 500 ng m^{-3} in Ruda Śląska, decreased several times during this decade. During the 90s, the concentrations of BaP dropped to less than 20 ng m^{-3} in each city of the Agglomeration. These concentrations have been still very high and hazardous to humans and the concentrations of PM_{10}-bound BaP, measured since 2001, were high (Fig. 4), like the concentrations of PM-bound one.

In general, the ambient particles with the aerodynamic diameter greater than 10 μm are not inhalable. Therefore, from the sanitary point of view, the concentrations of PM_{10} should be the measure of the hazard from ambient particles. In all cities of the Agglomeration, the yearly average concentrations of PM_{10} and PM_{10}-bound BaP exceed their permissible levels (40 μg m^{-3} and 1 ng m^{-3}, respectively, Fig. 2). The limit of 40 μg m^{-3} on the yearly average concentration of PM_{10} has been in effect in majority of the European countries for about 20 years (in Poland since 1998). Although the air quality has been improving for the latest thirty years, there is no city in the Silesian Agglomeration where this standard was not yearly exceeded in the period 1993-2010 (Fig. 2). Even the yearly average $PM_{2.5}$ concentrations, measured continuously since 2001 in Zabrze, exceed 25 μg m^{-3}, the standard for the $PM_{2.5}$ concentrations, every year (Rogula-Kozłowska et al., 2010; Klejnowski et al., 2007a, 2007b, 2009; PIEP, 2002-2011). Although the spectacular reduction of the industrial emissions, especially of coarse dust, caused significant drop of the concentrations of PM and PM-bound BaP the PM_{10} concentrations decreased only a little (Fig. 2).

The poor air quality conditions in the cities of the Agglomeration are due to bad spatial arrangement of industrial and urban infrastructure. The industrial objects are interspersed with living quarters. The differences in the level of industrialization, urbanization and land use cause spatial non-homogeneity of the air pollution from industrial, municipal or vehicular sources even over small areas. The area of the Agglomeration is also affected by periodically occurring episodes of very high concentrations of air pollutants (especially of PM_{10} in winter in city centers), which inflate the yearly PM concentrations.

The inventory of the Silesian air pollution sources from 2006 contains 107 business entities emitting PM_{10} and BaP within the Silesian Agglomeration (Air protection program [APP], 2010). In 2006, they emitted 4.9 Gg of PM_{10} and 1.3 Mg of benzo(a)pyrene into the air. In average, this industrial emission was 44% and 28% of, respectively, the total PM_{10} and BaP emissions in the Agglomeration. The municipal and household emission[2], mainly from local heating plants and domestic stoves, have even greater share in the PM_{10} and the BaP total

---

[2]Heating plants and furnaces fed with solid fuel (mainly coal) are main sources of PM_{10} and BaP. They include small (local) heating plants and domestic stoves. In the Silesian Agglomeration, the estimated average of the household heat demands covered by solid fuel combustion in domestic stoves is still greater than 34%. The emissions of PM_{10} and PM_{10}-bound BaP from the solid fuel furnaces are more than 94% of their totals in the surface emission. It is due to bad technical condition and age of the heating plants and stoves and also to the poor quality of the combusted coal.

concentrations. In 2006, it was 5.3 Gg and 3.2 Mg, what made 47% and 71% of the total PM_{10} and BaP emissions. This contribution is even greater in a heating season — in winter.

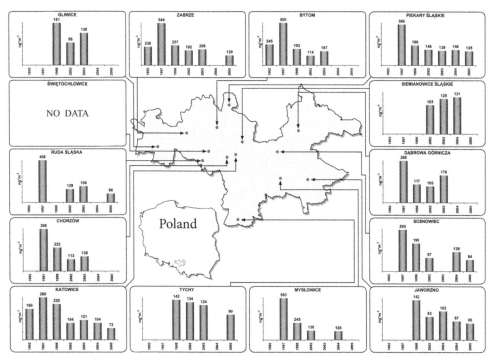

Fig. 3. The concentrations of the sum of 9 PM-bound PAH (Fl, BaA, Ch, BbF, BkF, BaP, DBA, BghiP, IP) in 14 cities of the Silesian Agglomeration in 1992-2005 (SSI, 1993-2006)

The elevated autumn and winter PAH concentrations in the Agglomeration my be linked with the emission from municipal sources. The seasonal dependence of the PM-bound PAH concentrations was observed in three Silesian cities: Katowice, Sosnowiec and Zawiercie in 2008. The seasonal PAH concentrations were 5.1-18.6 ng m^{-3} in spring, 5.4-7.6 ng m^{-3} in summer, 6.8-24.0 ng m^{-3} in autumn and 26.2-61.3 ng m^{-3} in winter (Zaciera et al., 2010).

Although the reduction of the industrial emission improved the air quality in the Silesian Agglomeration the hazard from PM_{10} (especially from the smallest particles) might grew owing to growth of vehicular emission. For the last two decades, the Poles have imported more than 920000 second-hand cars yearly. Majority of the cars were older than 10 years, they did not meet the emission standards and had high fuel-consumption.

In Gliwice, from April to June 2003, in a trafficked street canyon (1400 vehicles per hour), the average PM_{10} concentration was 94.0 μg m^{-3} and was higher by 40.0 μg m^{-3} than the one measured 100 m apart. The average concentration of total PAH, equal to 191.56 ng m^{-3}, was 1.5 times greater than the background concentration (Grynkiewicz-Bylina et al., 2005). In Zabrze, in summer 2005, the average concentration of the total PM_{10}- and the total $PM_{2.5}$-bound PAH at crossroads were 65.6 ng m^{-3} and 44.4 ng m^{-3}, and were 1.9 and 3.4 times greater than the background concentrations, respectively (Ćwiklak et al., 2009). In Bytom, on the turn of February and March 2007, the average concentrations of the total vehicular $PM_{2.5}$- and the total vehicular PM_{10}-bound PAH were between 56.2 and 73.4 ng m^{-3} and 75.1 and 91.0 ng m^{-3}, respectively. The significant in this city influence of the industrial emission and the emission from low sources was excluded by proper location of the measuring points (Kozielska et al., 2009).

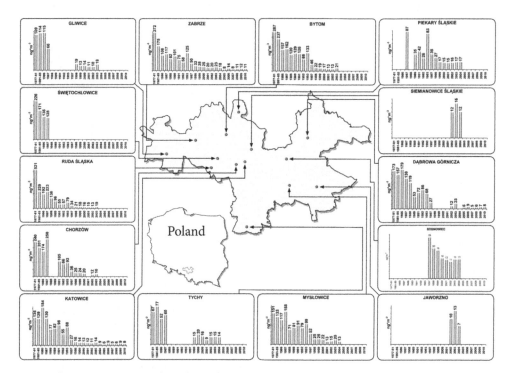

Fig. 4. The concentrations of PM-bound BaP in 1977-2001 and PM_{10}-bound BaP in 2002-2010 in 14 cities of the Silesian Agglomeration (SSI, 1982-2006; PIEP, 2005-2011)

The hazard from the air pollution in the neighborhood of trafficked roads is also elevated by the elevated toxicity of traffic PM_{10} containing allergens and carcinogens, also PAH (Kinney et al., 2000; Pakkanen et al., 2003; Vogt et al., 2003).

Now, in the Silesian Agglomeration, when the number of industrial and of the vast fugitive emission sources of PM (mine heaps, dumps) decreased, main PM and PAH sources are domestic furnaces, heat and power plants and road traffic. In the Agglomeration, the industrial and the municipal sources contribute to the air pollution much more than such sources in the Western Europe countries (EMEP, 2009), and the role of vehicular emission grows. As a result, the proportion of fine particles in PM increases. Ambient fine particles have relatively great surface to adsorb PAH (Ravindra et al., 2001; Sheu et al., 1997), so the fine dust is richer in PAH than the coarse one. Therefore, investigation of the particle size distribution of PM and the PAH content of PM fractions is crucial to abate the adverse effects of PM.

The work presents the method and the results of the investigations of the mass distribution of some PM-bound PAH among the PM fractions in the Silesian Agglomeration. The investigated PAH, three-ring: Acy, Ace, F, Ph, An, four-ring: Fl, Py, BaA, Ch, five-ring: BbF, BkF, BaP, DBA and six-ring: BghiP, and IP, are on the USEPA CWA list of the priority pollutants.

2. Methods

The site in Zabrze, selected for the experiment, is representative of the air pollution conditions typical in the Silesian Agglomeration – by the Directive 2008/50/EC definition, it is an urban background measuring point (Directive, 2008). The effects of the industrial and the municipal emissions on living quarters of the Agglomeration are represented and may be observed here very well.

Ambient dust was sampled with the use of a thirteen stage DEKATI low pressure impactor (DLPI), which collects thirteen PM fractions onto thirteen separate substrate filters (Table 1). The principle of DLPI operating may be found in (Klejnowski et al., 2010).

There were two periods of sampling: from 7 May to 2 August (summer) and from 26 October to 27 December (winter) 2007. One sample-taking lasted about one week. Seven such sample-takings were done in winter and nine in summer, the measurements covered the sampling periods in 98 and 92%, respectively, and whole summer (2nd and 3rd quarter of 2007) and winter (1st and 4th quarter of 2007) in 27% and 45%.

The mass of the dust collected on aluminum substrates was determined by weighing the substrates before and after exposure on a Mettler Toledo micro-balance. Before weighing, the substrates were kept in the weighing room for 48 hours (temperature 20 ± 2^0C, relative air humidity $48\pm5\%$). The concentrations of the fractions of PM were computed from the volume of air passed through the impactor and the masses of the dust collected on its stages. All the samples (substrates), till analyzing, were kept in a refrigerator in tight and lightproof containers.

The winter and summer samples were developed separately. For each of the thirteen PM fractions, all its samples (7 in winter and 9 in summer samples per fraction) were extracted together in an ultrasonic bath in dichloromethane (CH_2Cl_2). The extract was percolated, washed and dried by evaporating in the helium atmosphere. The dry residue was diluted in propane-2 ($CH_3CH(OH)CH_3$) and distilled water was added to receive the proportion 15/85 (v/v) of propanol-2 to water. For selective purification, the resulting samples were solidified (SPE) by extracting in columns filled with octadecylsilane (C_{18}, Supelco). PAH were eluted with the use of dichloromethane (CH_2Cl_2). The extract of a PAH fraction was condensed in the helium atmosphere to the volume of 0.5 cm^3. The samples were analysed on a Perkin Elmer Clarus 500 gas chromatograph with a Flame Ionization Detector (FID). An RTX 5 Restek capillary 30 m x 0.32 mm x 0.25 μm column was used to separate the sample components. The flow of the carrier gas, helium, was 1.5 cm^3 min^{-1}. Calibration curves for 15 PAH standards were used in quantitative determinations. The linear correlation of the surfaces of the peaks with the PAH concentrations was checked in the concentration range 1 – 4 ng μl^{-1}. The correlation coefficient ranged from 0.90 to 0.97. The time of the whole analysis was 40 min. FID was provided with hydrogen (45 cm^3 min^{-1}) and air (450 cm^3 min^{-1}). The recoveries of PAH were determined using a standard containing the 15 PAH. They ranged from 85% to 93%.

3. Discussion of the results

3.1 Concentrations of PM- and PM-bound PAH. Origin of PAH in Zabrze

The winter PM_{10} concentrations exceeded 46 µg m^{-3}, the summer ones reached almost 19 µg m^{-3} (Table 1). Such a difference is due to very high emission of PM from combustion of fossil fuels in winter, specific of the Silesian Agglomeration (Rogula-Kozłowska et al., 2008; Pastuszka et al., 2010). The average concentrations of PM_{10} and PM in the experimental period (33.5 µg m^{-3} and 32.5 µg m^{-3}) were lower than the yearly concentrations of PM_{10} in 2007 and PM in 2005 (Fig. 1 and 2). The explanation may by that the former were sampled with DLPI at about 5 m above the ground level during about 16 weeks, missing a part of winter when in Zabrze the highest PM concentrations occur (Klejnowski et al., 2007a, 2009), the latter were measured by SSI at the height 2.5 m during the whole year.

Fraction, µm	PM, µg m^{-3}	Acy	Ace	F	Ph	An	Fl	Py	BaA	Ch	BbF	BkF	BaP	DBA	BghiP	IP	ΣPAH
0.03-0.06 W	0.4	nd	nd	0.14	0.14	nd	0.03	0.04	nd	0.08	0.07	0.08	0.05	nd	nd	nd	0.62
S	0.09	0.01	0.02	0.02	0.05	nd	0.04	0.02	nd	0.11	0.14	0.14	0.04	0.01	0.05	0.06	0.7
0.06-0.108 W	0.59	nd	nd	0.11	0.1	nd	0.04	0.06	nd	nd	0.07	0.11	nd	nd	nd	nd	0.49
S	0.3	0.02	0.03	0.08	0.14	0.09	0.02	0.02	nd	0.05	0.1	0.07	0.04	nd	nd	nd	0.66
0.108-0.17 W	1.54	nd	nd	0.16	0.16	0.05	0.07	0.07	0.04	0.09	0.22	0.18	0.15	nd	nd	nd	1.19
S	0.64	0.01	0.03	0.14	0.17	0.03	0.03	0.09	nd	0.06	0.15	0.06	0.03	0.01	0.01	0.01	0.84
0.17-0.26 W	6.57	0	0.05	0.12	0.25	0.03	1.29	1.4	1.62	1.83	1.16	1.25	1.37	0.06	0.21	0.25	10.88
S	1.25	0.01	0.01	0.09	0.16	0.1	0.02	0.04	0.01	0.01	0.04	0.02	0.01	nd	0.01	nd	0.52
0.26-0.40 W	8.19	nd	0.1	0.28	0.99	0.33	5.08	6.75	5.79	5.51	4.17	4.35	5.39	0.24	0.03	1.96	40.97
S	2.08	nd	nd	0.04	0.1	0.01	0.02	0.02	0.01	0.03	0.02	0.04	0.02	nd	nd	nd	0.3
0.40-0.65 W	8.77	0.09	0.1	0.27	1.1	0.16	4.64	5.03	5.29	5.06	3.79	4.2	4.86	0.23	1.67	2.14	38.65
S	3.09	nd	nd	0.03	0.1	0.02	0.02	0.02	0.01	0.04	0.01	0.02	0.02	nd	nd	nd	0.31
0.65-1.0 W	7.59	0.1	0.08	0.28	1.33	0.31	4.65	4.68	5.04	4.54	3.36	3.59	4.27	0.2	1.1	1.77	35.3
S	2.2	nd	nd	0.01	0.02	nd	0.02	0.01	nd	0.05	0.01	0.03	0.01	nd	nd	nd	0.14
1.0-1.6 W	5	0.1	0.05	0.13	0.86	0.06	3.23	4.24	3.31	3.48	1.88	2.13	2.79	0.09	0.34	0.78	23.47
S	1.67	0.01	nd	nd	0.02	nd	0.02	0.01	0.01	0.03	0.05	0.06	0.01	nd	nd	0.01	0.24
1.6-2.5 W	2.66	nd	nd	0.04	0.09	0.22	0.26	0.26	0.26	0.3	0.21	0.28	0.32	nd	0.02	0.08	2.34
S	1.5	0.02	0.03	0.15	0.16	0.04	0.03	0.04	nd	nd	0.1	0.07	0.13	nd	nd	nd	0.77
2.5-4.4 W	2.19	nd	nd	0.1	0.17	0.12	0.04	0.05	0.05	nd	0.05	0.04	nd	nd	nd	nd	0.63
S	2.15	nd	0.01	0.02	0.04	0.01	0.02	0.01	nd	0.09	0.04	0.02	0.11	nd	nd	nd	0.37
4.4-6.8 W	1.55	nd	nd	0.14	0.21	0.09	0.02	0.04	nd	0.08	0.11	0.11	0.08	nd	nd	nd	0.88
S	1.99	0.02	0.04	0.09	0.06	0.02	0.03	0.03	nd	0.07	0.07	0.05	0.17	nd	nd	nd	0.62
6.8-10.0 W	1.24	nd	nd	0.14	0.13	0.04	0.02	0.03	nd	0.08	0.08	0.12	0.05	nd	nd	nd	0.69
S	1.94	nd	0.01	0.02	0.02	0.01	0.02	0.06	nd	0.04	0.07	0.07	0.06	nd	nd	nd	0.39
> 10.0 W	2.28	nd	nd	0.15	0.19	0.09	0.04	0.04	nd	0.04	0.1	0.11	0.08	nd	nd	nd	0.85
S	2.9	0.01	0.02	0.05	0.06	0.01	0.02	0.05	nd	0.08	0.02	0.03	0.13	nd	nd	nd	0.48

nd – not detected

Table 1. The concentrations of 13 PM fractions and the fraction-related PAH in Zabrze in winter (W) and summer (S) 2007

The winter concentration of total PAH (ΣPAH) in Zabrze in 2007 were only a little higher than the average concentration of the sum of 9 PAH (Σ9PAH) observed by SSI in 2005 (Table 2, Fig. 3). The summer concentrations of ΣPAH in Zabrze in 2007 were lower than 7 ng m^{-3}. The average concentrations of BaP, in 2007 (about 10 ng m^{-3} in the whole measuring period, 1 ng m^{-3} and 20 ng m^{-3} in summer and winter) very well agree with the BaP concentrations measured by SSI in the recent years (Fig. 4).

The winter concentrations of ΣPAH bound onto particular PM fractions in Zabrze in 2007 were from 0.5 to almost 41.0 ng m^{-3}. ΣPAH bound onto PM$_{0.26-0.40}$, PM$_{0.40-0.65}$, PM$_{0.65-1.0}$ had the greatest concentrations (Table 1) because of the high concentrations of Fl, Py, BaA, Ch, BbF, BkF, BaP.

The summer concentrations of ΣPAH bound onto PM fractions in Zabrze in 2007 were from 0.15 to 0.85 ng m^{-3}. ΣPAH bound onto PM$_{0.03-0.06}$, PM$_{0.06-0.108}$, PM$_{0.108-0.17}$, PM$_{0.17-0.26}$, PM$_{1.6-2.5}$ and PM$_{4.4-6.8}$ had higher concentrations than ΣPAH bound to other fractions.

In winter, the PM concentrations and the concentrations of majority the PM fractions were two or three times higher than in summer. For PM$_{0.03-0.06}$ it was 4.5, for PM$_{0.17-0.26}$ more than 5 and for PM$_{0.26-0.40}$ almost 4 times (Table 1). The PM$_{2.5-10}$ concentrations in summer were close to or higher than the PM$_{2.5-10}$ winter concentrations.

The masses of PM$_{2.5}$ and PM$_1$ were 89 and 72% in winter and 68 and 51% in summer of the mass of PM$_{10}$, respectively (Fig. 5). The concentrations of ΣPAH behaved similarly. In winter, the concentrations of PM$_{2.5}$ and PM$_1$-bound ΣPAH were 99 and 82% of the PM$_{10}$-bound ΣPAH concentrations; in summer it was 77 and 69%, respectively. Despite the disparities between the concentrations of PM$_{1-}$, PM$_{1-2.5}$ and PM$_{2.5-10}$-bound ΣPAH in both seasons (Fig. 5), the mass contributions of ΣPAH to PM$_1$, PM$_{1-2.5}$, and PM$_{2.5-10}$ was 0.02$-$0.04% in summer and 0.38, 0.34 and 0.04% in winter, respectively. In each season, PM$_1$ and PM$_{1-2.5}$ had almost equal PAH contents. It means that in each season, PM$_1$ and PM$_{1-2.5}$ as well as PM$_1$- and PM$_{1-2.5}$-bound PAH came from the same sources.

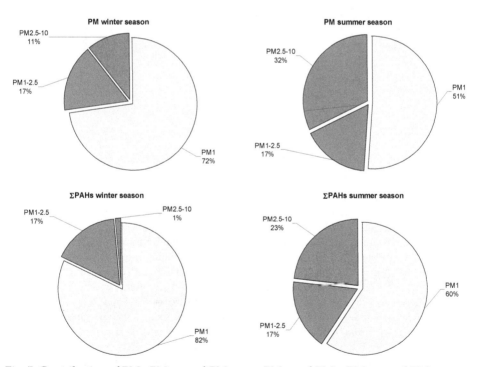

Fig. 5. Contribution of PM$_1$, PM$_{1-2.5}$ and PM$_{2.5-10}$ to PM$_{10}$ and PM$_{1-}$, PM$_{1-2.5}$- and PM$_{2.5-10}$-bound ΣPAH to PM$_{10}$-bound ΣPAH in Zabrze in winter and summer 2007

The differences between the summer and the winter concentrations of the fraction-bound ΣPAH are greater than the differences between the concentrations of the PM fractions. The concentrations of ΣPAH bound onto $PM_{0.17-0.26}$, $PM_{0.26-0.40}$, $PM_{0.40-0.65}$, $PM_{0.65-1.0}$, $PM_{1.0-1.6}$ are greater from 21 to 251 times in winter than in summer. Noticeably, these fractions are usually formed by primary particles originating from combustion (Chow, 1995; Zhao et al., 2008; Wingfors et al., 2011).

For each of $PM_{0.06-0.108}$, $PM_{0.108-0.17}$, $PM_{0.17-0.26}$, $PM_{0.26-0.40}$, $PM_{0.40-0.65}$ and $PM_{1.6-2.5}$, among the fifteen PAH bound onto each of these fractions, Ph, one of the markers of emission from car engines (Harrison et al., 1996), had the highest summer concentrations. The summer concentrations of F, Py, An and Ch bound onto these fractions were also high. These PAH are also attributed to combustion of gasoline and oil in car engines (Chang et al., 2006; Miguel et al., 1998). In winter, Py, BaA, Ch, BbF and BkF had the greatest concentrations among the 15 PAH in each of these fractions. They belong to CPAH (Prahl & Carpenter, 1983), the nine so called combustion PAH (Fl, Py, BaA, BbF, BkF, BaP, BeP, IP and BghiP; Rogge et al., 1993a, 1993b; Kavouras et al., 1999; Bi et al., 2002; Manoli et al., 2002, 2004; Sienra et al., 2005). In winter, about 80% of the mass of these PAH was in PM_1. In summer, almost 100% of DBA and BghiP and about 80% of IP were in PM_1. Each of the remaining PAH was contained in particles greather than 1 µm at least in several dozen percent (Fig. 6).

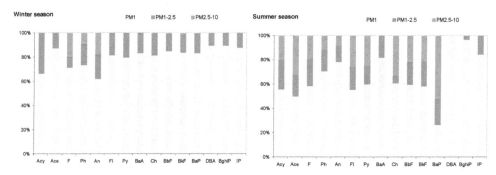

Fig. 6. Contribution of PM_1-, $PM_{1-2.5}$- and $PM_{2.5-10}$-bound PAH to PM_{10}-bound PAH in Zabrze in winter and summer 2007

In summer, the concentration of PM_1-bound BaP was a little more than 20% of the concentration of PM_{10}-bound BaP. The concentrations of BaP were greatest among the concentrations of $PM_{1-2.5}$- and $PM_{4.4-6.8}$-bound PAH.

The winter and the summer profiles of PM-bound PAH differ. In winter, five- and four-ring PAH were 87% of PM-bound ΣPAH. In summer, five- and four-ring PAH were only 58.5% of PM-bound ΣPAH. Three-ring PAH were 39.1% of PM-bound ΣPAH in summer, six-ring PAH were 6.6% and 2.4% of PM-bound ΣPAH in winter and summer, respectively (Table 1). In winter, $PM_{0.26-0.40}$, $PM_{0.40-0.65}$, $PM_{0.65-1.0}$, $PM_{1.0-1.6}$ were richest in PAH (Fig. 7). $PM_{0.26-1.6}$ contained more than 88% of each: ΣPAH, six-, five- and four-ring PAH; for three-ring PAH it was over 67%. In summer, the PM fractions of PAH between the fractions was uniform except for six-ring PAH, whose contribution to $PM_{0.03-0.06}$ was over 73%.

The mass size distribution of PM is multimodal. Usually, PM is represented by three subdistributions (modes). They are called the nucleation, accumulation and coarse modes. The nucleation mode covers the mass distribution of the population of particles with diameters up to approximately 0.1 μm, the accumulation mode – the mass distribution of the particles with diameters in the interval 0.1-2 μm, and the coarse mode is for the particles with diameters greater than 2 μm (Willeke & Whitby 1975; Sverdrup & Whitby, 1977; Hinds, 1998). In practice, the particles in the nucleation mode weigh very little and the mass distribution density may have only two maxima, representing the accumulation and the coarse modes.

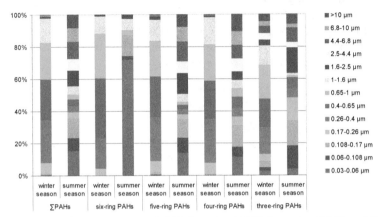

Fig. 7. Distributions of ΣPAH, six-ring PAH, five-ring PAH, four-ring-PAH and three-ring PAH among 13 PM fractions in Zabrze in winter and summer 2007

In Zabrze, in summer 2007, the PM mass distribution with respect to the particle aerodynamic diameter was bimodal. The probability density function had two maxima, one in the interval of particle diameters 0.4-0.65 μm (accumulation mode), and the second between 6.8 and 10 μm (coarse mode, Fig. 8). In winter, the mass distribution was unimodal, the density function had its only maximum between 0.26-0.4 μm.

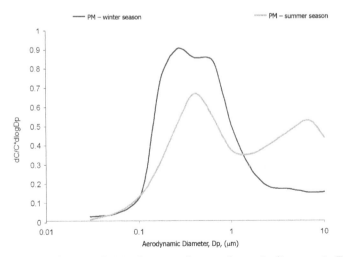

Fig. 8. Mass size distribution of PM relative to the aerodynamic diameter in Zabrze in winter and summer 2007

In winter, the PAH and ΣPAH mass distributions with respect to particle aerodynamic diameter of the particles they were adsorbed onto, except for the three-ring PAH, were bi- or trimodal, with one mode, like for PM, between 0.26 and 0.4 µm (Fig. 9 and 10). The second maximum occurred usually between 0.65 and 1.0 µm (except for BbF, BkF, BaP) and the third one between 4.4 and 6.8 µm or 6.8 and 10 µm (except for BaA, IP, DBA and BghiP). Three-ring PAH, in winter and in summer, were not detectable in some PM fractions (Table 1) because the lighter than Ph ambient species occur in the gaseous form (Guo et al., 2003; Fang et al., 2006; Akyüz & Çabuk, 2008). In summer, ΣPAH and, in general, four-, five-, and six-ring PAH (Fl, Py, Ch, BbF, BkF, BaP) had tri- or bimodal distributions. The distributions of IP and BghiP were bimodal, and of DBA — unimodal.

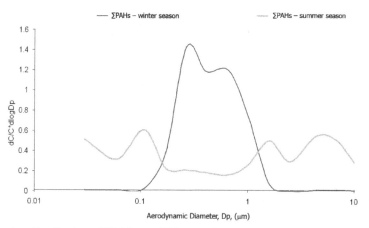

Fig. 9. Mass size distribution of PM-bound ∑PAH relative to the aerodynamic diameter of the particles PAH are adsorbed on in Zabrze in winter and summer 2007

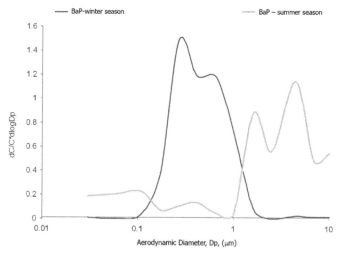

Fig. 10. Mass size distribution of PM-bound BaP relative to the aerodynamic diameter of the particles BaP is adsorbed on in Zabrze in winter and summer 2007

Like in winter, in summer too, majority of the probability functions of PAH and the function of ΣPAH had one maximum between 1.0 and 1.6 µm or 1.6 and 2.5 µm and the second between 4.4 and 6.8 µm or 6.8 and 10 µm (coarse particles).

The PAH diagnostic ratio (DR) is a proportion of ambient PAH concentrations. DR, in some way characterizes the origin of the involved PAH. In the present work, DR are used to determine the effect of the combustion sources in Zabrze on the concentrations of PAH bound to some PM fractions (PM$_1$, PM$_{2.5}$, PM$_{10}$ and fractions where the majority of the probability functions of size distributions of PAH or ΣPAH concentrations assume their maximum). Some DR were taken from literature (Table 2).

Emission source	[BaA]/[BaP]	[Fl]/([Py]+[Fl])	[BaA]/([Ch]+[BaA])	[BbF]/[BkF]	[Ph]/([Ph]+[An])	*ΣCPAHs/ΣPAHs
Vehicular emissions		0.4 – 0.5 [2, 3, 4] 0.44a) [5]	0.2 – 0.35 [4]			0.41e) [5] 0.51f) [5]
Gasoline emissions	0.5 [1]	0.4 ± 0.08 [6] < 0.5 [7, 8]		1.1 – 1.5 [13]	0.50 [15]	
Diesel emissions	1.0 [1]	> 0.5 [7, 8] 0.6 – 0.7 [5]			0.65 [15]	
Used motor oil		0.36 ± 0.08 [6]				
Crude oil			0.16 ± 0.12 [6]		> 0.70 [6]	
Combustion (stationary sources)						0.78 ± 0.16g) [2]
Coal combustion		> 0.5 [3, 4] 0.57 [10]	0.46 [10] 0.50b) [11] 0.17 – 0.36 [12]	3.5 – 3.9 [14]	0.76 [15]	
Wood combustion	1.0 [1]	> 0.5 [3, 4] 0.51 [9,10]	0.40 ± 0.09 [6]	0.8 – 1.1 [14]		
Natural gas combustion		0.49 [9,10]				
Other sources				2.5 – 2.9c) [14]	> 0.7d) [16]	

a) Vehicular emissions – converter equipped automobiles; b) Coal combustion for domestic heating; c) Smelters; d) Associated with lubricant oil and fossil fuels; e) Non-catalyst automobiles; f) Catalyst-equipped automobiles; g) General dominance of combustion sources
[1]: (Li & Kamens, 1993); [2]: (Kavouras et al., 1999); [3]: (Zencak et al., 2007); [4]: (Yunker et al., 2002); [5]: (Rogge et al., 1993a, 1993b); [6]: (Sicre et al., 1987); [7]: (Ravindra et al., 2006); [8]: (Ravindra et al. 2008); [9]: (Galarneau, 2008); [10]: (Tang et al., 2005); [11]: (Gschwend & Hites, 1981); [12]: (Dickhut et al., 2000); [13]: (Masclet et al., 1987); [14]: (Khalili et al., 1995); [15]: (Alves et al., 2001)
*ΣCPAHs/ΣPAHs=([Fl]+[Py]+[BaA]+[BbF]+[BkF]+[BaP]+[BeP]+[IP]+[BghiP])/([Acy]+[Ace]+[F]+[Ph]+ [An]+[Fl]+[Py]+[BaA]+[Ch]+[BbF]+[BkF]+[BaP]+[DBA]+[BghiP]+[IP])

Table 2. PAH diagnostic ratios (DR) from literature

The effect of stationary combustion sources on the ambient PAH concentrations is characterized by the proportion ΣCPAH/ΣPAH of the PM-bound ΣCPAH and ΣPAH concentrations (Rogge et al., 1993a, 1993b; Kavouras et al., 1999; Manoli et al., 2004; Sienra et al., 2005). In Zabrze, ΣCPAH/ΣPAH indicate that in winter PM$_{0.26-0.4}$-, PM$_{0.65-1.0}$- and PM$_{1.6-2.5}$-bound PAH came from stationary combustion sources (Table 2 and 3). In winter, these fractions contained half of the mass of the PM-bound PAH and, not surprisingly, the winter PM$_1$-, PM$_{2.5}$- and PM$_{10}$-bound PAH should in general come from stationary combustion.

Probably, $PM_{6.8-10}$-bound PAH came in part from vehicular sources ($\Sigma CPAH/\Sigma PAH=0.56$). In summer, PM_1- and $PM_{2.5}$-bound PAH might also have come from vehicular sources ($\Sigma CPAH/\Sigma PAH$ was lower than in winter) but the summer concentrations of $PM_{1.6-2.5}$-, $PM_{6.8-10}$- and PM_{10}-bound PAH were affected by stationary combustion sources as well. However, in summer, the degradation processes (oxidation, photochemical reactions) and in winter the variability of the parameters of combustion processes and the variety of used fuels limit the reliability of DR making the source apportionment uncertain (Kavouras et al., 1999; Hong et al., 2007). Nevertheless, other DR confirm these findings. [Ph]/([Ph]+[An]) indicates PAH from hard coal combustion and, in winter, from unidentified sources and crude oil (traffic). [BaA]/([BaA]+[Ch]) indicates the effect of crude oil combustion on the PAH concentrations also in summer (0.10, 0.11 and 0.08 for PM_1, $PM_{2.5}$ and PM_{10}, respectively). For $PM_{0.26-0.4}$-bound PAH, [BaA]/([BaA]+[Ch]) is 0.27 indicating car exhaust (Tables 2 and 3). The winter [BaA]/([BaA]+[Ch]) for $PM_{1.6-2.5}$-bound PAH is 0.46, for $PM_{1.6-2.5}$-bound PAH−0.53, they are higher than in summer and prove that hard coal combustion was the main source of ambient PAH in Zabrze.

The effect of vehicular sources and wood combustion on the ambient PAH concentrations is reflected by [BbF]/[BkF]. In summer [BbF]/[BkF] were slightly higher than in winter for majority of the PM fractions (Table 3). [BbF]/[BkF] allows to apportion PM_1-, $PM_{2.5}$- and PM_{10}-bound PAH to traffic in summer. In winter, except for $PM_{6.8-10}$, [BbF]/[BkF] were between 0.8 and 1.1. It indicates the contribution of wood combustion to the winter PAH concentrations. The values of [BaA]/[BaP] also indicate the contribution of wood combustion to the winter concentration of PAH in Zabrze (Tables 2 and 3).

Fraction, μm	[BaA]/[BaP]		[Fl]/([Py]+[Fl])		[BaA]/([Ch]+[BaA])		[BbF]/[BkF]		[Ph]/([Ph]+[An])		ΣCPAHs/ΣPAHs	
	S	W	S	W	S	W	S	W	S	W	S	W
PM$_{0.26-0.4}$	0.75	1.07	0.51	0.43	0.27	0.51	0.59	0.96	0.91	0.75	0.46	0.83
PM$_{0.65-1.0}$	-	1.18	0.58	0.50	-	0.53	0.34	0.94	1.00	0.81	0.50	0.82
PM$_{1.6-2.5}$	0.02	0.81	0.40	0.50	1.00	0.46	1.44	0.77	0.79	0.29	0.65	0.73
PM$_{6.8-10}$	-	-	0.31	0.43	-	-	1.03	0.67	0.82	0.77	0.81	0.56
PM$_{1.0}$	0.23	1.11	0.42	0.47	0.10	0.51	1.29	0.93	0.75	0.82	0.54	0.82
PM$_{2.5}$	0.16	1.11	0.43	0.46	0.11	0.51	1.25	0.92	0.76	0.81	0.57	0.82
PM$_{10}$	0.08	1.11	0.44	0.46	0.08	0.50	1.26	0.92	0.77	0.80	0.60	0.81

Table 3. PAH diagnostic ratios (DR) in summer (S) and winter (W) 2007 in Zabrze

In winter, [Fl]/([Fl]+[Py]) were close to the values characteristic of PAH coming from wood or natural gas combustion (Galarneau, 2008). In general, [Fl]/([Fl]+[Py]) were lower in summer than in winter. [Fl]/([Fl]+[Py]), oscillating about 0.4, suggest the effect of car exhaust. Relatively high summer [Fl]/([Fl]+[Py]) for $PM_{0.26-0.4}$ and $PM_{0.65-1.0}$ suggest diesel car exhaust as the source of PAH in these fractions (Rogge et al., 1993a, 1993b).

3.2 Comparison of the concentrations of PM- and PM-bound PAH in Zabrze and selected sites in the world

The ambient concentrations of PAH bound to various PM fractions, especially to $PM_{2.5}$ and PM_{10}, are investigated at many sites in the world (Kavouras et al., 1999; Odabasi et al., 1999; Panther et al., 1999; Takeshi & Takashi, 2004; Mantis et al., 2005). Most often, the investigations cover short periods, results of long-term experiments are not abundant. The direct comparison of the concentrations of PM-bound PAH from various urban areas should be done cautiously. The data may be affected by the method of sampling and the technique of determination of PAH in the dust samples (Takeshi & Takashi, 2004). The local conditions, meteorological (air temperature and relative air humidity, direction and velocity of wind, precipitation) and other, neighborhood of the sampling point, a season of a year are also important (Wang X.H. et al., 2007; Evagelopoulos et.al., 2010).

Table 4 is a collation of the concentrations of the sum of 15 PAH (ΣPAH) and PM_1-, $PM_{2.5}$- and PM_{10}-bound BaP in Zabrze in 2007 and in various places in the world, mainly in Europe. BaP was selected because it is used as an air pollution indicator. The data cover the period from 2005 to 2009. In the last column of Table 4, in parentheses, the number of analyzed compounds is given. The concentration of ΣPAH was computed by summing up the concentrations of these PAH that were investigated in Zabrze. The table allows for rough evaluation of differences in the concentrations of BaP and ΣPAH bound to several PM fractions in various countries.

The ranges of the concentrations of BaP and ΣPAH are wide because of the causes mentioned above and different numbers of PAH in ΣPAH. In general, the lowest concentrations (lower than 0.1 ng m^{-3} for BaP and than 2 ng m^{-3} for ΣPAH) occurred in clean areas (not affected by vehicular or industrial sources), such as Virolahti (regional background station) in Finland (Makkonen et al., 2010) and Chréa National Park (great forest area) in Algeria (Ladji et al., 2009). In most places the BaP concentration did not exceed 1 ng m^{-3}, the limit established by the European Commission. The exceptions were the urban-traffic site in Oporto (Portugal; Slezakova et al., 2011), road tunnel in Marseille (France; El Haddad et al., 2009), urban background station in Flanders (Belgium; Vercauteren et al., 2011), measuring point in Zonguldak (Turkey; Akyüz & Çabuk, 2008) and urban background station and crossroads in Zabrze (Ćwiklak et al., 2009).

Majority of ambient PAH, especially in urbanized areas, are anthropogenic (Kulkarni & Venkataraman, 2000; Hien et al., 2007; Wang X.H. et al., 2007), and come mainly from combustion of fossil fuels, wastes or biomass and also from industry and road traffic. Most of the greatest values of BaP concentrations shown in Table 4, reaching 50 ng m^{-3}, come from sites located in industrialized and densely populated areas in Asiatic countries, such as Fushun (residential-commercial site, Kong et al., 2010), Beijing (campus site, Wang H. et al., 2009), Guiyu (electronic waste recycling site, Deng et al., 2006) in the People's Republic of China or Chennai in the Republic of India (Mohanraj et al., 2011). The concentrations of ΣPAH in these regions were also high, some times, like in Funshun, China, where the ΣPAH concentration reached 1.9 μg m^{-3}, many times higher than elsewhere in the world. In Europe, the highest BaP and ΣPAH concentrations were at traffic station in Sweden and at traffic and urban sites in Zabrze (Table 4).

BaP and ΣPAH tend to accumulate in the finest particles of PM (Table 4). Like in Zabrze, this tendency may be observed in the road tunnel in Lisbon, Portugal (Oliveira et al., 2011) and at the urban background and traffic stations in Los Angeles, USA (Phuleria et al., 2007).

Location		Sampling period	Fraction	Concentration (ng m⁻³)	
				BaP [a]	∑PAH [b]
Zabrze (Poland) [1]		summer 2007	PM_1	0.17	3.48 (15)
			$PM_{2.5}$	0.31	4.49 (15)
			PM_{10}	0.65	5.86 (15)
		winter 2007	PM_1	16.08	128.10 (15)
			$PM_{2.5}$	19.19	153.10 (15)
			PM_{10}	19.32	156.11 (15)
Zabrze (Poland) [2]	urban background	summer 2005	$PM_{2.5}$	_0.90_	12.80 (15)
			PM_{10}	_1.20_	21.60 (15)
	crossroads		$PM_{2.5}$	_2.10_	43.40 (15)
			PM_{10}	_4.00_	63.80 (15)
Flanders (Belgium) [3]	urban background	Oct 2006 - Mar 2007	PM_{10}	1.18	82.24 (15)
		Apr - Sep 2007	PM_{10}	0.51	74.68 (15)
	rural background	Oct 2006 - Mar 2007	PM_{10}	0.81	45.90 (15)
		Apr - Sep 2007	PM_{10}	0.41	32.23 (15)
Virolahti (Finland), regional background [4]		summer 2006	PM_1	< 0.08	1.77 (13)
			$PM_{2.5}$	< 0.08	1.77 (13)
			PM_{10}	< 0.08	1.78 (13)
		winter 2006	PM_1	0.52	7.54 (13)
			$PM_{2.5}$	0.69	14.53 (13)
			PM_{10}	0.73	13.9 (13)
Marseille (France), tunnel [5]		winter 2008	$PM_{2.5}$	2.42	74.67 (8)
			PM_{10}	6.73	108.39 (8)
Toulouse (France) [6]	traffic	Apr 2006	PM_{10}	_0.10_	18.20 (15)
	urban		PM_{10}	_0.15_	11.45 (15)
	industrial		PM_{10}	_0.80_	20.40 (15)
Thessaloniki (Greece), kerbside [7]	street level	Jan/Feb 2006	$PM_{<0.95}$	0.47	11.94 (13)
		Jan/Feb 2006	$PM_{0.95-1.5}$	0.10	2.36 (13)
		Jan/Feb 2006	$PM_{1.5-3}$	0.04	1.41 (13)
		Jan/Feb 2006	$PM_{3-7.5}$	0.02	0.98 (13)
		Jan/Feb 2006	$PM_{>7.5}$	0.02	0.80 (13)
	rooftop level	Jan/Feb 2006	$PM_{<0.95}$	0.45	11.54 (13)
		Jan/Feb 2006	$PM_{0.95-1.5}$	0.10	2.45 (13)
		Jan/Feb 2006	$PM_{1.5-3}$	0.03	1.06 (13)
		Jan/Feb 2006	$PM_{3-7.5}$	0.02	0.79 (13)
		Jan/Feb 2006	$PM_{>7.5}$	0.02	0.63 (13)
Athens (Greece), indoor samples [8]	smoking area	16-27 Jul 2007	PM_1	0.10	0.80 (15)
			$PM_{2.5}$	0.14	1.11 (15)
			PM_4	1.73	21.33 (15)
	no-smoking area		PM_1	0.03	0.44 (15)
			$PM_{2.5}$	0.05	0.65 (15)
			PM_4	0.00	5.13 (15)
	passive smokers` area		PM_4	3.58	29.04 (15)
Kozani (Greece), urban area surrounded by opencast coal mining [9]		Dec 2005 – Oct 2006	$PM_{2.5}$	0.38	4.77 (15)
			PM_{10}	0.11	1.46 (15)
Rome (Italy), downtown [10]		Apr - Jul 2007	PM_1	-	2.10 (14)
			$PM_{2.5}$	-	2.29 (14)
			PM_{10}	-	2.37 (14)

Location	Sub-site	Period	Fraction		
		Oct 2007 - Feb 2008	PM_1	-	6.70 (14)
			$PM_{2.5}$	-	7.77 (14)
			PM_{10}	-	7.98 (14)
Lisbon (Portugal), roadway tunnel [11]		Oct 2008	$PM_{0.49}$	6.98	113.80 (15)
			$PM_{0.49-0.95}$	0.35	14.70 (15)
			$PM_{0.95-2.5}$	0.17	7.33 (15)
			$PM_{2.5-10}$	0.19	2.66 (15)
Oporto Metropolitan Area (Portugal), urban-traffic [12]		Dec 2008	PM_{10}	2.02	20.67 (15)
			$PM_{2.5}$	1.88	18.96 (15)
Umeä (Sweden), traffic [13]		autumn 2009	$PM_{2.5}$	25.00	196.45 (12)
Zonguldak (Turkey), industrial city [14]		summer 2007	$PM_{2.5}$	0.40	5.70 (14)
		winter 2007	$PM_{2.5}$	15.70	152.70 (14)
		summer 2007	$PM_{2.5-10}$	0.20	1.60 (14)
		winter 2007	$PM_{2.5-10}$	0.70	10.50 (14)
Kabul (Afganistan), urban [13]		autumn 2009	$PM_{2.5}$	6.70	55.97 (12)
Mazar-eSharif (Afganistan), urban [13]			$PM_{2.5}$	0.09	2.19 (12)
Beijing (China); campus site [15]		summer 2005-2007	$PM_{2.5}$	1.19	39.28 (15)
			$PM_{2.5-10}$	1.14	24.87 (15)
		winter 2005-2007	$PM_{2.5}$	19.82	360.71 (15)
			$PM_{2.5-10}$	5.09	102.00 (15)
Guiyu (China), electronic waste recycling site [16]		AugI - Sep 2004	TSP	15.40	144.85 (15)
			$PM_{2.5}$	8.85	99.26 (15)
Xiamen (China) [17]	industrial-traffic intersection site	summer 2005	PM_{10}	0.10	5.24 (15)
		autumn 2005	PM_{10}	1.40	37.10 (15)
	residential site	summer 2005	PM_{10}	0.00	1.69 (15)
		autumn 2005	PM_{10}	1.50	26.60 (15)
Fushun (China) [18]	urban background	2004-2005	$PM_{2.5}$	10.71	261.82 (13)
			$PM_{2.5-10}$	1.98	72.44 (13)
	residential-commercial		$PM_{2.5}$	48.44	1899.36 (13)
			$PM_{2.5-10}$	3.44	166.92 (13)
Jinzhou (China) [18]	urban background		$PM_{2.5}$	13.61	190.86 (13)
			$PM_{2.5-10}$	1.13	20.78 (13)
	residential		$PM_{2.5}$	6.61	106.94 (13)
			$PM_{2.5-10}$	0.77	16.40 (13)
Chennai City (India) [19]	urban-residential	Dec 2009 – Feb 2010	$PM_{2.5}$	6.50	365.30 (11)
		Apr - Aug 2009	$PM_{2.5}$	7.40	326.90 (11)
	industrial/traffic site	Dec 2009 – Feb 2010	$PM_{2.5}$	8.10	681.80 (11)
		Apr - Aug 2009	$PM_{2.5}$	10.60	456.60 (11)
	urban-commercial	Dec 2009 – Feb 2010	$PM_{2.5}$	16.20	448.00 (11)
		Apr - Aug 2009	$PM_{2.5}$	2.50	313.90 (11)
Ho Chi Minh (Vietnam), roadside [20]		dry season, Jan-Feb 2005	TSP	_2.00_	39.80 (9)
		rainy season, Jul 2005	TSP	_5.70_	58.70 (9)

Boumerdes (Algeria), urban traffic congestion [21]		PM_1	0.11	1.53 (14)
		PM_{1-10}	0.02	0.33 (14)
Rouiba-Réghaia (Algeria), industrial zone [21]	Oct 2006	PM_1	0.30	2.70 (14)
		PM_{1-10}	0.04	0.59 (14)
Chréa National Park (Algeria), forest ecosystem [21]		PM_1	0.02	0.28 (14)
		PM_{1-10}	0.002	0.10 (14)
Golden (British Columbia, Canada), residential [22]	spring 2006	$PM_{2.5}$	0.14	1.76 (15)
	winter 2007	$PM_{2.5}$	2.67	31.39 (15)
Long Beach, California (USA), coastal city [23]	Jan – Mar 2005	$PM_{0.25}$	0.17	0.94 (10)
		$PM_{0.25-2.5}$	0.03	0.43 (10)
		$PM_{2.5-10}$	0.00	0.01 (10)
Los Angeles (USA) [24] background	Jan 2005	$PM_{0.108-2.5}$	0.03	0.31 (9)
		$PM_{>2.5}$	0.16	1.45 (9)
freeway		$PM_{0.108-2.5}$	0.04	0.38 (9)
		$PM_{>2.5}$	0.17	1.82 (9)

[a] Underlined italics mark the values read from a chart; [b] The number of PAH taken to compute ΣPAH concentration is in parentheses

[1]: (this study); [2]: (Ćwiklak et al., 2009); [3]: (Vercauteren et al., 2011); [4]: (Makkonen et al., 2010); [5]: (El Haddad et al., 2009); [6]: (Dejean et al., 2009); [7]: (Chrysikou et al., 2009); [8]: (Saraga et al., 2010); [9]: (Evagelopoulos et al., 2010); [10]: (Di Filippo et al., 2010); [11]: (Oliveira et al., 2011); [12]: (Slezakova et al., 2011); [13]: (Wingfors et al., 2011); [14]: (Akyuz & Cabuk, 2008); [15]: (Wang et al., 2009); [16]: (Deng et al., 2006); [17]: (Hong et al., 2007); [18]: (Kong et al., 2010); [19]: (Mohanraj et al., 2011); [20]: (Hien et al., 2007); [21]: (Ladji et al., 2009); [22]: (Ding et al., 2009); [23]: (Krudysz et al., 2009); [24]: (Phuleria et al., 2007)

Table 4. Comparison of the PM and PM-bound PAH concentrations at various sites in the world

Like in Zabrze in 2007, at some other sites in the world the BaP and ΣPAH concentrations depend on a season of a year and are higher in cold seasons. One of the most obvious causes of the seasonal variability of the PAH concentrations is home heating, central or individual, that is important, if not the most important, source of air pollutants in winter. Moreover, the meteorological conditions in winter (shallow mixing layer) are favorable for local occurrences of high air pollution, like in Zabrze, Zonguldak (Turkey) and in Golden (Canada). Instead, in summer, higher air temperatures and solar radiation intensify desorption of PAH from PM particles and their photochemical decomposition (Odabasi et al., 1999; Hong et al., 2007). Therefore, lower in summer than in winter concentrations of PM-bound PAH may also be due to releasing of PAH from PM particles. To prove it in Zabrze, the gas phase of ambient PAH would have to be investigated. In summer, more favorable conditions for dispersion and dilution of air pollutants (Mantis et al., 2005) and washing out of particles (with adsorbed PAH) by precipitation occur.

The concentrations of BaP and ΣPAH in Zabrze in summer 2007 were lower than in 2005 (Table 4). It may be due to the differences in the sampling periods (long sampling periods in 2007, 24-hour sampling in 2005), in the method of sampling (cascade impactor in 2007, manual sampler with a separating head in 2005) and in the method of PAH determination (combining of extracts for a season and GC-FID in 2007, averaging of diurnal concentrations in a season and GC-MS in 2005) (Ćwiklak et al., 2009).

Nevertheless, the concentrations at two measuring points in Zabrze in summer 2005, several times lower than in winter 2007, suggest that the municipal emission in Silesian Agglomeration may be much greater problem than the traffic emission. Instead, in other European cities, the greatest problem is the traffic emission. The ΣPAH concentrations in Zabrze in 2007 were higher than in other European cities except for the traffic sites (road tunnels) in Marseille (France) and Lisbon (Portugal).

3.3 Health hazard from PAH in Zabrze

Ambient PAH endanger human health by their mutageneity and carcinogenicity. Their strong adverse biological effect is documented by numerous works (Grimmer et al., 1986; White, 2002; Yan et al., 2004).

The risk from the exposure to particular PAH is expressed in terms of the most cancerogenous PAH, BaP, as the toxicity equivalence factor (TEF). The carcinogenicity of a combination of PAH, the BaP equivalence (BEQ), is computed as the linear combination of the concentrations of PAH entering the PAH combination and their TEF (Nisbet & LaGoy, 1992). BEQ for the 15 PAH discussed in this paper is:

$$BEQ = [Acy] \times 0.001 + [Ace] \times 0.001 + [F] \times 0.001 + [Ph] \times 0.001 + [An] \times 0.01 +$$
$$+[Fl] \times 0.001 + [Py] \times 0.001 + [BaA] \times 0.1 + [Ch] \times 0.01 + [BbF] \times 0.1 + [BkF] \times 0.1+$$
$$+[BaP] \times 1 + [DBA] \times 5 + [BghiP] \times 0.1 + [IP] \times 0.1 \qquad (1)$$

In winter, BEQ for $PM_{0.17-0.26}$-, $PM_{0.26-0.40}$-, $PM_{0.40-0.65}$-, $PM_{0.65-1.0}$-, and $PM_{1.0-1.6}$-bound PAH were 2.85 ng m^{-3}, 7.09 ng m^{-3}, 14.65 ng m^{-3}, 11.07 ng m^{-3} and 5.28 ng m^{-3}, respectively (Table 5). The five fractions contain 78% of the mass of PM_{10} and 96% of PM_{10}-bound ΣPAH (Table 1). The summer BEQ of PAH in these fractions do not exceed 1 ng m^{-3}. BEQ for the remaining eight fractions do not exceed 1 ng m^{-3} in both seasons. PAH in the seven fractions contained in $PM_{0.108-2.5}$ have BEQ higher, often several times, in winter than in summer.

BEQ for PAH in $PM_{2.5}$ and PM_{10} in summer 2007 in Zabrze were 0.79 ng m^{-3} and 1.16 ng m^{-3}, respectively. They were two or almost three times lower than Ćwiklak et al. (2009) determined in summer 2005 at the urban background site in Zabrze. They are comparable with other, foreign, values (Xiamen: 0.85 ng m^{-3} and 0.92 ng m^{-3} for PM10 and PM2.5 bound PAH; Hong et al., 2007). Instead, the winter BEQ for PAH in PM2.5 and PM10 are very high, 41.72 ng m^{-3} and 41.91 ng m^{-3}, and they are higher than BEQ computed for PAH in these fractions in areas with very high PAH concentrations, such as in Shanghai (BEQ equal to 15.77 ng m^{-3}; Guo et al., 2004) or in some Japanese cities (BEQ about 2 ng m^{-3}; Takeshi & Takashi, 2004).

The closer $\Sigma PAH_{carc}/\Sigma PAH$ to 1 are the more hazardous to humans ambient PAH are (Hong et al., 2007). In Zabrze, in summer, the values of $\Sigma PAH_{carc}/\Sigma PAH$ were between 0.17 and 0.74 (Table 5). They are dispersed owing to differentiation of the PAH profiles (Table 1). The winter $\Sigma PAH_{carc}/\Sigma PAH$ are close to or higher than 0.5. In general, in Zabrze, the values of $\Sigma PAH_{carc}/\Sigma PAH$ were high, much higher than ones determined for other urban areas (Bourotte et al., 2005; Sienra et al., 2005). They are comparable with the values received by Chen et al. (2004) for dust emitted from coal combustion.

Fraction		BEQ, ng m^{-3}	$\sum PAH_{carc}/\sum PAH$
PM$_{0.03-0.06}$	W	0.07	0.44
	S	0.32	0.70
PM$_{0.06-0.108}$	W	0.02	0.37
	S	0.06	0.39
PM$_{0.108-0.17}$	W	0.20	0.57
	S	0.10	0.39
PM$_{0.17-0.26}$	W	2.85	0.69
	S	0.07	0.17
PM$_{0.26-0.40}$	W	7.09	0.67
	S	0.03	0.39
PM$_{0.40-0.65}$	W	14.65	0.66
	S	0.02	0.34
PM$_{0.65-1.0}$	W	11.07	0.65
	S	0.01	0.63
PM$_{1.0-1.6}$	W	5.28	0.62
	S	0.02	0.74
PM$_{1.6-2.5}$	W	0.50	0.62
	S	0.15	0.39
PM$_{2.5-4.4}$	W	0.02	0.23
	S	0.12	0.70
PM$_{4.4-6.8}$	W	0.10	0.43
	S	0.18	0.55
PM$_{6.8-10.0}$	W	0.07	0.48
	S	0.07	0.63
PM$_{>10.0}$	W	0.10	0.39
	S	0.14	0.54
PM$_1$	W	35.93	0.66
	S	0.62	0.43
PM$_{1-2.5}$	W	5.78	0.62
	S	0.17	0.47
PM$_{2.5-10}$	W	0.19	0.39
	S	0.37	0.61
PM$_{2.5}$	W	41.72	0.65
	S	0.79	0.44
PM$_{10}$	W	41.91	0.65
	S	1.16	0.48

*PAH$_{carc}$/$\sum PAH$=([BaA]+[BaP]+[BbF]+[BkF]+[Ch]+[DBA]+[IP])/([Acy]+[Ace]+[F]+[Ph]+[An]+[Fl]+[Py]+[BaA]+[Ch]+[BbF]+[BkF]+[BaP]+[DBA]+[BghiP]+[IP])

Table 5. BEQ and $\sum PAH_{carc}/\sum PAH$ for PM-bound PAH In the PM fractions in Zabrze in winter (W) and summer (S) 2007

4. Conclusion

In Zabrze, the winter PM-bound BaP concentrations are 19 times greater than the limit for the yearly average BaP concentrations (1 ng m^{-3}). Both PM-bound BaP and Σ15PAH concentrations are much greater than the concentrations in other European cities. Despite the general improvement of the air quality in the Silesian Agglomeration during the last thirty years (decrease of the concentrations of ambient coarse particles and PAH related with this fraction), the concentrations of PM-bound BaP and $\Sigma15PAH$ are high. On the local scale, in winter, the most important sources of fine particles and particle-bound PAH are municipal sources (hard coal, wood and garbage combustion), and/or electric power and heat production from coal; in summer it is vehicular emission. At the urban background measuring point in Zabrze, the vast differences in the seasonal ambient concentrations of PAH and in PAH profiles refute the supposition on industry affecting mostly the air quality in the Silesian Agglomeration. However, DR applied in determination of probable PAH sources are not reliable and their application may give contradictive results (Simcik et al., 1999; Sienra et al., 2005; Evangelopoulos et al., 2010; Dvorská et al., 2011). The exact apportionment of PAH to sources needs measuring of the diurnal PM-bound PAH concentrations and some statistical reasoning must be done (e.g. multivariate factor analysis to apportion combinations of PAH to sources). The data base containing the intervals of DR determined for the specific conditions in the Silesian Agglomeration in vicinities of real PAH sources would appear very helpful. Also, the measurements of ambient gaseous PAH concentrations are necessary.

The high ambient PAH concentrations and the high five- and six-ring PAH content of total PAH are hazardous to the Zabrze population, especially in winter. The concentrations of the carcinogenic PAH was never lower than 50% of the ΣPAH concentration in winter, and BEQ for PM_1, $PM_{2.5}$ and PM_{10} were 35.93 ng m^{-3}, 41.72 ng m^{-3} and 41.91 ng m^{-3}, respectively.

In winter, all the four-, five- and six-ring PAH and ΣPAH had two- or trimodal distributions with one maximum between 0.26 and 0.4 µm, the second usually between 0.65 and 1.0 µm and the third between 4.4 and 10 µm. The greatest BEQ and $\Sigma PAH_{carc}/\Sigma PAH$ found for $PM_{0.17-1.6}$ suggest elevated toxicity of very fine particles, which are the core mass of PM, in Zabrze in winter. In summer, the distribution of ΣPAH and particular PAH with respect to the aerodynamic diameter of particles they are bound to, the values of BEQ and $\Sigma PAH_{carc}/\Sigma PAH$ were similar to those from other sites in the world (Chen et. al., 2004; Akyüz & Çabuk, 2008). The contribution of PM_1- and $PM_{2.5}$-bound ΣPAH to ΣPAH was in Zabrze, like elsewhere (Chrysikou et al., 2009; Kong et al., 2010; Makkonen et al., 2010; Oliveira et al., 2011), very high (99 and 82% in winter and 77 in 69% summer, respectively).

5. Acknowledgment

The work was partially supported by grant No. N N523 421037 from the Polish Ministry of Science and Higher Education.

6. References

Air protection program for areas of Silesia, which were found oversize levels of substances in the air. Appendix to Resolution No. III/52/15/2010 Silesian Provincial Assembly of 16 June 2010, Katowice, 2010 (in polish)

Akyüz, M., Çabuk, H. (2008): Particle-associated polycyclic aromatic hydrocarbons in the atmospheric environment of Zonguldak, Turkey, *Science of the Total Environment* 405 (1-3), pp. 62-70

Alves, C., Pio, C., Duarte, A. (2001). Composition of extractable organic matter of air particles from rural and urban Portuguese areas, *Atmospheric Environment*, 35 (32), pp. 5485-5496

Bi, X., Sheng, G., Peng, P.A., Zhang, Z., Fu, J. (2002). Extractable organic matter in PM10 from LiWan district of Guangzhou City, PR China, *Science of the Total Environment*, 300 (1-3), pp. 213-228

Bourotte, C., Forti, M.-C., Taniguchi, S., Bícego, M.C., Lotufo, P.A. (2005). A wintertime study of PAHs in fine and coarse aerosols in São Paulo city, Brazil, *Atmospheric Environment* 39 (21), pp. 3799-3811

Central Statistical Office of Poland. Environment 1975- 2000. Information and statistical studies, Warsaw 1976-2001

Central Statistical Office of Poland. Environment 2000- 2010. Information and statistical studies, Warsaw 2001-2011

Chang, K.-F., Fang, G.-C., Chen, J.-C., Wu, Y.-S. (2006). Atmospheric polycyclic aromatic hydrocarbons (PAHs) in Asia: A review from 1999 to 2004, *Environmental Pollution*, 142 (3), pp. 388-396

Chen, Y., Bi, X., Mai, B., Sheng, G., Fu, J. (2004). Emission characterization of particulate/gaseous phases and size association for polycyclic aromatic hydrocarbons from residential coal combustion, *Fuel*, 83 (7-8), pp. 781-79

Chow, J.C. (1995): Measurement methods to determine compliance with ambient air quality standards for suspended particles, *Journal of the Air and Waste Management Association*, 45 (5), pp. 320-382

Chrysikou, L.P., Gemenetzis, P.G., Samara, C.A. (2009). Wintertime size distribution of polycyclic aromatic hydrocarbons (PAHs), polychlorinated biphenyls (PCBs) and organochlorine pesticides (OCPs) in the urban environment: Street- vs rooftop-level measurements. *Atmospheric Environment*, 43 (2), pp. 290-300

Ćwiklak, K., Pastuszka, J.S., Rogula-Kozłowska W. (2009). Influence of traffic on particulate-matter polycyclic aromatic hydrocarbons in Urban atmosphere of Zabrze, Poland, *Polish Journal of Environmental Studies*, 18 (4), pp. 579-585

Dejean, S., Raynaud, C., Meybeck, M., Della Massa, J.-P., Simon, V. (2009): Polycyclic aromatic hydrocarbons (PAHs) in atmospheric urban area: monitoring on various types of sites. *Environmental Monitoring and Assessment*, 148 (1-4), pp. 27-37

Deng, W.J., Louie, P.K.K., Liu, W.K., Bi, X.H., Fu, J.M., Wong, M.H. (2006). Atmospheric levels and cytotoxicity of PAHs and heavy metals in TSP and PM2.5 at an electronic waste recycling site in southeast China, *Atmospheric Environment*, 40 (36), 6955

Dickhut, R.M., Canuel, E.A., Gustafson, K.E., Liu, K., Arzayus, K.M., Walker, S.E., Edgecombe, G., Gaylor, M.O., MacDonald, E.H. (2000). Automotive sources of carcinogenic polycyclic aromatic hydrocarbons associated with particulate matter in the Chesapeake Bay region, *Environmental Science and Technology*, 34 (21), pp. 4635-4640

Di Filippo, P., Riccardi, C., Pomata, D., Gariazzo, C., Buiarelli, F. (2010). Seasonal abundance of particle-phase organic pollutants in an urban/industrial atmosphere. *Water, Air, and Soil Pollution* 211 (1-4), pp. 231-250

Ding, L.C., Ke, F., Wang, D.K.W., Dann, T., Austin, C.C. (2009): A new direct thermal desorption-GC/MS method: Organic speciation of ambient particulate matter collected in Golden, BC, *Atmospheric Environment* 43 (32), pp. 4894-4902

Directive 2008/50/EC of the European Parliament and of the Council of 21 May 2008 on ambient air quality and cleaner air for Europe

Dvorská, A., Lammel, G., Klánová, J. (2011). Use of diagnostic ratios for studying source apportionment and reactivity of ambient polycyclic aromatic hydrocarbons over Central Europe, *Atmospheric Environment*, 45 (2), pp. 420-427

El Haddad, I., Marchand, N., Dron, J., Temime-Roussel, B., Quivet, E., Wortham, H., Jaffrezo, J.L., Baduel, C., Viosin, D., Besombes, J.L., Gille, G. (2009). Comprehensive primary particulate organic characterization of vehicular exhaust emissions in France, *Atmospheric Environment*, 43 (39), pp. 6190-6198

EMEP. *Transboundary Partuculate Matter in Europe*, Status report 4/2009

Evagelopoulos, V., Albanis, T.A., Asvesta, A., Zoras, S. (2010). Polycyclic aromatic hydrocarbons (PAHs) in fine and coarse particles, *Global Nest Journal*, 12 (1), pp. 63-70

Fang, G.-C., Wu, Y.-S., Chen, J.-C., Chang, C.-N., Ho, T.-T. (2006). Characteristic of polycyclic aromatic hydrocarbon concentrations and source identification for fine and coarse particulates at Taichung Harbor near Taiwan Strait during 2004-2005, *Science of the Total Environment*, 366 (2-3), pp. 729-738

Galarneau, E. (2008). Source specificity and atmospheric processing of airborne PAHs: Implications for source apportionment, *Atmospheric Environment*, 42 (35), pp. 8139-8149

Grimmer, G., Abel, U., Brune, H., Deutsch-Wenzel, R., Emura, M., Heinrich, U., Jacob, J., Kemena, A., Misfeld, J., Mohr, U. (1986). Evaluation of environmental carcinogens by carcinogen-specific test systems, *Experimental Pathology*, 29 (2), pp. 65-76

Grynkiewicz-Bylina, B., Rakwic, B., Pastuszka, J.S.(2005). Assessment of exposure to traffic-related aerosol and to particle-associated PAHs in Gliwice, Poland, *Polish Journal of Environmental Studies*,14 (1), pp. 117-123

Gschwend, P.M., Hites, R.A. (1981). Fluxes of polycyclic aromatic hydrocarbons to marine and lacustrine sediments in the northeastern United States, *Geochimica et Cosmochimica Acta*, 45 (12), pp. 2359-2367

Guo, H., Lee, S.C., Ho, K.F., Wang, X.M., Zou, S.C. (2003): Particle-associated polycyclic aromatic hydrocarbons in urban air of Hong Kong, *Atmospheric Environment*, 37 (38), pp. 5307-5317

Guo, H.L., Lu, C.G., Yu, Q., Chen, L.M. (2004). Pollution characteristics of polynuclear aromatic hydrocarbons on airborne particulate in Shanghai. Journal of Fudan University (Natural Science), 43, pp. 1107

Harrison R.M., Smith D.J.T., Luhana L., (1996). Source apportionment of atmospheric polycyclic aromatic hydrocarbons collected from an urban location in Birmingham. UK, *Environmental Science and Technology*, 30(3), pp. 825-832

Hien, T.T., Thanh, L.T., Kameda, T., Takenaka, N., Bandow, H. (2007). Distribution characteristics of polycyclic aromatic hydrocarbons with particle size in urban aerosols at the roadside in Ho Chi Minh City, Vietnam, *Atmospheric Environment*, 41 (8), pp. 1575-1586

Hinds, W.C. (1998): Aerosol technology. Properties, behaviour, and measurement of airborne particles. Second Edition. John Wiley & Sons, Inc. New York

Hong, H.S., Yin, H.L., Wang, X.H., Ye, C.X. (2007). Seasonal variation of PM10-bound PAHs in the atmosphere of Xiamen, China. *Atmospheric Research*, 85 (3-4), pp. 429-441

http://www.epa.gov/

Inspection for Environmental Protection, Silesian Voivodship Inspectorate for Environmental Protection in Katowice, The environmental status in the Silesia region in 1999-2000, Library of Environmental Monitoring, Katowice 2001 (WIOŚ, 2001) (in polish)

Kavouras, I.G., Lawrence, J., Koutrakis, P., Stephanou, E.G., Oyola, P. (1999). Measurement of particulate aliphatic and polynuclear aromatic hydrocarbons in Santiago de Chile: Source reconciliation and evaluation of sampling artifacts, *Atmospheric Environment*, 33 (30), pp. 4977-4986

Khalili, N.R., Scheff, P.A., Holsen, T.M. (1995). PAH source fingerprints for coke ovens, diesel and gasoline engines, highway tunnels, and wood combustion emissions, *Atmospheric Environment*, 29 (4), pp. 533-542

Kinney, P.L., Aggarwal, M., Northridge, M.E., Janssen, N.A.H., Shepard, P. (2000). Airborne concentrations of PM2.5 and diesel exhaust particles on Harlem sidewalks: A community-based pilot study, *Environmental Health Perspectives*, 108 (3), pp. 213-218

Klejnowski, K., Krasa, A., Rogula, W. (2007a). Seasonal variability of concentrations of total suspended particles (TSP) as well as PM10, PM2.5 and PM1 modes in Zabrze, Poland, *Archives of Environmental Protection*, 33 (3), pp. 15-27

Klejnowski, K., Talik, E., Pastuszka, J., Rogula, W., Krasa, A. (2007b). Chemical composition of surface layer of PM1, PM1-2.5, PM2.5-10, *Archives of Environmental Protection*, 33 (3), pp. 89-95

Klejnowski, K., Rogula-Kozłowska, W., Krasa, A., (2009). Structure of atmospheric aerosol in Upper Silesia (Poland) - Contribution of PM2.5 to PM10 in Zabrze, Katowice and Częstochowa in 2005-2007, *Archives of Environmental Protection*, 35 (2), pp. 3-13

Kong, S., Ding, X., Bai, Z., Han, B., Chen, L., Shi, J., Li, Z. (2010). A seasonal study of polycyclic aromatic hydrocarbons in PM2.5 and PM2.5-10 in five typical cities of Liaoning Province, China *Journal of Hazardous Materials* 183 (1-3), pp. 70-80

Kozielska, B., Rogula-Kozłowska, W., Pastuszka, J.S. (2009). Effect of road traffic concentration of PM2.5, PM10 and PAHs in zones of high and low municipal emission, *Polska Inżynieria Środowiska pięć lat po wstąpieniu do Unii Europejskiej, Monografie Komitetu Inżynierii Środowiska PAN*, 58(1), pp. 129-137, Lublin, ISBN 978-83-89293-81-7 (in polish)

Kristensson, A., Johansson, C., Westerholm, R., Swietlicki, E., Gidhagen, L., Wideqvist, U., Vesely, V. (2004). Real-world traffic emission factors of gases and particles measured in a road tunnel in Stockholm, Sweden, *Atmospheric Environment*, 38 (5), pp. 657-673

Krudysz, M.A., Dutton, S.J., Brinkman, G.L., Hannigan, M.P., Fine, P.M., Sioutas, C., Froines, J.R. (2009). Intra-community spatial variation of size-fractionated organic compounds in Long Beach, California. *Air Quality, Atmosphere and Health*, 2 (2), pp 69-88

Kulkarni, P., Venkataraman, C. (2000). Atmospheric polycyclic aromatic hydrocarbons in Mumbai, India. *Atmospheric Environment* 34 (17), pp. 2785-2790

Ladji, R., Yassaa, N., Balducci, C., Cecinato, A., Meklati, B.Y. (2009). Distribution of the solvent-extractable organic compounds in fine (PM1) and coarse (PM1-10) particles in urban, industrial and forest atmospheres of Northern Algeria, *Science of the Total Environment*, 408 (2), pp. 415-424

Li, C.K., Kamens, R.M., (1993). The use of polycyclic aromatic hydrocarbons as sources signatures in receptor modeling. *Atmospheric Environment*, 27A, pp. 523–532

Makkonen, U., Hellén, H., Anttila, P., Ferm, M. (2010): Size distribution and chemical composition of airborne particles in south-eastern Finland during different seasons and wildfire episodes in 2006, *Science of the Total Environment* 408, (3), pp. 644-651

Manoli, E., Voutsa, D., Samara, C. (2002). Chemical characterization and source identification/apportionment of fine and coarse air particles in Thessaloniki, Greece, *Atmospheric Environment*, 36 (6), pp. 949-961

Manoli, E., Kouras, A., Samara, C. (2004). Profile Analysis of Ambient and Source Emitted Particle-Bound Polycyclic Aromatic Hydrocarbons from Three Sites in Northern Greece, *Chemosphere*, 56 (9), pp. 867-878

Mantis, J., Chaloulakou, A., Samara, C. (2005). PM10-bound polycyclic aromatic hydrocarbons (PAHs) in the Greater Area of Athens, Greece, *Chemosphere* 59 (5), pp. 593-604

Masclet, P., Bresson, M.A., Mouvier, G. (1987). Polycyclic aromatic hydrocarbons emitted by power stations, and influence of combustion conditions, *Fuel*, 66 (4), pp. 556-562

Miguel, A.H., Kirchstetter, T.W., Harley, R.A., Hering, S.V. (1998). On-road emissions of particulate polycyclic aromatic hydrocarbons and black carbon from gasoline and diesel vehicles, *Environmental Science and Technology*, 32 (4), pp. 450-455

Mohanraj, R., Solaraj, G., Dhanakumar, S. (2011). Fine particulate phase PAHs in ambient atmosphere of Chennai metropolitan city, India. *Environmental Science and Pollution Research*, 18 (5), pp. 764-771

Nisbet, I.C.T., LaGoy, P.K.: (1992) Toxic equivalency factors (TEFs) for polycyclic aromatic hydrocarbons (PAHs), *Regulatory Toxicology and Pharmacology*, 16 (3), pp. 290-300

Odabasi, M., Vardar, N., Sofuoglu, A., Tasdemir, Y., Holsen, T.M. (1999). Polycyclic aromatic hydrocarbons (PAHs) in Chicago air, *Science of the Total Environment*, 227 (1), pp. 57-67

Oliveira, C., Martins, N., Tavares, J., Pio, C., Cerqueira, M., Matos, M., Silva, H., Oliveira, C., Camões, F. (2011). Size distribution of polycyclic aromatic hydrocarbons in a roadway tunnel in Lisbon, Portugal, *Chemosphere*, 83 (11), pp. 1588-1596

Pakkanen, T.A., Kerminen, V.-M., Loukkola, K., Hillamo, R.E., Aarnio, P., Koskentalo, T., Maenhaut, W. (2003). Size distributions of mass and chemical components in street-level and rooftop PM1 particles in Helsinki, *Atmospheric Environment*, 37 (12), pp. 1673-1690

Panther, B.C., Hooper, M.A., Tapper, N.J. (1999). A comparison of air particulate matter and associated polycyclic aromatic hydrocarbons in some tropical and temperate urban Environments, *Atmospheric Environment*, 33 (24-25), pp. 4087-4099

Pastuszka, J.S., Rogula-Kozłowska, W., Zajusz-Zubek, E. (2010). Characterization of PM10 and PM2.5 and associated heavy metals at the crossroads and urban background site in Zabrze, Upper Silesia, Poland, during the smog episodes, *Environmental Monitoring and Assessment*, 168 (1-4), pp. 613-627

Phuleria, H.C., Sheesley, R.J., Schauer, J.J., Fine, P.M., Sioutas, C. (2007). Roadside measurements of size-segregated particulate organic compounds near gasoline and diesel-dominated freeways in Los Angeles, CA, *Atmospheric Environment*, 41 (22), pp. 4653-4671

Prahl, F.G., Carpenter, R. (1983). Polycyclic aromatic hydrocarbon (PAH)-phase associations in Washington coastal sediment, *Geochimica et Cosmochimica Acta*, 47(6), pp.1013-1023

Program ochrony powietrza dla stref województwa śląskiego, w których stwierdzone zostały ponadnormatywne poziomy substancji w powietrzu. Załącznik do uchwały Nr III/52/15/2010 Sejmiku Województwa Śląskiego z dnia 16 czerwca 2010 r. Katowice, 2010 r. (in polish)

Querol, X., Alastuey, A., Rodriguez, S., Plana, F., Mantilla, E., Ruiz, C.R. (2001). Monitoring of PM10 and PM2.5 around primary particulate anthropogenic emission sources, *Atmospheric Environment*, 35 (5), pp. 845-858

Ravindra, Mittal, A.K., Van Grieken, R. (2001). Health risk assessment of urban suspended particulate matter with special reference to polycyclic aromatic hydrocarbons: A review, *Reviews on Environmental Health*, 16 (3), pp. 169-189

Ravindra, K., Bencs, L., Wauters, E., De Hoog, J., Deutsch, F., Roekens, E., Bleux, N., Berghmans, P., Van Grieken, R. (2006). Seasonal and site-specific variation in vapour and aerosol phase PAHs over Flanders (Belgium) and their relation with anthropogenic activities, *Atmospheric Environment*, 40 (4), pp. 771-785

Ravindra, K., Sokhi, R., Van Grieken, R. (2008). Atmospheric polycyclic aromatic hydrocarbons: Source attribution, emission factors and regulation, *Atmospheric Environment*, 42 (13), pp. 2895-2921

Rogge, W.F., Hildemann, L.M., Mazurek, M.A., Cass, G.R., Simonelt, B.R.T. (1993a). Sources of fine organic aerosol. 3. Road dust, tire debris, and organometallic brake lining dust: Roads as sources and sinks, *Environmental Science and Technology*, 27 (9), pp. 1892-1904

Rogge, W.F., Hildemann, L.M., Mazurek, M.A., Cass, G.R., Simoneit, B.R.T. (1993b). Sources of fine organic aerosol. 2. Noncatalyst and catalyst-equipped automobiles and heavy-duty diesel trucks, *Environmental Science and Technology*, 27 (4), pp. 636-651

Rogula-Kozłowska, W., Pastuszka, J.S., Talik, E. (2008). Influence of vehicular traffic on concentration and particle surface composition of PM10 and PM2.5 in Zabrze, Poland, *Polish Journal of Environmental Studies*, 17 (4), pp. 539-548

Rogula-Kozłowska W., Klejnowski K., Krasa A., Szopa S.(2010). Concentration and elemental composition of atmospheric fine particles in Silesia Province, Poland. [In] Environmental Engineering III, Pawłowski, Dudzińska. & Pawłowski (eds.), Taylor & Francis Group, London, pp. 75-81

Saraga, D.E., Maggos, T.E., Sfetsos, A., Tolis, E.I., Andronopoulos, S., Bartzis, J.G., Vasilakos, C. (2010): PAHs sources contribution to the air quality of an office environment: Experimental results and receptor model (PMF) application. *Air Quality, Atmosphere and Health*, 3 (4), pp. 225-234

Sheu, H.-L., Lee, W.-J., Lin, S.J., Fang, G.-C., Chang, H.-C., You, W.-C. (1997). Particle-bound PAH content in ambient air, *Environmental Pollution*, 96 (3), pp. 369-382

Sicre, M.A., Marty, J.C., Saliot, A., Aparicio, X., (1987). Aliphatic and aromatic hydrocarbons in the Mediterranean aerosol. *International Journal of Environmental Analytical Chemistry*, 29, pp. 73–94

Sienra, M.del.R., Rosazza, N.G., Préndez, M. (2005). Polycyclic Aromatic Hydrocarbons and Their Molecular Diagnostic Ratios in Urban Atmospheric Respirable Particulate Matter. *Atmospheric Research*, 75 (4), pp. 267-281

Simcik, M.F., Eisenreich, S.J., Lioy, P.J. (1999). Source apportionment and source/sink relationships of PAHs in the coastal atmosphere of Chicago and Lake Michigan, *Atmospheric Environment*, 33 (30), pp. 5071-5079

Slezakova, K., Castro, D., Begonha, A., Delerue-Matos, C., Alvim-Ferraz, M.D.C., Morais, S., Pereira, M.D.C. (2011). Air pollution from traffic emissions in Oporto, Portugal: Health and environmental implications, *Microchemical Journal*, 99 (1), pp. 51-59

Smith, D.J.T., Harrison, R.M. (1998) Atmospheric Particles. Wiley, New York

State Sanitary Inspection, Department in Katowice. Atmospheric pollution in the Katowice Province in 1979. Katowice 1980, (for official use, in polish)

State Sanitary Inspection, Department in Katowice. Atmospheric pollution in the Katowice Province in the years 1983 and 1984. Katowice 1984/85, (for official use, in polish)

State Sanitary Inspection, Department in Katowice. Atmospheric pollution in the Katowice Province in the years 1985 – 1987. Katowice 1988, (in polish)

State Sanitary Inspection, Department in Katowice. Atmospheric pollution in the Katowice Province in the years 1988 – 1990, Katowice 1991, (in polish)

State Sanitary Inspection, Department in Katowice. Atmospheric pollution in the Katowice Province in the years 1991 – 1993, Katowice 1994, (in polish)

State Sanitary Inspection, Department in Katowice. Atmospheric pollution in the Katowice Province in the years 1996 – 1997, Katowice 1998, (in polish)

State Sanitary Inspection, Department in Katowice. Atmospheric pollution in the Katowice Province in the years 1997 – 1998 and in the former provinces Bielsko and Czestochowa in 1998, Katowice 1999, (in polish)

State Sanitary Inspection, Department in Katowice. Atmospheric pollution in the Silesian Province in the years 2001 - 2002, Katowice 2003, (in polish)

State Sanitary Inspection, Department in Katowice. Atmospheric pollution in the Silesian Province in the years 2002 – 2003, Katowice 2004, (in polish)

State Sanitary Inspection, Department in Katowice. Atmospheric pollution in the Silesian Province in the years 2003 – 2004, Katowice 2005, (in polish)

State Sanitary Inspection, Department in Katowice. Atmospheric pollution in the Silesian Province in 2005, Katowice 2006, (in polish)

Sverdrup, G.M., Whitby, K.T. (1977). Determination of submicron atmospheric aerosol size distributions by use of continuous analog sensors, *Environmental Science and Technology*, 11 (13), pp. 1171-1176

Takeshi, O., Takashi, A. (2004).Spatial distributions and profiles of atmospheric polycyclic aromatic hydrocarbons in two industrial cities in Japan, *Environmental Sciences and Technology*, 38 (1), pp. 49-55

Tang, N., Hattori, T., Taga, R., Igarashi, K., Yang, X., Tamura, K., Kakimoto, H., Mishukov, V.F., Toriba, A., Kizu, R., Hayakawa, K. (2005). Polycyclic aromatic hydrocarbons and nitropolycyclic aromatic hydrocarbons in urban air particulates and their relationship to emission sources in the Pan-Japan Sea countries, *Atmospheric Environment*, 39 (32), pp. 5817-5826

The Provincial Inspectorate for Environmental Protection, The environmental status in the Silesia region in 2001, Library of Environmental Monitoring, Katowice 2002, (in polish)

The Provincial Inspectorate for Environmental Protection. The environmental status in the Silesia region in 2002, Katowice 2003, (in polish)

The Provincial Inspectorate for Environmental Protection. The environmental status in the Silesia region in 2003, Katowice 2004, (in polish)

The Provincial Inspectorate for Environmental Protection. The environmental status in the Silesia region in 2004, Katowice 2005, (in polish)

The Provincial Inspectorate for Environmental Protection. The environmental status in the Silesia region in 2005, Katowice 2006, (in polish)

The Provincial Inspectorate for Environmental Protection. The environmental status in the Silesia region in 2006, Katowice 2007, (in polish)

The Provincial Inspectorate for Environmental Protection. The environmental status in the Silesia region in 2007, Katowice 2008, (in polish)

The Provincial Inspectorate for Environmental Protection. The environmental status in the Silesia region in 2008, Katowice 2009, (in polish)

The Provincial Inspectorate for Environmental Protection. The environmental status in the Silesia region in 2009, Katowice 2010, (in polish)

The Provincial Inspectorate for Environmental Protection. The environmental status in the Silesia region in 2010, Katowice 2011, (in polish)

Vercauteren, J., Matheeussen, C., Wauters, E., Roekens, E., van Grieken, R., Krata, A., Makarovska, Y., Maenhaut, W., Chi, X., Geypens, B. (2011). Chemkar PM10: An extensive look at the local differences in chemical composition of PM10 in Flanders, Belgium, *Atmospheric Environment*, 45 (1), pp. 108-116

Vogt, R., Kirchner, U., Scheer, V., Hinz, K.P., Trimborn, A., Spengler, B. (2003). Identification of diesel exhaust particles at an Autobahn, urban and rural location using single-particle mass spectrometry, *Journal of Aerosol Science,*34 (3), pp. 319-337

Wang, X.H., Ye, C.X., Yin, H.L., Zhuang, M.Z., Wu, S.P., Mu, J.L., Hong, H.S. (2007). Contamination of Polycyclic Aromatic Hydrocarbons Bound to PM10/PM2.5 in Xiamen, China, *Aerosol and Air Quality Research*, 7(2), pp.260–276

Wang, H., Zhou, Y., Zhuang, Y., Wang, X., Hao, Z. (2009). Characterization of PM2.5/PM2.5-10 and source tracking in the juncture belt between urban and rural areas of Beijing, *Chinese Science Bulletin*, 54 (14), pp. 2506-2515

Willeke, K., Whitby, K.T. (1975). Atmospheric aerosols: size distribution interpretation, *Journal of the Air Pollution Control Association*, 25 (5), pp. 529-534

Wingfors, H., Hägglund, L., Magnusson, R. (2011). Characterization of the size-distribution of aerosols and particle-bound content of oxygenated PAHs, PAHs, and n-alkanes in urban environments in Afghanistan, *Atmospheric Environment*, 45 (26), pp. 4360-4369

White, P.A. (2002). The genotoxicity of priority polycyclic aromatic hydrocarbons in complex mixtures, *Mutation Research - Genetic Toxicology and Environmental Mutagenesis*, 515 (1-2), pp. 85-98

Yan, J., Wang, L., Fu, P.P., Yu, H. (2004). Photomutagenicity of 16 polycyclic aromatic hydrocarbons from the US EPA priority pollutant list, *Mutation Research - Genetic Toxicology and Environmental Mutagenesis*, 557 (1), pp. 99-108

Yunker, M.B., Macdonald, R.W., Vingarzan, R., Mitchell, R.H., Goyette, D., Sylvestre, S. (2002). PAHs in the Fraser River basin: A critical appraisal of PAH ratios as indicators of PAH source and composition, *Organic Geochemistry*, 33 (4), pp. 489-515

Zaciera M., Kurek J., Dzwonek L., Feist B., Jędrzejczak A. (2011) Seasonal variability of PAHs and nitro-PAHs concentrations in total suspended particulate matter in ambient air of cities of silesian voivodeship, [In] Modern achievements in the protection ambient air, Musialik-Piotrowska A. & Rutkowski J.D., Polish Association of Sanitary Engineers and Technicians Wroclaw, pp. 401-406; (in polish)

Zencak, Z., Klanova, J., Holoubek, I., Gustafsson, Ö. (2007). Source apportionment of atmospheric PAHs in the western balkans by natural abundance radiocarbon analysis, *Environmental Science and Technology*, 41 (11), pp. 3850-3855

Zhao, Y., Wang, S., Duan, L., Lei, Y., Cao, P., Hao, J. (2008). Primary air pollutant emissions of coal-fired power plants in China: Current status and future prediction. *Atmospheric Environment*, 42 (36), pp. 8442-8452

Zou, L.Y., Zhang, W., Atkiston, S. (2003). The characterisation of polycyclic aromatic hydrocarbons emissions from burning of different firewood species in Australia, *Environmental Pollution*, 124 (2), pp. 283-289

9

Rapid Detection and Recognition of Organic Pollutants at Trace Levels by Surface-Enhanced Raman Scattering

Zhengjun Zhang, Qin Zhou and Xian Zhang

Tsinghua University

P. R. China

1. Introduction

Organic pollutants are harmful even at trace level in the environment, and they are difficult to detect at that concentration. Our work provides a rapid and sensitive method — surface-enhanced Raman scattering — to detect and distinguish isomers of organic pollutants.

2. Application of surface-enhanced Raman scattering to organic pollutant detection

In the modern world, environmental problems have attracted more and more attention, for environmental pollutants are extremely harmful to human beings' health. Environmental pollutants, such as persistent organic pollutants, are widely separated in the environment and difficult to detect at trace level.

Within persistent organic pollutants, polychlorinated biphenyls (PCBs), due to their excellent dielectric properties, had been widely used since the 1920s in transformers, heat transfers, capacitors, etc., and had polluted nearly everywhere in the world.[1] In recent years, however, they have been found to be very harmful to human beings. They may cause serious diseases, such as cancers and gene distortion, when exceeding the critical dose in human bodies, and more seriously, PCBs can be accumulated in plants and animals from the environment and yield higher doses in human bodies, making PCBs very dangerous to human beings even in trace amounts.[1-3] Therefore, the detection of PCBs in trace amounts is crucial. Currently, the mostly applied detection technique for PCBs is the combination of high-resolution gas chromatography and mass spectrometry. It requires, however, very sophisticated devices, standard samples, complicated pretreatments of samples, favourable experimental environments and experienced operators.[4-7] Thus, new methods are demanded especially for the rapid detection of trace amounts of PCBs.

Surface-enhanced Raman scattering (SERS) has been proven to be an effective way to detect some organics.[8] With the great progress of nanoscale technology in recent years, SERS has attracted enormous attention due to its excellent performance and potential applications in the detection of molecules in trace amounts, even single molecule detection. For instance, using Ag nanorods as SERS substrates Rhodamine 6G with concentration of

10^{-14} M (dissolved in water) was detected;[9] with the alumina-modified AgFON substrates, bacillus subtilis spores were detected to 10^{-14} M;[10, 11] Vo-Dinh reported even the detection of specific nucleic acid sequences by the SERS technique.[12-14] In spite of the numerous studies on the application as a chemical and biological sensor,[15-17] the SERS technique has not yet been employed to detect PCBs as they are hardly dissolved in water.

While lots of researchers investigate SERS in the detection of biological and medical molecules,[12-17] SERS has also proved an excellent method in environment pollutants' detection, such as trace amounts of polychlorinated biphenyls (PCBs).[45, 47]

3. Fingerprint character of SERS: Understanding and simulating Raman spectrum of organic pollutants

SERS is an excellent method to detect and recognize trace amounts of organic pollutants as organic pollutants have different Raman spectrums due to their different molecule vibration modes, even when they have similar physical and chemical properties. These different Raman spectrums have peaks with different Raman shifts and different peak heights. These peaks are sharp and unique, which show a fingerprint character of SERS spectra and make SERS spectra easily distinguishable from each other.

The Raman spectrum shows detailed structure information of organic pollutants. Therefore, one can detect and recognize organic pollutants via the SERS spectrum even at trace level, just like one can recognize crystal structures via X-ray diffraction.

The relationship between Raman spectrum and molecule vibration modes can be analyzed by density functional theory. We performed a simulation using the Gaussian 03 programme package with the density functional theory. The simulations were carried out with the Becke's three-parameter hybrid method using the Lee-yang-Parr correlation functional (B3LYP) and the LANL2DZ basis set.[46] The Gaussian View was used to input investigated compounds data visually.

4. Fabrication high sensitive silver nanorods SERS substrates

The detection sensitivity of SERS depends considerably on the surface property of the SERS substrate. High aspect ratio, nanostructured Ag, Au, Cu substrates are proved to be good SERS substrates. For instance, using ordered arrays of gold particles prepared through a porous alumina template as the SERS substrate, Rhodamine 6G (R6G) molecules were detected to a concentration limit of 10^{-12} M; arrays of silicon nanorods coated with thin films of Ag served as good SERS substrates for R6G molecule detection, etc.[12, 16] Thus the preparation of SERS substrates with preferred surface property is of great importance. There are several methods to prepare these kinds of SERS substrates and in this chapter we take glancing angle deposition as an example.

Glancing angle deposition (GLAD) technique is a simple but powerful means of producing thin films with pre-designed nanostructures, such as nanopillars, slanted posts, zigzag columns and spirals. Silver nanorod arrays prepared by GLAD are excellent SERS substrates.

In addition, the SERS properties are related to the optical properties of the nanorod arrays. Both SERS properties and optical properties depend on the structure of the nanorods, such as the shape, length, separation, tilting angle and so on, which can be tuned by the deposition conditions.

4.1 Fabrication of sensitive SERS substrates by GLAD

The detection sensitivity of the SERS technique depends greatly on the surface property of the SERS substrate.[40, 41] Among the approaches so far available to prepare nanostructured materials, the glancing angle deposition (GLAD) technique is a simple but powerful means of producing thin films with pre-designed nanostructures,[42, 43] such as nanopillars, slanted posts, zigzag columns, spirals,[18] [19] etc. [20-23] For example, arrays of Ag nanorods were found to be good SERS substrates for the detection of trans-1,2-bis(4-pyridyl)ethane molecules, with a SERS enhancement factor greater than 10^8.[16] It is therefore of great interest to investigate the growth of metal nanostructures by the GLAD technique.[44]

Pristine Si wafers with (001) orientation were used as substrates. These were supersonically cleaned in acetone, ethanol and de-ionized water baths in sequence, and were fixed on the GLAD substrate in an e-beam deposition system. The system was pumped down to a vacuum level of 3×10^{-5} Pa and then the thin Ag film was deposited on the substrate with a depositing rate of 0.5 nm/s, with the thickness monitored by a quartz crystal microbalance. To produce films of aligned Ag nanorods, the incident beam of Ag flux was set at ~ 85 º from the normal of the silicon substrate, at different substrate temperatures. The morphology and structure of the thin Ag films was characterized by scanning electron microscope (SEM), transmission electron microscope (TEM) and high-resolution TEM, selected area diffraction (SAD) and X-ray diffraction (XRD), respectively. The performance of the nanostructured Ag films as SERS substrates was evaluated with a micro-Raman spectrometer using R6G as the model molecule.

It is well known that the major factors influencing the growth morphology of the films by GLAD are the incident direction of the depositing beam flux, the temperature and the movement of the substrate, and the deposition rate, etc. When fixing the incident Ag flux at ~85 º from the normal of the substrate and the deposition rate at ~0.5 nm/s, the growth morphology of the Ag films was greatly dependent on the temperature and movement of the substrate. Figure 1 shows the growth morphology of thin Ag films versus the temperature and movement of the substrate. The SEM micrographs were taken by a FEI SEM (QUANTA 200FEG) working at 20 kV.

Figure 1(a) and (b) shows typical SEM images of the surface morphology of thin Ag films deposited at 120 ºC, without substrate rotation and with substrate rotation at a speed of 0.2 rpm, respectively. One sees from the images that at this temperature, Ag nanorods formed in two films with a length of 500 nm, yet they were not well separated - most nanorods were joined together. A major difference between the two is the growth direction of the joined nanorods, i.e. without rotation the nanorods grew at a glancing angle on the substrate, while with substrate rotation the nanorods grew vertically aligned. Another difference noticeable is the size of the nanorods, i.e. nanorods grown with substrate rotation have a slightly larger diameter.

Figure 1(c) and (d) shows respectively the surface morphology of thin Ag films deposited at -40 ºC, without substrate rotation and with rotation at a speed of 0.2 rpm. Comparing with figures 1(a) and (b), it can be seen that the decrease in the deposition temperature led to the separation of Ag nanorods in the two films, while the rotation of the substrate also determined the growth direction and diameter of the nanorods, as observed from figures 1(a) and 1(b). The Ag nanorods grown at this temperature are 20-30 nm in diameter, ~ 800 nm in length and are well separated. Therefore, through adjusting the temperature and movement of the substrate one can grow well separated and aligned Ag nanorods on planar silicon substrates.

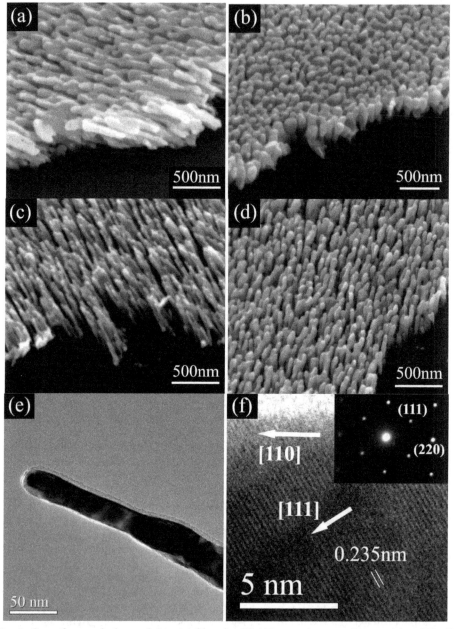

Fig. 1. Growth morphology of thin Ag films by GLAD at various conditions. (a) at 120 °C without substrate rotation; (b) at 120 °C and substrate rotation at 0.2 rpm; (c) at -40 °C without substrate rotation; and (d) at -40 °C and substrate rotation at 0.2 rpm. (e) and (f) shows respectively a bright-field TEM and a HRTEM image of the nanorods shown by figure 1(c); inset of (f) is the corresponding SAD pattern.

Figure 1(e) and 1(f) shows respectively a bright-field TEM and a HRTEM image of Ag nanorods shown by figure 1(c); inset of figure 1(f) is the corresponding SAD pattern. The images and the SAD pattern were taken with a JEM-2011F working at 200 kV. One sees from the figures that the Ag nanorod is ~ 30 nm in diameter and its micro-structure is single crystalline. By indexing the SAD pattern it is noticed that during the growth process the {111} plane of the nanorod was parallel to the substrate surface, with its axis along the <110> direction. This was confirmed by XRD analysis. Figure 2 shows a XRD pattern of the Ag nanorods shown by figure 1(c). The pattern was taken with a Rigaku X-ray diffractometer using the Cu k$_\square$ line, working at the θ -2 θ coupled scan mode. From the figure, a very strong (111) texture is observed, indicating that the {111} plane of the Ag nanorods was parallel to the substrate surface. These suggest that one can produce arrays of aligned, single crystalline Ag nanorods by the GLAD technique even at a low substrate temperature, i.e. -40 °C.

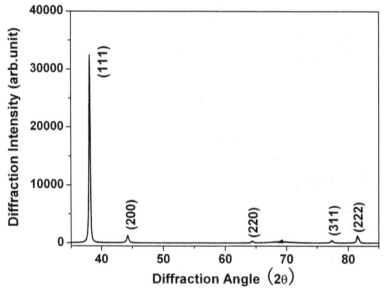

Fig. 2. A XRD pattern of the Ag film consisting of well separated, single crystalline nanorods shown by figure 1(c).

By using Rhodamine 6G as the model molecule, the performance of thin Ag films shown by figures 1(a)-(d) is examined as the SERS substrates. These samples were dipped in a 1x10^{-6} mol/L solution of R6G in water for 30 minutes and dried with a continuous gentle nitrogen blow. Figure 3(a) and 3(b) show Raman spectra of R6G obtained on the four nanostructured Ag films by a Reinshaw 100 Raman spectrometer using a 514 nm Ar$^+$ laser as the excitation source. It is observed that with the thin Ag films as the SERS substrate, all spectra exhibit clearly the characteristic peaks of R6G molecules, at 612, 774, 1180, 1311, 1361, 1511, 1575 and 1648 cm^{-1}, respectively.[12] However, the intensity of the Raman peaks was dependent on the morphology of the films. It is noticed that on Ag films consisting of well separated nanorods, see figures 3(b), the Raman peaks of R6G are much stronger than those on films of joined nanorods, see figures 3(a). This suggests that arrays of aligned but well separated Ag nanorods represent excellent SERS performance.

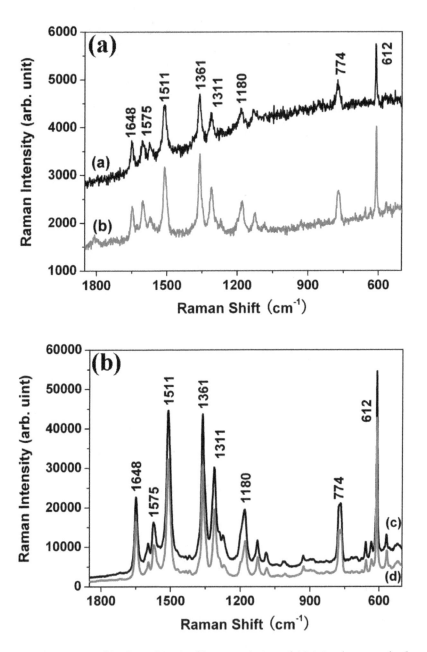

Fig. 3. Raman spectra of R6G on thin Ag films consisting of (a) joined nanorods shown by figures 1(a) (black line) and 1(b) (grey line); and (b) separated Ag nanorods shown by figures 1(c) (black line) and 1(d) (grey line), respectively, at a concentration of 1x10⁻⁶ mol/L.

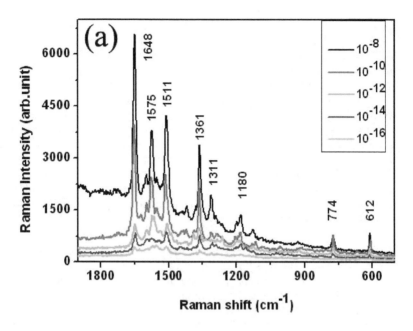

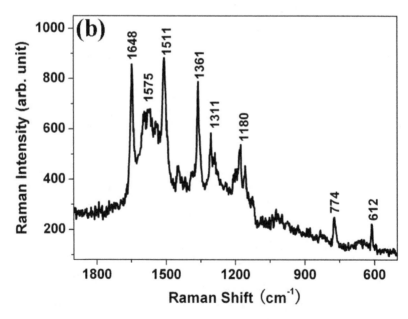

Fig. 4. (a) Raman spectra of R6G at concentrations ranging from 1x10⁻⁸ to 1x10⁻¹⁶ mol/L; and (b) the Raman spectrum of R6G at a concentration of 1x10⁻¹⁴ mol/L, on the thin Ag film consisting of well separated, single crystalline Ag nanorods.

Using arrays of aligned Ag nanorods shown by figures 1(c) and 1(d) as SERS substrates, we examined the detection limit of R6G molecules in water by the SERS technique. Figure 4(a) shows Raman spectra of R6G obtained on Ag nanorods shown by figure 1(c), as a function of the concentration of R6G in water ranging from 1×10^{-8} to 1×10^{-16} mol/L. Similar results were also obtained for Ag nanorods shown by figure 1(d). The Raman spectra were obtained by one scan with an accumulation time of 10 s, at a laser power of 1 % to avoid decomposition of R6G. It is found that characteristic peaks of R6G were all observed at all concentrations. To clearly show this, we plot the Raman spectrum at 10^{-14} mol/L in figure 4(b). It is noticed that although the intensity of the peaks is almost two orders lower than that at 10^{-6} mol/L, the spectrum contains the clear characteristic peaks of R6G.[12] These suggest that Ag films consisting of aligned and well separated Ag nanorods with single crystalline could serve as excellent SERS substrate for trace amount detection of R6G molecules. However, in the Raman spectrum at 10^{-16} mol/L in figure 4(a), some of the peaks of R6G disappear. That suggests the concentration limit of this method is 10^{-14} mol/L in the authors' work.[9]

4.2 Enhancing the sensitivity of SERS substrates via underlayer films

Although the Ag nanorod arrays present sensitive SERS performance, it is still necessary to enable the substrate to detect organic pollutants at trace amount with adequate sensitivity. There are several ways to promote the sensitivity of Ag nanorods as SERS substrates.

Much effort has been devoted to achieving highly sensitive SERS substrates. In particular, multilayer structures can improve SERS enhancement, such as "sandwich" structures with silver oxide or carbon inside and Ag or Au as both underlayer and overlayer.[24-28] Other researchers found that multilayer structures of Ag/Au nanostructures on the smooth metallic underlayer exhibited better SERS sensitivity compared to those without metallic underlayer (EF = 5×10^8).[29-31] However, the factor that governs the enhancement for multilayer structures is not very clear. Recently, Misra et al. obtained remarkably high SERS sensitivity using a micro-cavity with a radius of several micrometers.[32] Shoute et al. obtained high SERS signals (EF = 6×10^6) for molecules adsorbed on the silver island films supported by thermally oxidized silicon wafers and declared that the additional enhancement was due to the optical interference effect.[33] All the above experiments and those conducted by Driskell et al. suggested that the underlayer reflectivity could play an important role in the multilayer SERS substrates.[29]

We have investigated in detail the relationship of underlayer reflectivity and the SERS enhancement of Ag nanorod substrates prepared by oblique angle deposition. We use thin films of different materials with different thicknesses as underlayers to modulate the reflectivity systematically. With the coating of the same Ag nanorods, we find that the SERS intensity increases linearly with the underlayer reflectivity. This conclusion can be explained by a modified Greenler's model we recently developed.[34]

To change the reflectivity of the underlayer films, one can vary the dielectric constant and the thickness of the films systematically. We proposed to use Ag, Al, Si and Ti films, since they have different dielectric constants and can be fabricated easily. With a transfer matrix method, we can calculate the reflectivity of those films.[35, 36] Figure 5(a) shows the calculated reflectivity spectra of 100 nm Ag, Al, Si, and Ti films. In general, the reflectivity, $R_{Ag} > R_{Al} > R_{Ti} > R_{Si}$, except that at $\lambda \sim 600$ nm where the Si film has a large constructive interference. Figure 5(b) plots the film thickness dependent reflectivity for Ag, Al, Si and Ti at a fixed wavelength $\lambda_0 = 785$ nm. The reflectivity of Ag, Al, Ti, e.g. metals, increases monotonically with the film thickness d. The reflectivity R of Ag, Al and Ti thin films increases sharply when $d < 100$ nm and almost remains unchanged when 100 nm $\leq d \leq 400$ nm, while R_{Si} shows an oscillative behaviour due to the interference effect of a dielectric layer.

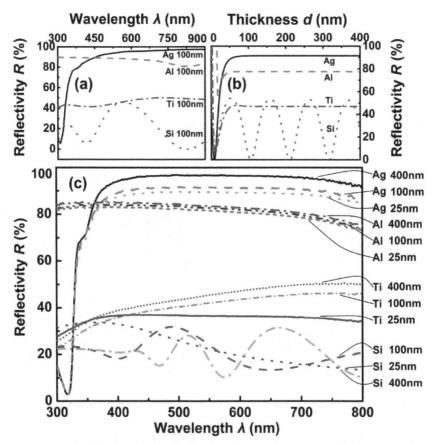

Fig. 5. (a) Calculated reflectivity R of thin Ag, Al, Si and Ti films at different wavelengths λ with film thickness of 100 nm; b) calculated reflectivity R of thin Ag, Al, Si and Ti films with different thicknesses d at $\lambda_0 = 785$ nm; c) experimentally obtained reflectivity spectra of thin Ag, Al, Si and Ti films with different thicknesses.

We deposited thin Ag, Al, Si and Ti films, all with thickness $d = 25$, 100, and 400 nm, respectively, to achieve different reflectivity. All depositions were carried out in a custom-designed electron-beam deposition system.[16] Before the deposition, the glass slide substrates were cleaned by piranha solution ($H_2SO_4 : H_2O_2 = 4:1$ in volume). The pellets of source materials, Ag, Al, Ti, with 99.99% purity, were purchased from Kurt J. Lesker Company, and Si with 99.9999% purity was purchased from Alfa Aesar Company. The film thickness was monitored *in situ* by a quartz crystal microbalance (QCM) facing toward the vapour source. After the deposition, the reflectivity of the deposited thin films was measured by an Ultraviolet-Visible Spectrophotometer (UV-Vis) double beam spectrophotometer with an integrating sphere (Shimadzu UV-Vis 2450). Figure 1(c) shows the reflectivity spectra of the twelve thin films obtained. The shapes of the reflection spectra are qualitatively consistent with those predicted by the calculations, as shown in Figs. 1(a) and (b). At the same wavelength, in general, $R_{Ag} > R_{Al} > R_{Ti} > R_{Si}$. In the visible wavelength region, the reflectivity of Ag, Al and Ti increases with the thickness d, while Si demonstrates an oscillating behaviour.

The twelve deposited planar thin film samples were then loaded into another custom-designed electron-beam evaporation system for Ag nanorod deposition through the so-called oblique angle deposition (OAD).[16, 29, 37] In this deposition, the background pressure was 1×10^{-7} Torr and the substrate holder was rotated so that the deposition flux was incident onto the thin films with an angle $\theta = 86°$ with respect to the surface normal of the substrate holder. The Ag nanorod arrays were formed through a self-shadowing effect.[16, 29, 37] During the deposition, the Ag deposition rate was monitored by a QCM directly facing the incident vapour. The deposition rate was fixed at 0.3 nm/s and the deposition ended when the QCM read 2000 nm (our optimized condition).

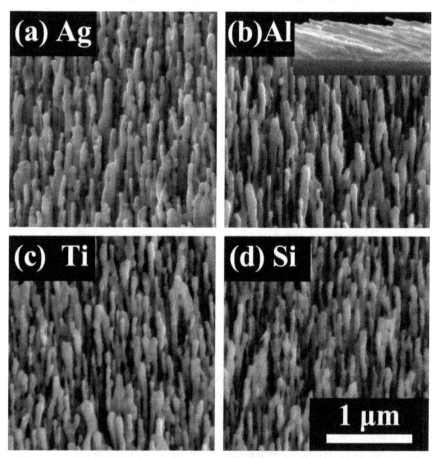

Fig. 6. Representative SEM images of Ag nanorod arrays on 100 nm underlayer thin films with different materials: (a) Ag; (b) Al; (c) Ti; (d) Si. All the figures have the same scale bar.

The morphologies of the Ag nanorod arrays on different thin film substrates were characterized by a scanning electron microscope (SEM, FEI Inspect F). The typical top-view SEM images are shown in Fig. 6 and they all look very similar. From the cross-section and top-view SEM images, the length L, diameter D and separation S of these Ag nanorods on different planar thin films are obtained statistically: $L_{Ag} = 940 \pm 70$ nm, $D_{Ag} = 90 \pm 10$ nm, $S_{Ag} = 140 \pm 30$ nm; $L_{Al} = 950 \pm 50$ nm, $D_{Al} = 90 \pm 10$ nm, $S_{Al} = 140 \pm 30$ nm; $L_{Si} = 900 \pm 50$ nm, $D_{Si} = 80 \pm 10$ nm, $S_{Si} = 130 \pm 20$ nm; and $L_{Ti} = 930 \pm 60$ nm, $D_{Ti} = 90 \pm 10$ nm, $S_{Ti} = 130 \pm 20$ nm, respectively. The Ag nanorod tilting angles β were measured to be about 73°with respect to substrate normal, which are consistent with our previous results.[16, 29, 37] These structure parameters are very close to one another, implying that the Ag nanorod arrays deposited on different thin film substrates are statistically the same. The SERS response of these Ag nanorod substrates were evaluated under identical conditions: A 2 µL droplet of a Raman probe molecule, trans-1, 2- bis (4-pyridyl) ethylene (BPE) with a concentration of 10^{-5} M, was uniformly dispersed onto the Ag nanorod substrates. The SERS spectra were recorded by the HRC-10HT Raman Analyzer from Enwave Optronics Inc., with an excitation wavelength of $\lambda_0 = 785$ nm, a power of 30 mW and an accumulation time of 10 s.

Figure 7(a) shows the representative BPE SERS spectra obtained at $\lambda_0 = 785$ nm from the Ag nanorod arrays on thin Ag, Al, Si and Ti film underlayers (thickness $d = 100$ nm). Each spectrum is an average of at least 15 different spectra taken at different spots on the substrates. All of them show the three main Raman bands of BPE, $\Delta\nu = 1639$, 1610, and 1200 cm^{-1}, which can be assigned to the C=C stretching mode, aromatic ring stretching mode and in-plane ring mode, respectively.[38] The SERS intensity of the Ag nanorods grown on thin Ag film are higher than others and the SERS intensity of the Ag nanorods on Al film is larger than that on Ti film. The Ag nanorods on Si film show the smallest SERS intensity. According to Fig. 5(c), this seems to follow a trend: the larger the underlayer reflectivity, the larger the SERS intensity. To quantitatively compare the SERS response of these substrates, the Raman peak intensity I_{1200} at $\Delta\nu = 1200$ cm^{-1} is analyzed.

Figure 7(b) plots the SERS intensity I_{1200} versus the reflectivity R of the underlayer thin films at $\lambda_0 = 785$ nm. The error bar for the Raman intensity is the standard deviation from 15 or more measurements from multiple sampling spots on the same substrates and the error bar for the reflectivity data is calculated from multiple reflectivity measurements at $\lambda_0 = 785$ nm. In Fig. 7(b), the SERS intensity and reflectivity follow a linear relationship: when the reflectivity of the underlayer increases, the SERS enhancement factor increases. This linear relationship of the underlayer reflectivity and SERS intensity can be explained by a modified Greenler's model developed by Liu et al.[34] Greenler's model is proposed through classical electrodynamics to explain the effects of the incident angles and polarization, and the collecting angle on the Raman scattering from a molecule adsorbed on a planar surface.[32] The modified Greenler's model extended the Greenler's model from a planar surface to Ag nanorod substrates and considered the effect of the underlying substrate.[34] The main point of the modified Greenler's model is to consider the conditions of both the incident and scattering fields near the molecule absorbed on a nanorod to calculate the enhancement.

Assuming that relative Raman intensity η is the ratio of the total Raman scattering power to incident light power, according to the modified Greenler's model, η excited by an unpolarized light can be explicitly expressed as[34]

$$\eta = <E^2_{Raman}> / <E^2_{incident}>$$

$$= \frac{1}{2}\{[1+R_p+n_2^4R'_p\cos\delta_p\cos2(\varphi-\beta)+2n_2^2R'^{\frac{1}{2}}_p\cos(\delta'_p+2\pi\Delta/\lambda)\sin2\beta$$

$$+2n_2^2R_p^{\frac{1}{2}}R'^{\frac{1}{2}}_p\sin2\varphi\cos(\delta'_p+2\pi\Delta/\lambda-\delta_p)](1+R_p+2\sqrt{R_p}\cos\delta_p)\cos^2(\varphi-\beta)$$

$$+[1+n_2^4R'_s+2n_2^2R'^{\frac{1}{2}}_s\cos(\delta'_s+2\pi\Delta/\lambda)]\}$$

where R_p and R_s are the reflectivity of p- and s-polarized lights by the Ag nanorod surface, and R'_p and R'_s are the reflectivity of p- and s-polarized components by the underlayer thin film; n_2 is complex refractive index of Ag, and $n_2 = 0.03 + 5.242i$ (for $\lambda_0 = 785$ nm); φ is the light incident angle, and β is the Ag nanorod tilting angle; $\Delta = d (1 + \cos 2\varphi) / \cos \varphi$, where d is the thickness of Ag nanorod layer; δ_p, δ_s, δ'_p, and δ'_s are the reflectivity phase shifts of p- and s-polarization E-fields from Ag nanorods and underlayer thin film, defined as

$$\delta_p = \tan^{-1}[Im(r_p) / Re(r_p)], \quad \delta_s = \tan^{-1}[Im(r_s) / Re(r_s)],$$

$$\delta'_p = \tan^{-1}[Im(r'_p) / Re(r'_p)], \quad \delta'_s = \tan^{-1}[Im(r'_s) / Re(r'_s)]$$

By setting the light incident angle $\varphi = 0°$, the Ag nanorod tilting angle $\beta = 73°$, the thickness of Ag layer $d = 300$ nm, the relative Raman intensity η as a function of the underlayer reflectivity R at $\lambda_0 = 785$ nm is calculated and plotted in Fig. 7(c). It shows that the η indeed increases linearly with R, which is in very good agreement with our experimental data shown in Fig. 7(b). Therefore, the underlayer reflectivity is one significant parameter to consider for improving the SERS response of multilayer substrates.

Both our experiments and the modified Greenler's model demonstrate that the higher the underlayer reflectivity, the higher the SERS intensity for the Ag nanorod based SERS substrates. Accordingly, in order to further improve the SERS response of the Ag nanorod substrates, one can further increase the reflectivity of the underlayers through a proper surface coating such as multilayer dielectric coating.[39]

5. Detecting trace amount PCBs by SERS method

With the highly sensitive SERS substrates described before, one can detect trace amount organic molecules by the SERS method.

SERS is extremely sensitive in water solutions, for water does not have any Raman peaks. When detecting organic pollutants in nonaqueous systems, we use volatile organic solvents, such as acetone, to dilute pollutants. As the organic solvent shows high Raman background, we need to make the solvent volatilizated completely before SERS measurement.

The powders of 2, 3, 3', 4, 4'-pentachlorinated biphenyl used in this study were commercially available from the AccuStandard Company. Since there is no Raman data of 2, 3, 3', 4, 4'- pentachlorinated biphenyl reported, we first measured its Raman spectrum and that of acetone, for comparison, see Figure 8(a). To clearly show most characteristic peaks of 2, 3, 3', 4, 4'-pentachlorinated biphenyl, the Raman spectrum was plotted in two regions of 300 to 1000 cm⁻¹ and 1000 to 1700 cm⁻¹ respectively, see Figures 11(b) and 11(c). From the figures one sees that the strongest peaks are located at 342, 395, 436, 465, 495, 507, 517, 598, 679, 731, 833, 891, 1032, 1136, 1179, 1254, 1294, 1573, and 1591 cm⁻¹, respectively; while for acetone the characteristic peaks are at 530, 786, 1065, 1220, 1428 and 1709 cm⁻¹, respectively. It is suggested that 2, 3, 3′, 4, 4′-pentachlorinated biphenyl is distinguishable from acetone and that acetone can be used as the solvent for the SERS measurements, as 2, 3, 3′, 4, 4′-pentachlorinated biphenyl is not soluble in water.

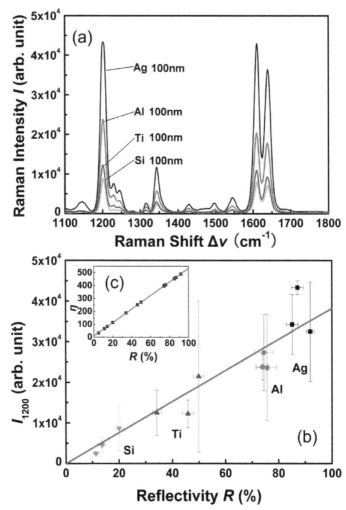

Fig. 7. (a) BPE SERS spectra obtained from Ag nanorod arrays deposited on 100 nm thin Ag, Al, Si and Ti film underlayers; (b) the plot of experimental Raman intensity as a function of underlayer reflectivity at λ_0 = 785 nm. Different symbol groups represent different kinds of substrates. (c) The plot of the enhanced Raman intensity ratio η as a function of underlayer reflectivity calculated by the modified Greenler's model.

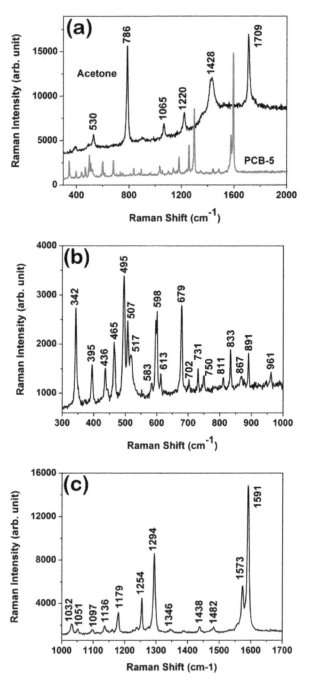

Fig. 8. (a) Comparison of Raman spectra of PCB-5 powders and acetone; (b) and (c) show details of the Raman spectrum of PCB-5 powders.

Because the SERS sensitivity is also dependent on the sample treatment, we employed in this study a very simple method to prepare SERS samples, i.e. dropping a small volume (~ 0.5 uL) of solutions of 2, 3, 3', 4, 4'-pentachlorinated biphenyl in acetone on Ag nanorods using a top single channel pipettor and then blowing away the acetone with a continuous, gentle nitrogen blow. Figure 9(a) shows the Raman spectra of 2, 3, 3', 4, 4'-pentachlorinated biphenyl dissolved in acetone at concentrations of 10^{-4} to 10^{-10} mol/L, respectively. The accumulation time of each Raman spectrum was 50 seconds and we used only 1% laser power to avoid changing of pentachlorinated biphenyl. When the small volume (~ 0.5 uL) of solutions of 2, 3, 3', 4, 4'-pentachlorinated biphenyl was dropped on Ag nanorods, it became a circular spot with diameter of about 4 mm. The Raman spectrum was accumulated from a 2 um diameter circular area on the substrates. Therefore, for the solution at concentration of 10^{-10} mol/L, only about ten 2, 3, 3', 4, 4'-pentachlorinated biphenyl molecules ($2*10^{-23}$ mol) would be accumulated in SERS; if that was at concentration of 10^{-8} mol/L, about 1000 molecules ($2*10^{-21}$ mol) would be accumulated and so on.

One sees that the Raman peaks of 10^{-4} mol/L PCB-5 solution located at 342, 495, 598, 679, 1032, 1136, 1179, 1254, 1294, 1573 and 1591 cm^{-1} march the Raman peaks of powder PCB-5 very well, this is quite different from the characteristic peaks of acetone. The peak around 1390 cm^{-1} represents disordered and amorphous carbon on the substrates. Figure 9(b) shows the SERS spectra of 10^{-8} mol/L PCB-5. Peaks located at 495, 1032, 1294, 1573 and 1591 cm^{-1} can march the Raman peaks of powder PCB-5. It indicates that the peaks shown in Figure 9(a) and (b) are the characteristic peaks of dissolved PCB-5, and PCB-5 with a concentration of 10^{-8} mol/L can be detected by the SERS method in the authors' work.

Large scale arrays of aligned and well separated single crystalline Ag nanorods on planar silicon substrate can be fabricated by GLAD method and these Ag films can be used as SERS substrates. With these substrates 2, 3, 3', 4, 4'- PCB-5 molecules were detected even at a concentration of 10^{-8} mol/L by the SERS method, which indicates that trace amount of PCBs can be detected by the SERS method with Ag nanorods as SERS substrates.[45]

6. Rapid recognition of isomers and homologues of PCBs at trace levels by SERS

Detecting trace amount PCBs by the SERS method is introduced in section 4, but simple detection is not enough for organic pollutant detection, one also needs to distinguish kinds of organic pollutants from each other.

Furthermore, isomers and homologues of organic pollutants are hard to distinguish – especially in trace amounts – due to the similarities in their physical and chemical properties. The SERS method with silver nanorods as a substrate can be used to identify the Raman characteristics of isomers of monochlorobiphenyls and recognize these compounds, even at trace levels.

The Raman spectra of biphenyl, 2-, 3- and 4-chlorobiphenyls were measured by a Renishaw Raman 100 spectrometer using a 633 nm He-Ne laser as the excitation source at room temperature. Powders of these compounds are commercially available from the AccuStandard Company. Simulation of these Raman spectra was performed using the Gaussian 03 programme package with the density functional theory, to better understand the vibrational modes observed and figure out fingerprints of these compounds. For the SERS measurements, powders of chlorobiphenyl were dissolved in acetone to concentrations from 10^{-4} to 10^{-10} mol/L. The substrates were Ag nanorods prepared by electron beam deposition. The deposition of the Ag nanorods was described as aforementioned. A small volume of the solutions (~ 0.5 μL) was dropped on the surface of Ag nanorods and acetone was blown away using a nitrogen flow.

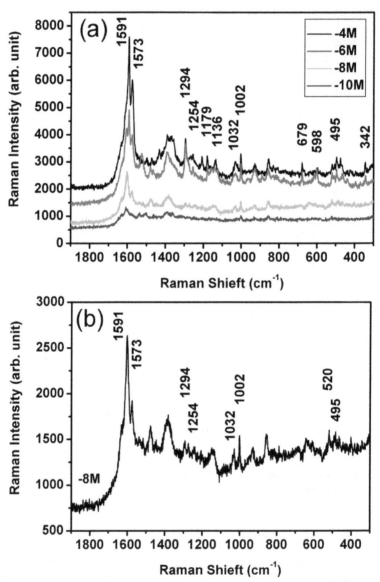

Fig. 9. (a) SERS spectra of PCB-5 dissolved in acetone with various concentrations; (b) SERS spectrum of PCB-5 in acetone at a concentration of 10^{-8} mol/L.

Figures 10 (a), (b), (c) and (d) show the measured Raman spectrum of biphenyl, 2-, 3- and 4-chlorobiphenyl, respectively. One sees that the four derivatives have strong peaks at ~ 3065, 1600, 1280, 1030 and 1000 cm⁻¹, demonstrating the common feature of biphenyl and its derivatives. One may also notice the differences among the Raman spectra of the four derivatives. For example, (1) biphenyl, 3- and 4-chlorobiphenyl have strong Raman peaks around 1276 cm⁻¹, while the peak for 2-chlorobiphenyl was at ~ 1297 cm⁻¹; (2) biphenyl has a strong peak at 738 cm⁻¹, 2- and 4-chlorobiphenyl have strong peaks at ~ 760 cm⁻¹, but the peak for 3-chlorobiphenyl was negligible; (3) both 2- and 3-chlorobiphenyl have strong peaks around ~ 680 cm⁻¹, while biphenyl and 4-chlorobiphenyl have no visible peak nearby; (4) only 2-chlorobiphenyl has a strong peak at ~ 432 cm⁻¹. The above features might be used to detect and distinguish biphenyl, 2-, 3- and 4-chlorobiphenyl.

To gain a clear understanding of these features, we performed simulations using the Gaussian 03 programme package with the density functional theory. The simulations were carried out with the Becke's three-parameter hybrid method using the Lee-yang-Parr correlation functional (B3LYP) and the LANL2DZ basis set.[46] The Gaussian View was used to input investigated compounds' data visually. The π bond length of the benzene ring was set to be 1.409 Å, the σ bond length between C and H atoms was set to be 1.088 Å and the σ bond length between C and Cl atoms was set to be 1.760 Å.

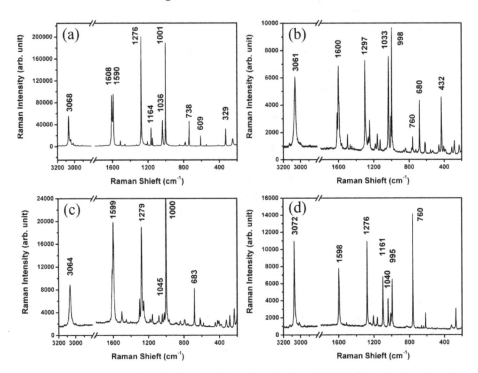

Fig. 10. Raman spectra of (a) biphenyl; (b) 2-chlorobiphenyl; (c) 3-chlorobiphenyl; and (d) 4-chlorobiphenyl, measured using powders commercially available from the AccuStandard Company.

Table I lists major vibrational modes of the four derivatives obtained by the above simulations. The common features of their Raman spectra at ~ 3065, 1600, 1280, 1030 and 1000 cm⁻¹, seen in Figure 10, can be attributed to the C-H stretching mode (~ 3100 cm⁻¹), the ring CCC stretching mode (~ 1650 cm⁻¹), the C-C bridge bond stretching mode (~ 1280 cm⁻¹), the C-H bending in-plane mode (1050 to 1100 cm⁻¹) and the CCC trigonal breathing mode(~ 1000 cm⁻¹), respectively.

Vibrational Model	Raman Shift /cm-1	Raman Int.of 2-chlorobiphenyl	Raman Int.of 3-chlorobiphenyl	Raman Int.of 4-chlorobiphenyl
CH stretching	3100	511	359	351
CCC stretching	1650	31	34	38
CC-bridge stretching	1280	50	50	62
CH bending in-plane	1050	46	22	15
trigonal breathing	1000	34	44	27
CCC bending in-plane (1-4 direction)	760	3	4	23
CCC bending in-plane (3-6 direction)	680	11	8	7
CCC bending in-plane (2-5 direction)	460	13	0	0

Table 1. Major simulated vibrational modes for 2-, 3- and 4-chlorobiphenyl.

One sees that due to the replacement of the H by Cl atom, the CCC bending (ring deformation) in-plane modes (~760 cm⁻¹, 680 cm⁻¹ and 460 cm⁻¹) changed differently for 2-, 3- and 4-chlorobiphenyl. Figures 11 (a), (b) and (c) show respectively the strong CCC bending in-plane mode for 2-, 3- and 4-chlorobiphenyl. It is seen that for the 2-chlorobiphenyl, the 2-5 direction CCC bending at 460 cm⁻¹ is the strongest one, the 3-6 direction CCC bending at 680 cm⁻¹ has a similar intensity, while the 1-4 direction bending at 760 cm⁻¹ is weak. For 3-chlorobiphenyl, the 2-5 direction bending mode is negligible, the 3-6 direction bending mode is strong, while the 1-4 direction bending mode is weak. For 4-chlorobiphenyl, the 2-5 direction bending mode is negligible, the 3-6 direction bending mode is weak, while the 1-4 direction bending mode is strong. These results are in agreement with the experimental measurements and suggest that these features in the CCC bending in-plane modes can be used to recognize the three homologues.

The substrate used in the SERS measurements was Ag nanorods prepared by the electron beam deposition technique. Powders of 2-, 3- and 4-chlorobiphenyl were dissolved in acetone and diluted into solutions with a concentration ranging from 10⁻⁴ to 10⁻¹⁰ mol/L. A small volume (0.5 µL) of these solutions was dropped on Ag nanorods and the acetone was blown away using a gentle nitrogen flow. Figures 12 (a), (b) and (c) show respectively the SERS spectra of the 2-, 3- and 4-chlorobiphenyl, at various concentrations. The accumulation time of each spectrum was fixed at 30 seconds per 100 cm⁻¹ and we used only 10% laser power (0.47mW) to avoid radiation damage. From these figures we notice that the characteristic Raman peaks are all clearly observed for three homologues, even at a concentration of 10⁻⁸ mol/L, suggesting that the SERS technique is able to detect chlorobiphenyls even at such a low concentration.

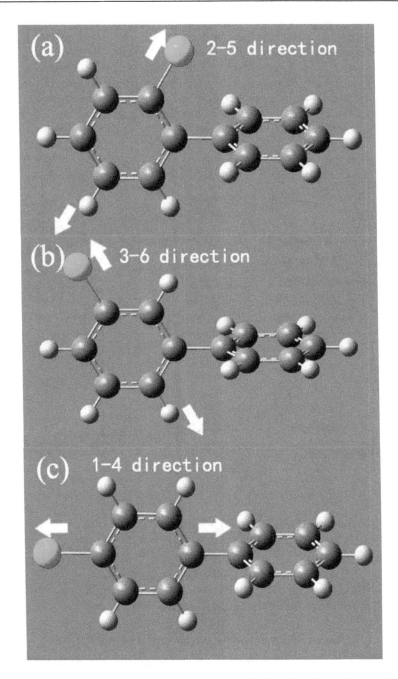

Fig. 11. The strong ring deformation in-plane modes for the three homologues of chlorobiphenyl. (a) 2-chlorobiphenyl; (b) 3-chlorobiphenyl; and (c) 4-chlorobiphenyl.

Figure 12 (d) compares SERS spectra of the three homologues at a concentration of 10^{-6} mol/L. One sees that at this concentration the three spectra show clearly the common features as mentioned above around 1600, 1280, 1030 and 1000 cm^{-1} (the 3100 cm^{-1} was not measured). The spectra also clearly show the characteristic Raman peaks for three homologues, i.e. the difference in the CCC bending (ring deformation) in-plane modes caused by the Cl atom replacement. These suggest that by using Ag nanorods as substrates, the SERS technique is capable of detecting chlorobiphenyls at trace amounts and is capable of recognizing the homologues at small concentrations.

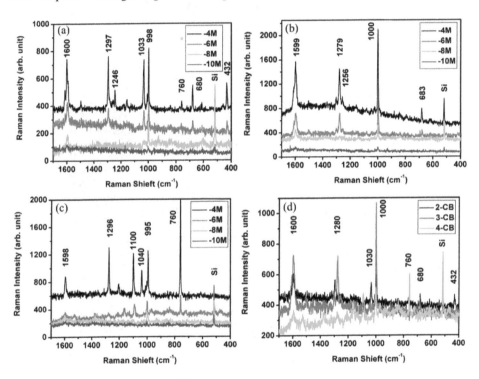

Fig. 12. SERS spectra of (a) 2-chlorobiphenyl; (b) 3-chlorobiphenyl; and (c) 4-chlorobiphenyl at concentrations from 10^{-4} to 10^{-10} mol/L in acetone. (d) Compares the SERS spectra of the three homologues at a concentration of 10^{-6} mol/L.

In summary, based on the understanding of Raman characteristics of these compounds, one can detect and recognize the homologues of chlorobiphenyls, even at the trace amount, by using the SERS technique with Ag nanorods as substrates.[47]

7. Detecting trace POPs in real environmental samples

In sections 4 and 5 we introduced the SERS method to detect and distinguish trace amount PCBs and their isomers and homologues. In those experiments, the PCBs are in acetone solutions, as fundamental study. In this section we introduce some examples in practical trace POPs detection.

7.1 Detecting trace PCBs in dry soil samples

The polluted soil samples were dried and made into small powers which were acquired from the Nanjing Institute of Soil (China). With a combination of the high-resolution gas chromatography and mass spectrometry techniques, sample I proved to contain about 5 µg/g PCBs and sample II proved to contain about 300 µg/g PCBs. 0.2 g soil sample I was put into 20 mL acetone and was agitated uniformly for about 5 minutes. This suspension was precipitated for 30 minutes and the transparent acetone solution in the upper layer was taken as solution sample A. 0.2 g soil sample I was put into 200 mL acetone and solution sample B was obtained through the aforementioned process. 0.2 g soil sample II was put into 20 mL acetone to obtain solution sample C and was put into 200 mL acetone to obtain solution sample D.

The Ag nanorods SERS substrates were put into the solution samples A, B, C and D, respectively. After 30 minutes, the Ag nanorods substrates were taken out of the solutions and the acetone on the substrates was blown away using a nitrogen flow. The Raman spectra of these substrates dipped into solution samples were measured by a Renishaw Raman 100 spectrometer using a 633 nm He-Ne laser as the excitation source at room temperature.

Figures 13 (a), (b), (c) and (d) show the measured Raman spectrum of the Ag substrates dipped into sample A, B, C and D, respectively. From Figures 13 (a), (b) and (c), one sees peaks at ~1600, 1280, 1240, 1150, 1030 and 1000 cm^{-1} clearly, demonstrating the common feature of PCBs. The peaks around 1590~1600cm^{-1} present benzene stretching vibration mode; the peak around 1280cm^{-1} presents CC bridge stretching vibration mode; the peak around 1030cm^{-1} presents CH bending in-plane mode; the peak around 1000cm^{-1} presents trigonal breathing vibration mode; and peaks around 1240~1250cm^{-1} and 1140~1200cm^{-1} present the vibration peaks induced by Cl substituent. These characteristic peaks suggest that PCBs in dry soil can be detected by the SERS method by dissolving into acetone. The most widely used PCBs are trichlorobiphenyls and pentachlorobiphenyls, we assumed that the molecular weight of the PCBs in the soil samples is 300, then the concentration of the PCBs acetone solution in solution sample A, B, C and D are about 10^{-5} mol/L, 10^{-5} mol/L, 10^{-6} mol/L, 10^{-7} mol/L, and 10^{-8} mol/L, respectively. Thus, with silver nanorod substrates, 5ug/g PCBs in dry soil samples can be detected by the SERS method.

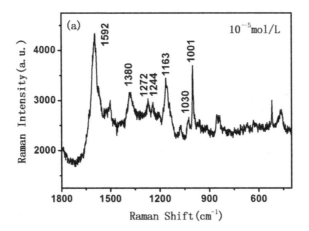

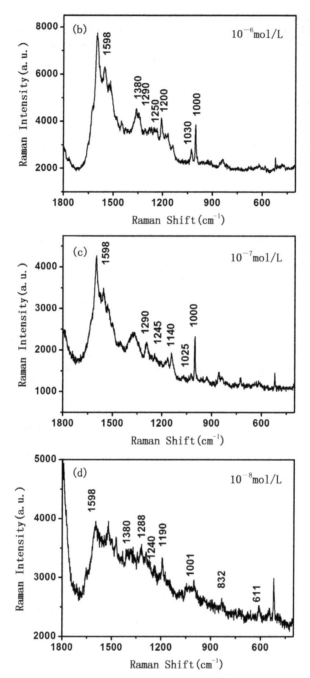

Fig. 13. SERS spectra of PCBs in dry soil samples after being treated by acetone: (a) ~10⁻⁵ mol/L; (b) ~10⁻⁶ mol/L; (c) ~10⁻⁷ mol/L; (d) ~10⁻⁸ mol/L.

7.2 Detecting PCBs in white spirit

PCBs in white spirit can also be detected by the SERS method with silver nanorod substrates. The concentration of PCBs in white spirit is about 10^{-4} mol/L. We put a drop of PCBs "polluted" white spirit on the silver nanorod substrates and made the white spirit volatilized away. Then, we found PCBs Raman signal with the SERS method described before. Figures 14 (a) and (b) show the SERS spectra of pure white spirit and white spirit with 10^{-4} mol/L PCBs, respectively. One can recognize characteristic Raman peaks of PCBs around 1590, 1290, 1240, 1030 and 1000 cm^{-1} in Figure 14 (b).

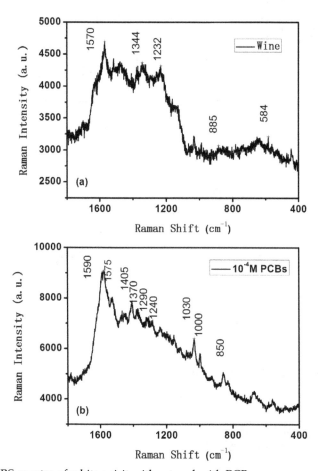

Fig. 14. SERS spectra of white spirit without and with PCBs.

7.3 Detecting melamine in milk

In the year 2009, milk produced by Sanlu Co. (China) was found to contain amounts of Melamine in much higher concentrations than usual. Milk with Melamine seems to contain more protein when detecting nitrogen concentration, but it is poisonous to children. With the SERS method with silver nanorods as substrates, we detected Melamine in milk. Figures 15 (a) and (b) show the SERS spectra of pure Melamine and milk with Melamine.

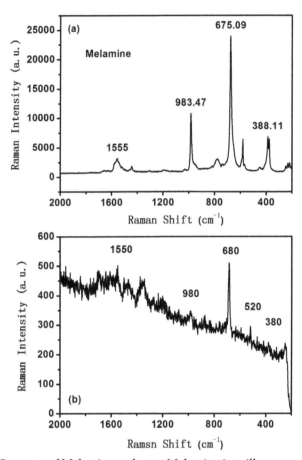

Fig. 15. SERS spectra of Melamine and trace Melamine in milk.

8. Summary

Although persistent organic pollutants such as PCBs are difficult to detect at trace amount, they can be detected and recognized rapidly via the SERS technique. Ag nanostructured SERS substrates prepared by the glancing angle deposition method are excellent at detection and their sensitivity can be further improved by tuning the thin underlayer films and Ag nanorod structures. With well designed and prepared Ag nanostructured SERS substrates, pentachlorinated biphenyl molecules are detected and isomers of chlorobiphenyl molecules are recognized at trace level using the SERS method. These series of studies provide a potential method for trace pollutant detection via nano technology.

9. References

[1] Ross G. The public health implications of polychlorinated biphenyls (pcbs) in the environment. *Ecotoxicology and Environmental Safety*, 2004, 59(3): 275~291.

[2] Ohtsubo Y, Kudo T, Tsuda M, et al. Strategies for bioremediation of polychlorinated biphenyls. *Applied Microbiology and Biotechnology*, 2004, 65(3): 250~258.

[3] Cicchetti D V, Kaufman A S, Sparrow S S. The relationship between prenatal and postnatal exposure to polychlorinated biphenyls (pcbs) and cognitive, neuropsychological, and behavioral deficits: a critical appraisal. *Psychology in the Schools*, 2004, 41(6): 589~624.

[4] Hong J E, Pyo H, Park S J, et al. Determination of hydroxy-pcbs in urine by gas chromatography/mass spectrometry with solid-phase extraction and derivatization. *Analytica Chimica Acta*, 2005, 531(2): 249~256.

[5] Namiesnik J, Zygmunt B. Selected concentration techniques for gas chromatographic analysis of environmental samples. *Chromatographia*, 2002, 56Suppl. S: S9~S18.

[6] Pitarch E, Serrano R, Lopez F J, et al. Rapid multiresidue determination of organochlorine and organophosphorus compounds in human serum by solid-phase extraction and gas chromatography coupled to tandem mass spectrometry. *Analytical and Bioanalytical Chemistry*, 2003, 376(2): 189~197.

[7] Barra R, Cisternas M, Suarez C, et al. Pcbs and hchs in a salt-marsh sediment record from south-central Chile: use of tsunami signatures and cs-137 fallout as temporal markers. *Chemosphere*, 2004, 55(7): 965~972.

[8] Moskovits M. Surface-enhanced spectroscopy. *Reviews of Modern Physics*, 1985, 57(3): 783~826.

[9] Zhou Q, Li Z C, Yang Y, et al. Arrays of aligned, single crystalline silver nanorods for trace amount detection. *Journal of Physics D-Applied Physics*, 2008, 41(15200715).

[10] Kudelski A. Analytical applications of raman spectroscopy. *Talanta*, 2008, 76(1): 1~8.

[11] Zhang X Y, Zhao J, Whitney A V, et al. Ultrastable substrates for surface-enhanced raman spectroscopy: al2o3 overlayers fabricated by atomic layer deposition yield improved anthrax biomarker detection. *Journal of the American Chemical Society*, 2006, 128(JA063876031): 10304~10309.

[12] Tan R Z, Agarwal A, Balasubramanian N, et al. 3d arrays of sers substrate for ultrasensitive molecular detection. *Sensors and Actuators a-Physical*, 2007, 139(1-2Sp. Iss. SI): 36~41.

[13] Isola N R, Stokes D L, Vo-Dinh T. Surface enhanced raman gene probe for hiv detection. *Analytical Chemistry*, 1998, 70(7): 1352~1356.

[14] Vodinh T, Houck K, Stokes D L. Surface-enhanced raman gene probes. Analytical Chemistry, 1994, 66(20): 3379~3383.

[15] Tripp R A, Dluhy R A, Zhao Y P. Novel nanostructures for sers biosensing. *Nano Today*, 2008, 3(3-4): 31~37.

[16] Chaney S B, Shanmukh S, Dluhy R A, et al. Aligned silver nanorod arrays produce high sensitivity surface-enhanced raman spectroscopy substrates. *Applied Physics Letters*, 2005, 87(0319083).

[17] Zhang Z Y, Zhao Y P. Tuning the optical absorption properties of ag nanorods by their topologic shapes: a discrete dipole approximation calculation. *Applied Physics Letters*, 2006, 89(0231102).

[18] Malac M, Egerton R F, Brett M J, et al. Fabrication of submicrometer regular arrays of pillars and helices. *Journal of Vacuum Science & Technology B*, 1999, 17(6): 2671~2674.

[19] Dick B, Brett M J, Smy T. Investigation of substrate rotation at glancing-incidence on thin-film morphology. *Journal of Vacuum Science & Technology B*, 2003, 21(6): 2569~2575.

[20] Dick B, Brett M J, Smy T J, et al. Periodic magnetic microstructures by glancing angle deposition. *Journal of Vacuum Science & Technology A-Vacuum Surface and Films*, 2000, 18(4Part 2): 1838~1844.

[21] Alouach H, Fujiwara H, Mankey G J. Magnetocrystalline anisotropy in glancing angle deposited permalloy nanowire arrays. *Journal of Vacuum Science & Technology A*, 2005, 23(4): 1046~1050.

[22] Singh J P, Tang F, Karabacak T, et al. Enhanced cold field emission from 100 oriented beta-w nanoemitters. *Journal of Vacuum Science & Technology B*, 2004, 22(3): 1048~1051.

[23] Hawkeye M M, Brett M J. Glancing angle deposition: fabrication, properties, and applications of micro- and nanostructured thin films. *Journal of Vacuum Science & Technology A*, 2007, 25(5): 1317~1335.

[24] Leverette C L, Shubert V A, Wade T L, et al. Development of a novel dual-layer thick ag substrate for surface-enhanced raman scattering (sers) of self-assembled monolayers. *Journal of Physical Chemistry B*, 2002, 106(34): 8747~8755.

[25] Li H G, Cullum B M. Dual layer and multilayer enhancements from silver film over nanostructured surface-enhanced raman substrates. *Applied Spectroscopy*, 2005, 59(4): 410~417.

[26] Li H G, Baum C E, Sun J, et al. Multilayer enhanced gold film over nanostructure surface-enhanced raman substrates. *Applied Spectroscopy*, 2006, 60(12): 1377~1385.

[27] Yang Y A, Bittner A M, Kern K. A new sers-active sandwich structure. *Journal of Solid State Electrochemistry*, 2007, 11(2): 150~154.

[28] Mulvaney S P, He L, Natan M J, et al. Three-layer substrates for surface-enhanced raman scattering: preparation and preliminary evaluation. *Journal of Raman Spectroscopy*, 2003, 34(2): 163~171.

[29] Driskell J D, Shanmukh S, Liu Y, et al. The use of aligned silver nanorod arrays prepared by oblique angle deposition as surface enhanced raman scattering substrates. *Journal of Physical Chemistry C*, 2008, 112(4): 895~901.

[30] Driskell J D, Lipert R J, Porter M D. Labeled gold nanoparticles immobilized at smooth metallic substrates: systematic investigation of surface plasmon resonance and surface-enhanced raman scattering. *Journal of Physical Chemistry B*, 2006, 110(35): 17444~17451.

[31] Addison C J, Brolo A G. Nanoparticle-containing structures as a substrate for surface-enhanced raman scattering. *Langmuir*, 2006, 22(21): 8696~8702.

[32] Misra A K, Sharma S K, Kamemoto L, et al. Novel micro-cavity substrates for improving the raman signal from submicrometer size materials. *Applied Spectroscopy*, 2009, 63(3): 373~377.

[33] Shoute L, Bergren A J, Mahmoud A M, et al. Optical interference effects in the design of substrates for surface-enhanced raman spectroscopy. *Applied Spectroscopy*, 2009, 63(2): 133~140.

[34] Liu Y J, Zhao Y P. Simple model for surface-enhanced raman scattering from tilted silver nanorod array substrates. *Physical Review B*, 2008, 78(0754367).

[35] Mitsas C L, Siapkas D I. Generalized matrix-method for analysis of coherent and incoherent reflectance and transmittance of multilayer structures with rough surfaces, interfaces, and finite substrates. *Applied Optics*, 1995, 34(10): 1678~1683.

[36] Fu J X, Park B, Zhao Y P. Nanorod-mediated surface plasmon resonance sensor based on effective medium theory. *Applied Optics*, 2009, 48(23): 4637~4649.

[37] Abell J L, Driskell J D, Dluhy R A, et al. Fabrication and characterization of a multiwell array sers chip with biological applications. *Biosensors & Bioelectronics*, 2009, 24(12): 3663~3670.

[38] Yang W H, Hulteen J, Schatz G C, et al. A surface-enhanced hyper-raman and surface-enhanced raman scattering study of trans-1,2-bis(4-pyridyl)ethylene adsorbed onto silver film over nanosphere electrodes. Vibrational assignments: experiment and theory. *Journal of Chemical Physics*, 1996, 104(11): 4313~4323.

[39] Zhou Q, Liu Y J, He Y P, et al. The effect of underlayer thin films on the surface-enhanced raman scattering response of ag nanorod substrates. *Applied Physics Letters*, 2010, 97(12190212).

[40] Michaels A M, Jiang J, Brus L. Ag nanocrystal junctions as the site for surface-enhanced raman scattering of single rhodamine 6g molecules. *Journal of Physical Chemistry B*, 2000, 104(50): 11965~11971.

[41] Mcfarland A D, Young M A, Dieringer J A, et al. Wavelength-scanned surface-enhanced raman excitation spectroscopy. *Journal of Physical Chemistry B*, 2005, 109(22): 11279~11285.

[42] Qin L D, Zou S L, Xue C, et al. Designing, fabricating, and imaging raman hot spots. *Proceedings of the National Academy of Sciences of the United States of America*, 2006, 103(36): 13300~13303.

[43] Liu Y J, Zhang Z Y, Zhao Q, et al. Surface enhanced raman scattering from an ag nanorod array substrate: the site dependent enhancement and layer absorbance effect. *Journal of Physical Chemistry C*, 2009, 113(22): 9664~9669.

[44] Gansel J K, Thiel M, Rill M S, et al. Gold helix photonic metamaterial as broadband circular polarizer. *Science*, 2009, 325(5947): 1513~1515.

[45] Zhou Q, Yang Y, Ni J, et al. Rapid detection of 2, 3, 3', 4, 4'-pentachlorinated biphenyls by silver nanorods-enhanced raman spectroscopy. *Physica E-Low-Dimensional Systems & Nanostructures*, 2010, 42(5): 1717~1720.

[46] Fleming G D, Golsio I, Aracena A, et al. Theoretical surface-enhanced raman spectra study of substituted benzenes i. Density functional theoretical sers modelling of benzene and benzonitrile. *Spectrochimica Acta Part a-Molecular and Biomolecular Spectroscopy*, 2008, 71(3): 1049~1055.

[47] Zhou Q, Yang Y, Ni J E, et al. Rapid recognition of isomers of monochlorobiphenyls at trace levels by surface-enhanced raman scattering using ag nanorods as a substrate. Nano Research, 2010, 3(6): 423~428.

Permissions

The contributors of this book come from diverse backgrounds, making this book a truly international effort. This book will bring forth new frontiers with its revolutionizing research information and detailed analysis of the nascent developments around the world.

We would like to thank Dr. Tomasz Puzyn and M.Sc. Eng Aleksandra Mostrag-Szlichtyng, for lending their expertise to make the book truly unique. They have played a crucial role in the development of this book. Without their invaluable contribution this book wouldn't have been possible. They have made vital efforts to compile up to date information on the varied aspects of this subject to make this book a valuable addition to the collection of many professionals and students.

This book was conceptualized with the vision of imparting up-to-date information and advanced data in this field. To ensure the same, a matchless editorial board was set up. Every individual on the board went through rigorous rounds of assessment to prove their worth. After which they invested a large part of their time researching and compiling the most relevant data for our readers. Conferences and sessions were held from time to time between the editorial board and the contributing authors to present the data in the most comprehensible form. The editorial team has worked tirelessly to provide valuable and valid information to help people across the globe.

Every chapter published in this book has been scrutinized by our experts. Their significance has been extensively debated. The topics covered herein carry significant findings which will fuel the growth of the discipline. They may even be implemented as practical applications or may be referred to as a beginning point for another development. Chapters in this book were first published by InTech; hereby published with permission under the Creative Commons Attribution License or equivalent.

The editorial board has been involved in producing this book since its inception. They have spent rigorous hours researching and exploring the diverse topics which have resulted in the successful publishing of this book. They have passed on their knowledge of decades through this book. To expedite this challenging task, the publisher supported the team at every step. A small team of assistant editors was also appointed to further simplify the editing procedure and attain best results for the readers.

Our editorial team has been hand-picked from every corner of the world. Their multi-ethnicity adds dynamic inputs to the discussions which result in innovative outcomes. These outcomes are then further discussed with the researchers and contributors who give their valuable feedback and opinion regarding the same. The feedback is then collaborated with the researches and they are edited in a comprehensive manner to aid the understanding of the subject.

Apart from the editorial board, the designing team has also invested a significant amount of their time in understanding the subject and creating the most relevant covers. They scrutinized every image to scout for the most suitable representation of the subject and create an appropriate cover for the book.

The publishing team has been involved in this book since its early stages. They were actively engaged in every process, be it collecting the data, connecting with the contributors or procuring relevant information. The team has been an ardent support to the editorial, designing and production team. Their endless efforts to recruit the best for this project, has resulted in the accomplishment of this book. They are a veteran in the field of academics and their pool of knowledge is as vast as their experience in printing. Their expertise and guidance has proved useful at every step. Their uncompromising quality standards have made this book an exceptional effort. Their encouragement from time to time has been an inspiration for everyone.

The publisher and the editorial board hope that this book will prove to be a valuable piece of knowledge for researchers, students, practitioners and scholars across the globe.

List of Contributors

Alenka Majcen Le Marechal, Simona Vajnhandl and Julija Volmajer Valh
University of Maribor, Faculty of Mechanical Engineering, Maribor, Slovenia

Boštjan Križanec
Environmental Protection Institute, Maribor, Slovenia

Zaharia Carmen and Suteu Daniela
Gheorghe Asachi' Technical University of Iasi, Faculty of Chemical Engineering and Environmental Protection, Romania

Radim Vácha
Research Institute for Soil and Water Conservation, Prague, Czech Republic

Guillermo Espinosa-Reyes, Donaji J. González-Mille, César A. Ilizaliturri-Hernández, Fernando Díaz-Barríga Martínez and Jesús Mejía-Saavedra
Universidad Autónoma de San Luis Potosí, Facultad de Medicina-Departamento de Toxicologia Ambiental, Mexico

Bernard Clément
LEHNA-IPE, Université de Lyon-ENTPE, France

Mahyar Sakari
Water Research Unit & School of Science and Technology, Universiti Malaysia Sabah, Malaysia

Ross Sadler and Des Connell
Griffith University, Australia

Wioletta Rogula-Kozłowska, Barbara Błaszczak and Krzysztof Klejnowski
Institute of Environmental Engineering of the Polish Academy of Sciences, Zabrze, Poland

Barbara Kozielska
Faculty of Energy and Environmental Engineering, Department of Air Protection, Silesian University of Technology, Gliwice, Poland

Zhengjun Zhang, Qin Zhou and Xian Zhang
Tsinghua University, P. R. China

Printed in the USA
CPSIA information can be obtained
at www.ICGtesting.com
JSHW011435221024
72173JS00004B/811